Understanding and Implementing ISO 9000:2000

Second Edition

David L. Goetsch

Stanley B. Davis

Upper Saddle River, New Jersey
Columbus, Ohio

Library of Congress Cataloging-in-Publication Data

Goetsch, David L.
Understanding and implementing ISO 9000:2000 / David L. Goetsch, Stanley B. Davis.
p. cm.
Includes bibliographical references and index.
ISBN 0-13-041106-X
1. ISO 9000 Series Standards. I. Davis, Stanley, 1931-

TS156.6 .G62 2002
658.5′62—dc21

2001034005

Editor in Chief: Stephen Helba
Executive Editor: Debbie Yarnell
Media Development Editor: Michelle Churma
Production Editor: Louise N. Sette
Copy Editor: Ben Shriver
Design Coordinator: Robin G. Chukes
Cover Designer: Mark Shumaker
Production Manager: Brian Fox
Marketing Manager: Jimmy Stephens

This book was set in Clearface by The Clarinda Company. It was printed and bound by R. R. Donnelley & Sons Company. The cover was printed by The Lehigh Press, Inc.

Pearson Education Ltd., *London*
Pearson Education Australia Pty. Limited, *Sydney*
Pearson Education Singapore Pte. Ltd.
Pearson Education North Asia Ltd., *Hong Kong*
Pearson Education Canada, Ltd., *Toronto*
Pearson Educación de Mexico, S. A. de C. V.
Pearson Education—Japan, *Tokyo*
Pearson Education Malaysia Pte. Ltd.
Pearson Education, *Upper Saddle River, New Jersey*

10 9 8 7 6 5 4 3 2 1
ISBN: 0-13-041106-X

Preface

> *In December, 2000, ISO released the new version of ISO 9000. Its departure from the earlier version is dramatic, especially in its adoption of the eight quality management principles (see Chapter One) and emphasis on customer satisfaction and continual improvement. All of these are part of the fundamental philosophy of Total Quality Management (TQM) too long ignored by ISO 9000. Even though these changes dictated that virtually everything in the original book be replaced by totally new material for this second edition, the authors are in complete accord with ISO's move toward TQM. We salute the members of TC 176, the committee that has brought a remarkable new ISO 9000 to organizations around the world.*

WHY WAS THIS BOOK WRITTEN AND FOR WHOM?

Understanding and Implementing ISO 9000:2000 was written in response to the need for a practical teaching resource and a how-to guide that would provide a step-by-step model for understanding the year 2000 version of the ISO 9000 standard and its supporting documents, and for implementation and registration to the standard.

It is the authors' intent in the writing of this book that it might serve a triple role.

- First, it is intended to be a primary text in courses based on ISO 9000, the international standard for quality management systems, and a supplemental text in courses dealing with management, industrial management, and total quality management.
- Second, for private and public sector organizations, the book was designed to be a practical hands-on manual for implementing a quality management system conforming to ISO 9000, and doing so at an affordable cost.
- Third, the book is intended for organizations in the private and public sectors that are currently registered to the 1994 version of ISO 9000, to guide them through the preparations necessary to retain registration under the far different 2000 version.

Since its first release in 1987, more than 350,000 organizations worldwide have registered under the ISO 9000 International Standard for Quality Management Systems, with tens of thousands more registering each year. ISO 9000 has truly become a force in the world's markets for products and services. The latest version of the standard was released at the end of 2000, and has dramatically raised the quality management system bar for excellence. ISO 9000:2000 has all but eliminated the gap with Total Quality Management, and can now present the opportunity for registered organizations to offer customers consistently high quality products and services while becoming more competitive. These benefits, however, will not come without considerable additional effort on the part of organizations, whether currently registered to ISO 9000:1994 or not yet registered. *Understanding and Implementing ISO 9000:2000* shows all organizations what they must do to achieve conformance to the new standard, and helps prepare students for effective roles in private or public sector organizations.

ORGANIZATION OF THIS BOOK

This book begins with a comprehensive background on the purpose of standardization, the International Organization for Standardization, and the ISO 9000 standard and why it came into being. Chapter 2 discusses ISO 9000's value and applicability to the service sector, something too long ignored. Chapter 3 leads the reader through the component parts of ISO 9000:2000 and their relationships with each other and with its environmental management system counterpart, ISO 14000, and provides clarification for the sometimes confusing language of the standard. Chapter 4 establishes and explains all of the requirements of the standard in a way that is easily understood. Chapter 5 defines, and develops an understanding of the concept of a quality management system (QMS), its elements, structure, and supporting activities. Chapter 6 clearly defines the documentation and documentation system required, and elaborates on what must be and what need not be documented. Chapter 7 explains and clarifies the registration and audit processes, and offers recommendations for the selection of registrars. Chapter 8 expands on the concept of continual improvement now required by ISO 9000. Chapter 9 provides a step-by-step process for taking the organization through preparation for registration, and finally through the registration audit—and doing it at minimum cost. Chapter 10 explores the new relationship between ISO 9000:2000 and Total Quality Management. Chapter 11 briefly discusses other related ISO, and ISO 9000-based, standards that already exist or are anticipated. Five appendices provide interesting and useful information for the study and application of ISO 9000, including a comprehensive checklist that will assist organizations in assessing their readiness for registration. With the checklist, it is possible not only to determine areas that already meet the requirements of ISO 9000:2000, but also the specific actions that must be taken for conformance or registration.

ABOUT THE AUTHORS

David L. Goetsch is provost of the joint campus of the University of West Florida and Okaloosa-Walton Community College in Fort Walton Beach, Florida. He also administers

Florida's Center for Manufacturing Competitiveness that is located on the campus, and is president of the Institute for Corporate Competitiveness, a private company. Dr. Goetsch is cofounder of The Quality Institute, a partnership of the University of West Florida, Okaloosa-Walton Community College, and the Okaloosa Economic Development Council. He currently serves on the executive board of the Institute.

Stanley B. Davis was a manufacturing executive with Harris Corporation until his retirement in 1992. He was founding managing director of The Quality Institute and is a well-known expert in the areas of implementing Total Quality Management, statistical process control, just-in-time manufacturing, benchmarking, quality management systems, and environmental management systems. He currently serves as Professor of Quality at the Institute, and heads his own consulting firm, Stan Davis Consulting, which is dedicated to assisting private industry and public organizations throughout North America achieve world-class performance and competitiveness.

Contents

CHAPTER FOUR
Requirements of ISO 9000 53

CHAPTER FIVE
The Quality Management System (QMS) 173

CHAPTER SIX
QMS Documentation 223

CHAPTER SEVEN
Registration and the Audit Process 243

CHAPTER ONE

Background of ISO 9000: 2000 and Why It Exists

MAJOR TOPICS

- Standards and Standardization
- International Organization for Standardization
- Evolution of ISO 9000
- Objective of ISO 9000
- Scope of ISO 9000
- Applicability of ISO 9000
- Rationale for ISO 9000 Certification

STANDARDS AND STANDARDIZATION

This book is about ISO 9000, an international **standard** for quality management systems. Throughout the book we discuss standards and **standardization.** Consequently it is important that the reader understand the terms. Webster's defines *standard* as, ". . . 3: something established by authority, custom, or general consent as a model or example . . . 4: something set up and established by authority as a rule for the measure of quantity, weight, extent, value, or quality. . . ." Further, Webster's defines *standardize* as "1: to compare with a standard. 2: to bring into conformity with a standard."[1]

We use standards all the time without giving them so much as a thought; in fact, life would be difficult without them. Consider language, for instance. We are able to communicate because of standardization within a common language. Words and phrases have common (standard) meanings. Therefore, the words and phrases may be easily transferred from a speaker to a listener or from a writer to a reader who, as a result of standardization, can understand what is being said. When there is no standard—no common meaning as between English and Mandarin—we have great difficulty communicating our thoughts and ideas.

On another level, when buying a package of cookies, you don't give a thought to whether a one-pound package of cookies baked in New York has the same weight as a package baked in California. Nor do you worry about whether a one-gallon milk container really holds one gallon, regardless of the dairy of origin. These things are taken for granted because we live in a society with a common set of standard weights and measures.

When you turn on a television anywhere in North America, you are confident that no matter what station you choose, it will be compatible with the TV's operating parameters. This is because all broadcasters and television receiver manufacturers adhere to a common set of technical standards. However, in Europe or Japan, if you tried to use the same television to receive local broadcasts, the result would be different. Why? The television standards used in Europe or Japan differ from the standards used in North America. In fact, there is no common international broadcast standard. Imagine what it would be like if every state in the United States had its own standards, or even worse, if there were no standards. Standards are pervasive, and if they are planned well and adopted widely, they are of great benefit to mankind.

In the business world, standardization is imperative so that a manufacturer in one country can sell its products worldwide without requiring hundreds of location-specific models. For example, a size eight dress made in Malaysia will fit just like a size eight dress made in South Carolina. Likewise, a disk manufacturer's product must be compatible with any computer, regardless of brand. A system of accepted standards is essential for the widespread marketability, efficient production, and convenient use of products—from complex mechanical systems like automobiles, and even the nuts and bolts that hold them together, to purely intellectual products such as computer programs.

Perhaps the most visible example of the absence of a common standard involves the automobile. In most of the world cars are driven on the right side of the road, and configured with driving controls and instruments on the left. In some countries, notably the United Kingdom and Japan, cars are driven on the left side of the road, so driving controls and instruments are configured on the right. This not only creates interesting situations as international travelers maneuver in traffic in rental cars, but also makes exporting cars more expensive, since the driver's controls must conform to the importing country's standard, thus requiring the manufacturer to produce both left- and right-hand drive models.

The most common kind of standard relates to some type of measurement (e.g., cycles per second, threads per inch or centimeter, dimensions, fit, shape, counts, weights, and measures). Another kind of standard has to do with processes, how things are done. One example is a quality management system conforming to the **ISO 9000** standard. Another is an environmental management system conforming to ISO 14000. These standards deal not with absolutes, but with how the quality or environmental management system is established and executed. Unlike the far more common absolute standards, ISO 9000 and ISO 14000 are generic standards, usable by virtually any organization, large or small, public or private, regardless of product or service provided.

The subject of this book, ISO 9000, is about standardizing the approach organizations everywhere take to managing and improving the processes that ultimately result in their products and services. As will be seen, this is different from simply conforming to a specification for a specific level of quality.

ISO INFO

The fact that we have a high degree of standardization has made life simpler for us in ways so basic and so obvious that we do not even realize they exist. It has given us the free national market which we take so casually. To you as end man, the American consumer, it has given lower prices and better quality, more safety, greater availability, prompter exchange and repair service, and all the other material advantages of mass production. Is this something to be taken for granted?[2]

W. Edwards Deming

INTERNATIONAL ORGANIZATION FOR STANDARDIZATION

In the previous section we saw that standards are important to modern society. The more *standard* the standard is, that is, the more widely accepted and utilized, the better. Widely accepted standards lead to more efficient use of resources for producers, more equitable international competition, and lower cost to consumers. The world has too many competing standards, such as standards for electrical power generation and distribution (50 Hz vs. 60 Hz), units of measure (metric vs. the English System), television broadcast standards,[3] and many others. For every difference that exists between national or regional standards, some competitor is put at a disadvantage and consumers ultimately pay higher prices.

The present situation is much improved from what it might have been without the worldwide movement that followed World War II to rationalize the thousands of conflicting standards of the various nations. The **International Organization for Standardization (ISO),** based in Geneva, Switzerland, has been the standard bearer for that effort. ISO was established in 1947 to promote standards in international trade, communications, and manufacturing. ISO is a nongovernmental organization, and, contrary to widely held beliefs, is not an arm of the European Union or the United Nations. It has no power to impose its standards. Rather, ISO is a worldwide federation of national standards organizations from 130 nations. Since in most countries standardization is a function of government, nearly all of these 130 member bodies are government organizations.

ISO INFO

ISO is not a European organization, although it is based in Geneva, Switzerland. Nor is ISO an agent of any government or federation of governments. When ISO was established in 1947, the American National Standards Institute (ANSI) was a founding member. ANSI is one of five permanent members of ISO's governing council, and one of four permanent members of ISO's Technical Management Board. ISO is a worldwide organization.

ISO INFO

About ANSI, The U.S. Representative to ISO

ANSI, the American National Standards Institute, was founded in 1918 by five engineering societies and three government agencies. Since then it has been the administrator and coordinator of the United States voluntary standardization system. ANSI's primary goal is "... the enhancement of global competitiveness of U.S. business and the American quality of life by promoting and facilitating voluntary consensus standards and conformity assessment systems and promoting their integrity."[4] *ANSI's members include more than 1300 United States businesses, professional societies, trade associations, and government agencies. Although the ANSI Federation includes such United States government entities as the Department of Energy, Occupational Safety and Health Administration, Department of Commerce, and the Environmental Protection Agency, these entities are conferred no special status relative to members from the private sector.*

The principal exception to this is the United States, whose ISO representative is the **American National Standards Institute (ANSI),** a private-sector organization.

ISO membership falls into three categories. A **full member** is a national body designated by its respective country as the "most representative of standardization." Most nations are represented by full members. A nation that does not have a standardization body may be represented by a **correspondent member.** A small nation with a very small economy may become a **subscriber member,** at a reduced membership rate. Although only full members may participate in the development of standards, ISO keeps correspondent and subscriber members informed about activities of interest. A list of ISO member bodies (full members) may be found in Appendix B.

EVOLUTION OF ISO 9000

In response to the need to harmonize dozens of national and international quality standards that existed throughout the world in the 1970s and 1980s, ISO formed **Technical Committee 176 (TC 176),** an international team representing 75 nations (refer to Appendix C). The United States is represented on TC 176 by the American National Standards Institute. ANSI is assisted by the American Society for Quality (ASQ), its affiliate responsible for quality management and related standards.

TC 176's objective was to develop a universally accepted set of quality standards. This became the ISO 9000 series of standards, first released in 1987. ISO 9000 was subsequently given a mild update by TC 176 and was rereleased in 1994. The term "family of standards" was attached to the 1994 versions. All organizations certified to ISO 9000 through December of 2000 were certified to the 1994 version. After the December 2000 publication of ISO 9000:2000, certifications gradually switched to the new standard.

ISO INFO

A three-year transition period went into effect with the December 2000 publication of ISO 9000:2000. Although the intent was for organizations to "upgrade" to the new standard as quickly as possible, ISO permitted new registrations to the 1994 standard during the transition period to accommodate organizations who were nearing readiness for certification. Existing (pre December, 2000) ISO 9000:1994 certifications retain validity until their next recertification date, no later than December 2003. ISO 9000:1994 certifications issued any time after December 2000 expire December 2003. Thereafter, all certifications will be to ISO 9001:2000.

The subject of this book is the year 2000 release of ISO 9000. ISO 9000:2000 represents a fundamental change in approach, and is a major, and needed, improvement over the two earlier versions. TC 176 has aligned ISO 9000 much closer—very close in fact—with the Total Quality Management approach. Although still used, the term "family of standards" is superfluous since there is now only one real standard (ISO 9001:2000) instead of three, and one guidance document, or manual, (ISO 9004:2000) instead of eight. See Figure 1-1.

ISO 9000's evolution has aligned it more closely with the Total Quality Management philosophy. It seemed to many observers, including the authors, that the 1987 and 1994 versions deliberately shied away from association with TQM, or from acknowledging its existence. Even in the 2000 version, which borrows heavily from TQM, there is no specific acknowledgment of it. The fact is, of course, that with the tutelage of Deming and Juran, the Japanese started the development of the management system we now know as TQM in 1950. Over the years several Japanese experts—Ishikawa, Shingo, Ohno, and others—emerged, and by the early 1970s TQM had been widely accepted in Japan. By 1980 the western world began taking note. By the time ISO 9000:1987 was developed, TQM was a mature management system, well understood by many in the West. It is clear TC 176 borrowed some elements of TQM for ISO 9000:1987, most notably its documentation requirements. ISO 9000:1994 moved a bit closer to TQM, at least mentioning (though not requiring) continual improvement. But any acknowledgment of TQM's superiority seemed to be deliberately avoided. ISO 9000:2000 has made a giant leap in comparison, especially in the area of continual improvement, which has gone from just cursory treatment to being a firm requirement. In addition, the standard now incorporates eight quality management principles that come directly from TQM.[5] They are:

1. **Customer focus**—understanding their needs, striving to exceed their expectations.
2. **Leadership**—establishing direction, unity of purpose, and a supporting work environment.

1987 Release	1994 Release	2000 Release	Purpose
ISO 9000:1987	ISO 9000-1:1994		Guidelines for Selection and Use
None	ISO 9000-2:1993		Guidelines for Application of the Standards
None	ISO 9000-3:1991		Guidelines for Application of ISO 9001 to Software Producers
None	ISO 9000-4:1993		Guide to "Dependability" Program Management
		ISO 9000:2000	Fundamentals and Vocabulary
ISO 9001:1987	ISO 9001:1994		Standard—Model for "Full House" Producers
ISO 9002:1987	ISO 9002:1994		Standard—Model for Production
ISO 9003:1987	ISO 9003:1994		Standard—Model for Final Inspection and Test Only
		ISO 9001:2000	Quality Management Systems—Requirements*
ISO 9004:1987	ISO 9004-1:1994		Guidelines for Quality System Elements and Management
None	ISO 9004-2:1991		Guidelines for Services
None	ISO 9004-3:1993		Guidelines for Processed Materials
None	ISO 9004-4:1993		Guidelines for Quality Improvement
		ISO 9004:2000	Guidelines for Organizational Performance Improvements
			*Certifications are to ISO 9001:2000 only.

Figure 1–1
ISO 9000 Evolution: 1987–2000

3. **Involvement of people**—ensuring that all employees at all levels are able to fully use their abilities for the organization's benefit.
4. **Process approach**—recognizing that all work is done through processes, and managed accordingly.
5. **System approach to management**—expands on the previous principle in that achieving any objective requires a system of interrelated processes.
6. **Continual improvement**—as a permanent organizational objective, recognizing and acting on the fact that no process is so good that further improvement is impossible.
7. **Factual approach to decision making**—acknowledging that sound decisions must be based on analysis of factual data and information.
8. **Mutually beneficial supplier relationships**—synergy can be found in such relationships.

Five of these principles (1, 2, 3, 6, and 7) are also principles listed in the primary eleven TQM principles.[6] The other three (4, 5, and 8) are also part of the TQM philosophy, the whole of which is now embedded to some degree in ISO 9000:2000. ISO considers the major changes in the revised standard to be:[7]

- Increased focus on top management commitment
- Customer satisfaction
- Emphasis on processes
- Continual improvement

All of these are fundamental to Total Quality Management. ISO and TC 176 are to be commended for these positive changes in the standard. Understand, however, that an organization can be certified to ISO 9000 without fully adopting TQM. But for those elements of the organization subject to the standard, the TQM alignment will be quite close, especially if ISO 9004:2000, which contains much of the TQM philosophy, is followed. (*Note:* Organizations are certified and audited to ISO 9001:2000 only. ISO 9004 is designed to show the organization how its performance can be further and continually improved.)

As a result of ISO 9000, any organization supplying products or services is able to develop and employ a quality management system that is recognized by customers worldwide. Customers around the world who deal with ISO 9000 registered organizations can expect that purchased goods and services will conform to a set of standards they recognize.

OBJECTIVE OF ISO 9000

Originally, the reason for creating ISO 9000 was to replace dozens of national and international quality standards with one single family of standards universally recognized and used worldwide. The objective of the original ISO 9000 standards was to enable organizations to consistently produce products (including services) that met the requirements of customers and lived up to the organization's stated intentions. Since its initial release in 1987, however, the objectives have been broadened to the point that with the 2000 release, the objectives include consistency in products, meeting customer and regulatory requirements, and having systems that address customer satisfaction, continual improvement, and prevention of nonconformity. The emphasis on customer satisfaction and continual improvement are new with ISO 9000:2000. The aims of ISO 9000:2000 also include the adoption by its registered organizations of a management system that is approaching Total Quality Management. In ISO's own words:

> The primary aim of the "consistent pair" [ISO 9001 and 9004] is to relate modern quality management to the processes and activities of an organization, including the promotion of continual improvement and achievement of customer satisfaction.[8]

ISO lists an ISO 9000 objective as being a "natural stepping stone toward Total Quality Management."[9] (We will explore the relationship between ISO 9000 and Total Quality Management in Chapter 10.)

Ultimately the objective of ISO 9000 is that any organization, whether large or small, public or private, regardless of what it does, can, by utilizing the management practices represented by the standard, (1) consistently deliver products and services that meet the requirements of its customers, and (2) continually improve the processes that affect its product or service quality.

SCOPE OF ISO 9000

ISO describes the scope of ISO 9000 as follows:

> . . . [ISO 9001] specifies requirements for a quality management system where an organization a) needs to demonstrate its ability to consistently provide product that meets customer and applicable regulatory requirements, and b) aims to enhance customer satisfaction through the effective application of the system, including processes for continual improvement of the system and the assurance of conformity to customer and applicable regulatory requirements.[10]

The scope of ISO 9000 includes any organization wishing to be certified to the standard. Within the organization the scope extends to any department or activity that can have an impact on the quality of the product or service. In a practical sense, this includes all departments and activities.

The standard requires the certified organization to structure its quality management system so that it addresses and conforms to all of the ISO 9001 clauses from clause 4 through clause 8. These clauses are explained in detail in Chapter 4.

More than 300,000 organizations are registered to ISO 9000. At least 60 percent of them are in Europe. The United States accounts for about 10 percent.[11] North American organizations hold more than 40,000 certificates, distributed approximately as follows.[12]

USA	30,000
Canada	10,000
Mexico	1,000

APPLICABILITY OF ISO 9000

ISO 9000:2000 applies to any organization that is certified, or seeks certification to the standard. All types and sizes of organizations from any sector may be included in the standard's applicability. In terms of North American registrations, actual applicability of ISO 9000 to the various sectors is as follows[13]:

Manufacturing	73%
Services	13%
Wholesale trade	8%
Transportation	4%
Construction	1%
All others	1%

RATIONALE FOR ISO 9000 CERTIFICATION

The rationale for adoption of TQM is survival. Many organizations must embrace TQM or cease to exist because their competitors are so much more efficient. While the same case might be made for ISO 9000, the most often stated rationale for certification is "keeping our customers."

No organization is forced to implement ISO 9000, although customer pressure, the major impetus for certification in North America, can be sufficient to cause some organizations to feel they must conform to the standard. To develop a conforming quality management system requires considerable effort and expense. The amount of effort and expense will depend upon many factors, including the size of the organization and its current quality management status. Why would an organization voluntarily take on the work and expense of certification to ISO 9000? There are several possible answers to this question. Some are appropriate and some are not. For example, if an organization seeks registration merely to use "ISO 9000" in its advertising, its motive is inappropriate and ISO 9000 will probably receive little more than lip service. Customer pressure may be legitimate motivation if the organization finds itself in a do or die situation. However, for most organizations, the rationale for implementing ISO 9000 should include one or more of the following:

- To improve product or service quality and consistency
- To improve organizational performance through better management of processes and resources
- To have a quality management system that will be recognized by customers worldwide

CASE STUDY

ISO 9000 in Action

This is the first installment of a serialized case study that appears throughout this book, with one installment at the end of each chapter. The case study describes how a hypothetical company, Anderson Manufacturing, Inc. (AMI), goes about preparing for ISO 9000 registration, but in the wrong way and for the wrong reasons. The simulated problems experienced by AMI and the solutions adopted will give readers an understanding of what they might face in a real setting.

Jake Butler had been hired by AMI to fill the position of Quality Systems Director. During the interview process, it became clear to Butler that his first task would be to prepare AMI for ISO 9000 registration. It also became clear to Butler that AMI's CEO knew plenty about the 1994 ISO registration standard, but very little about ISO 9000:2000. He had little knowledge of or interest in the "continual improvement" aspects of the updated standard. Nor did he appear to see the importance of pursuing ISO 9000 registration within the broader framework of implementing the TQM philosophy company wide. Butler became concerned that the CEO was more interested in the marketing benefits of registration than the actual performance-improvement aspects.

He hoped this was not the case. Quality management is Butler's profession and he knows it well. Butler knows that ISO 9000 registration, if undertaken for the right reasons, can be an important component of an organization's overall quality program. But he also knows that by itself ISO 9000, if undertaken in a cursory manner just to score marketing points, will not make a company a consistent winner in the global marketplace, a goal AMI's CEO says he wants to achieve.

Tomorrow morning Butler will have an opportunity to state his views to AMI's executive management team. He is scheduled to give a briefing on what he thinks the company needs to do in the area of quality in order to be competitive in the global marketplace. During the briefing Butler intends to explain that ISO 9000 registration should be undertaken within the broader framework of implementing the TQM philosophy and not as a marketing gimmick.

Jake Butler is new to AMI. Consequently, he does not know how his briefing will be received by the company's executive managers. Nonetheless he intends to press on. Butler's belief is that he might as well lay all of his cards face up on the table and find out from the beginning where AMI's executives stand. He knows that in order to succeed he will need a full commitment from AMI's executives. Better to learn at the outset whether or not he will have it.

SUMMARY

1. A standard is a criterion or set of criteria established by authority for the uniform measuring of such characteristics as quantity, weight, extent, value, quality, and the like, or as a model or example. Standardizing involves bringing about conformity with the standard's criteria. ISO 9000 is an international standard for managing and improving the processes that ultimately govern the quality of products and services.

2. The International Standards Organization (ISO) is a nongovernmental organization established in 1947 to promote standardization in international trade, communications, and manufacturing. It consists of national standards organizations from 130 countries. The United States is represented by the American National Standards Institute (ANSI).
3. ISO 9000:2000 differs from its predecessor, ISO 9000:1994 in several ways. One of the most important is the new standard's adherence to the TQM philosophy, which is built on the following concepts: customer focus, leadership, involvement of people, process approach, systems approach to management, continual improvement, factual approach to decision making, and mutually beneficial supplier relationships. ISO considers the major changes in the new standard to be increased focus on commitment from top management, customer satisfaction, emphasis on processes, and continual improvement.
4. The aim of ISO 9000:2000 is to enable organizations to consistently produce products (including services) that meet the requirements of customers and live up to the organization's stated intentions.
5. The scope of ISO 9000 is as follows: It specifies the requirements for a quality management system where an organization needs to consistently provide products or services that meet customer expectations and/or regulatory requirements and where the organization aims to enhance customer satisfaction through continual improvement.
6. ISO 9000 applies to any organization that wishes to seek certification to the standard. The manufacturing, services, wholesale-trade, transportation, and construction industries all have ISO 9000 registered companies.
7. The rationale for implementing ISO 9000 should include one of the following: improve product or service quality and consistency; improve organizational performance; establish a quality management system that will be recognized by customers internationally.

KEY TERMS AND CONCEPTS

American National Standards Institute (ANSI)
Continual improvement
Correspondent member
Customer focus
Factual approach to decision making
Full member
International Organization for Standardization (ISO)
Involvement of people
ISO 9000
Leadership
Mutually beneficial supplier relationships
Process approach
Standard
Standardize
Subscriber member
Systems approach to management
TC 176

REVIEW QUESTIONS

1. Define the term *standard* and give an example of how you use a standard or standards in your daily life.
2. Describe the International Standards Organization as if explaining it to a colleague who has never heard of it.
3. Explain the evolution of ISO 9000 from the beginning to present.
4. List and explain the eight fundamental principles of quality management.
5. What are the major changes—according to ISO—in ISO 9000:2000 when compared to the previous version?
6. Explain the aim of ISO 9000.
7. Explain the scope of ISO 9000.
8. What are the major points to be made when explaining the rationale for ISO 9000 certification?

APPLICATION ACTIVITIES

1. Undertake a research project to determine how W. Edwards Deming came to be the leader of the Japanese Total Quality movement. Your research should answer the following questions as a minimum:
   a. Why did Deming go to Japan rather than help manufacturers in the United States?
   b. Why did Japan need help from Deming?
   c. What is the Deming Prize in Japan and what are its requirements?
2. Develop a presentation that you, as a newly hired quality manager, could make to the executive management team of your company to convince its members of the benefits and costs of pursuing ISO 9000 registration. Assume yours will be a skeptical audience.

ENDNOTES

1. *Merriam Webster's Collegiate Dictionary*, 10th ed. (Springfield, MA: Merriam-Webster, Incorporated, 1993), pp. 1145–1146.
2. Dr. W. Edwards Deming, *Quality, Productivity, and Competitive Position* (Cambridge, MA: Massachusetts Institute of Technology, Center for Advanced Engineering Study, 1982), p. 345.
3. Standards related to electrical and electronic engineering are handled by another organization, also based in Geneva, the International Electrotechnical Commission (IEC).
4. American National Standards Institute, *An Introduction to ANSI,* 1998.
5. ISO 9000:2000, clause 0.2.

6. David L. Goetsch and Stanley B. Davis *Quality Management: Introduction to Total Quality Management for Production, Processing, and Services,* 3rd ed. (Upper Saddle River, NJ: Prentice Hall, 2000), p. 50.
7. ISO, *The New Year 2000 ISO 9000 Standards—An Executive Summary,* p. 3.
8. Ibid.
9. Ibid.
10. ISO 9001:2000, clause 1.1.
11. ISO press release, 27 August 1999.
12. Stewart Anderson, article in *ISO 9000 & ISO 14000 News* (ISO, 2/2000).
13. Ibid.

CHAPTER TWO

The Service Sector and ISO 9000

MAJOR TOPICS

- Application of ISO 9000 to the Service Sector
- Definition of "Service"
- Clarifying the ISO Language for the Service Sector
- Examples of Service-Providing Organizations
- Service Customers: External and Internal
- Relationship of "Service" to "Product"
- Relationship of Processes to Services
- Quality Characteristics of Services
- Inspecting Services
- Documentation Requirements for Service Providers
- Which Clauses Apply to Service Providers?

APPLICATION OF ISO 9000 TO THE SERVICE SECTOR

There is a widespread misconception that ISO 9000 is applicable exclusively to the manufacturing sector. However, ISO's intent is that the standard apply to "all organizations, regardless of type, size and product provided."[1] The question is, what is a "product"? In ISO 9000:2000, clause 3.4.2 defines *product* as "the result of a process," and goes on to list four generic product categories as:

Services (e.g., transport)
Software (e.g., computer program, dictionary)
Hardware (e.g., engine mechanical part)
Processed materials (e.g., lubricants)

There is no question that manufacturing organizations have become heavily involved with ISO 9000, but the fact is, 25 percent of registered firms in North America are from

ISO INFO

Throughout the text of this international standard, wherever the term "product" occurs, it can also mean "service."

ISO 9001, clause 3 and ISO 9004, clause 3

the **service** sector. Applicability to the service sector is greatly enhanced with the 2000 release of the standard. ISO 9000 now incorporates the eight quality management principles listed in ISO 9000:2000, clause 0.2, ISO 9004:2000, clause 4.3, and (by reference) ISO 9001:2000, clause 0.1:

1. Customer focus
2. Leadership
3. Involvement of people (employee involvement)
4. Process approach
5. System approach to management
6. Continual improvement
7. Factual approach to decision making
8. Mutually beneficial supplier relationships

All of these apply as much to organizations that provide services for their customers as to those that manufacture tangible products. In addition, this version of ISO 9000 promotes the adoption of a *process approach* for the quality management system and the enhancement of customer satisfaction.[2] The process approach emphasizes the importance of:

- Understanding and fulfilling requirements
- Considering processes in terms of added value
- Obtaining results of process performance and effectiveness
- Continual improvement of processes based on objective measurement

All of these items are applicable to organizations operating in the service sector.

DEFINITION OF "SERVICE"

According to *Webster's New World Dictionary of the American Language*, *service*, in the present context, is defined as follows:

- "work done or duty performed for another or others: as, professional services, repair service, a life devoted to public service."

- "the act or manner of serving food: as, the food was of even worse *quality* than the service."
- "an activity carried on to provide people with the use of something, as electric power, water, transportation, mail delivery, telephones, etc."
- "the act or method of providing these."
- "the *quality* of that which is provided: as, our electric service is poor."
- "anything useful, as maintenance, supplies, installation, repairs, etc., provided by a dealer or manufacturer for people who have bought things from him."

The italic in the bulleted items was added to note that quality is considered an inherent characteristic of service. From the definitions given we can say that *service is work performed for someone else*. It may be in the form of providing something (e.g., electric service), or the whole process of supplying, installing, maintaining, repairing, and so on. In addition, the method of providing the service, as well as the quality of the service are important characteristics of the service.

As noted above, a product is the result of one or more processes, and a *product* may be a *service*. ISO has overlooked the fact that when dealing with services, it is more than the *result* of the activity or process that is important. The very *act of doing* the activity or process is frequently observed and evaluated by the customer. For example, when you order a cup of coffee at the lunch counter the waitress may be efficient and gracious in providing you with the coffee, or she may be curt and offensive. The result of the process in both instances is that you got the cup of coffee, but the *way in which it was done* is as important to you as the coffee itself. In other words, the quality of the service will affect the size of the gratuity, or whether or not you will return to this place of business in the future. Not all services are performed in view of the customer, but those that are need special care given to the manner in which they are performed. Customer focus becomes critically important here.

CLARIFYING THE ISO LANGUAGE FOR THE SERVICE SECTOR

Interest in ISO 9000 on the part of the service sector has not been as keen as in manufacturing, perhaps because of the following:

1. ISO 9000 was widely heralded as a quality standard for manufactured products.
2. The language of the standards was oriented to manufacturing.
3. The service sector was not subjected to the same pressure as manufacturing to comply with quality standards.
4. The perceived notion that it is difficult to determine the quality required or achieved in a service setting.

There is no question that the committee that developed the ISO 9000 standards was slanted toward the production of tangible products, as demonstrated by the language of the standards. Since the late 1980s, however, pressure has increased steadily for service

ISO INFO

People pay others to do work that falls into one of the following categories:

- *Work they are not—or feel they are not—competent to do themselves*
- *Work they do not want to do themselves*
- *Work they do not have time to do themselves*

All such work falls under the broad umbrella of services. Regardless of why people pay others to perform services, they typically have high expectations concerning the quality of those services.

providers to improve their quality, or suffer the same fate as their manufacturing counterparts who failed to get the message. Since service has become such a major element of the world's economy, and since competition in the service sector is rapidly growing, in many parts of the world poor or inconsistent quality of services is no longer tolerated. Additionally, the notion that it is more difficult to determine the level of quality desired by the service customer, and whether it has been achieved or not, is pure fiction. Service customers make that determination relentlessly. There is no reason why service providers cannot make the same sort of determinations, and there is every reason why they should.

More and more service organizations of all types are becoming certified to ISO 9000. One of the early examples was Federal Express, the overnight package deliverer. Even though FedEx is a TQM practitioner, their ISO certification was no small task since all of their facilities around the world had to register individually. The motivating force behind the decision to undertake the work and expense necessary to achieve global certification was not competition, as one might expect. Rather, the motivation stemmed from correctly perceiving ISO 9000 as the right thing to do in a company that operates in every corner of the world and that has built its reputation on dependability and customer satisfaction.

In reviewing the standards, one finds words that do not seem to fit with the service sector. In Figure 2-1 some of these terms are listed. The chart also provides a translation more in keeping with the language of the service industry.

Once you get past these few terms which seem to be more akin to the production of tangible goods than to providing services, the rest should fall into place without difficulty. Service providers are required to develop the same kinds of documentation discussed in Chapters 4, 5, and 6. Consequently, you will become familiar with the language through application.

EXAMPLES OF SERVICE-PROVIDING ORGANIZATIONS

ISO 9004-2-1991, *Quality Management and Quality System Elements—Guidelines for Services* listed the following as examples of service-providing organizations: (*Note:* This is an obsolete document, but the list is still valid.)

ISO 9000	Translation
Design	Think in terms of *inventing* or *originating* your services or processes (Some service firms do provide design services. For these, no translation is necessary.)
Development	Think in terms of *improving* or *maturing* your services.
Conforming product	Services that conform to customer requirements.
Requirements	May be something ISO 9000 requires you to do (such as develop a quality manual), or something customers have specified as needed in your services.
Product	The service(s) you provide, or knowledge you impart to customers for remuneration.
Production	Everything you do in providing your services to your customers. The result of your processes.
Processes	Various kinds of processes include: the means by which you determine which services to provide, the means by which you design or develop them, the means by which you provide the services, and the means by which you verify that they were properly executed.
Procedures	Documented instructions for operating your processes, and for complying with the requirements of ISO 9000.

Figure 2–1
Translating Manufacturing Language to Service Language

- **Hospitality services**—catering, hotels, tourism, entertainment, radio, television, leisure.
- **Communications**—airports and airlines, road, rail, and sea transport, telecommunications, postal, data.
- **Healthcare services**—medical staff/doctors, ambulances, medical laboratories, dentists, opticians.
- **Maintenance**—electrical, mechanical, vehicles, heating systems, air conditioning, buildings, computers.
- **Utilities**—cleansing, waste management, water supply, grounds maintenance, electricity, gas, energy supply, fire, police, public services.
- **Trading**—wholesale, retail, distribution, marketing, packaging.
- **Financial**—banking, insurance, pensions, property services, accounting.

- **Professional**—building design (architects), surveying, legal, law enforcement, security, engineering, project management, quality management, consultancy, training and education.
- **Administration**—personnel, computing, office services.
- **Technical**—consultancy, photography, test laboratories.
- **Purchasing**—contracting, inventory management and distribution.
- **Scientific**—research, development, studies, decision aids.

This list is not inclusive of every type of service that may be offered, but it is comprehensive enough to provide a general idea of what ISO 9000 means by services.

SERVICE CUSTOMERS: EXTERNAL AND INTERNAL

The list of typical service organizations in the previous section is by no means complete. In the developed nations, the service industry is expanding far faster than any other economic sector, and, in fact, is considered to be the major industry in the United States. If the service industry is where more and more external customers spend ever larger proportions of their money, and if service represents a major segment of the economies of developed countries, then the opportunity for benefit through elimination of the waste associated with poor quality services is substantial and becoming more so. This means that ISO 9000 may be even more usefully applied to the service industry than to manufacturing. Certainly the imperative for ISO 9000 should be no less in the service industry than in manufacturing.

Beyond this, if one applies the definition of *service* as being "work done, or duty performed, for others," service is easily translated into any organization. Using this definition, and the concept of the **internal customer** it is easy to see that services are continually being performed in every office, manufacturing plant, and so on. Much of the work of business turns out to be services under this definition, and the importance of eliminating waste from these services, and infusing the highest possible quality into them takes on a greater urgency. This is ISO 9000's mission, and it is as applicable to the service sector as to manufacturing.

ISO INFO

The total quality movement introduced organizations to the concept of the internal *customer. If you and I work in the same organization, and if the quality of your work depends in part on how well I do mine, you are an internal customer to me. This concept alone should get service organizations interested in ISO 9000.*

RELATIONSHIP OF "SERVICE" TO "PRODUCT"

Some services are closely identified with tangible products, while others are not. Figure 2-2 illustrates the point.

In Figure 2-2, the further left on the continuum line, the higher the tangible product content or identification, as illustrated by a vehicle dealership. Moving to the right there is less tangible product content or identification, as indicated in the center by a restaurant. Clearly the restaurant delivers both products and services, but presumably the service itself takes on as much importance as the meal itself. Continuing to the right, tangible product content virtually disappears, as with legal services. What the customer is paying for in this case is a service that is based on the specialized knowledge of the lawyer, and there may be nothing more tangible than advice received.

In those elements of the service sector that are closely tied to physical products, the line between the service and the product is easily blurred. For example, a new car dealership is identified with the make of automobile it sells. Clearly the dealership has nothing to do with the production of the car. It is in the service business on two counts:

1. It provides sales-related services to customers.
2. It provides maintenance services for customers after the purchase.

If the services provided in either case are of poor quality, both the dealership and the manufacturer lose credibility with customers. Saturn was the first automobile manufacturer in the United States to employ a new customer-friendly approach to selling and servicing its cars. Saturn insisted that the sales personnel avoid even a semblance of high pressure tactics, baiting, or haggling. Sales personnel were there to provide information about the cars. Saturn used a similar philosophy on the vehicle maintenance side of the business. Many customers seem to like this approach—so much so that Saturn enjoyed a consistently high customer loyalty rating for several years. Although Saturn seems to have lost some of its focus on the customer recently, the example demonstrated that the sales and maintenance services provided can have an important effect on the customer's satisfaction with the associated product, and that the effort put into services to make them more customer-friendly can reflect favorably on the product brand.

The point here is that if the organization is in manufacturing, and also sells and/or services the goods it produces, the service processes of the organization are just as important to have under ISO 9000 as the manufacturing processes. In fact, if the organization manufactures a product that is subsequently sold and serviced through authorized

Figure 2–2
Product Content in a Service Continuum[3]

agents, it should make every effort to have the agents certified to ISO 9000. It probably makes no difference to the consumer whether her hair dryer is serviced by the original manufacturer or by some local authorized repair shop. But it will make a difference if the authorized repair shop is staffed by discourteous people and looks cluttered and unkempt, or if the needed part and expertise are not available. A negative service experience will turn away customers faster than almost any other factor.

Service organizations that are closely associated with branded products include retail stores and dealerships; repair shops, whether owned by the original equipment manufacturer (OEM) or authorized by the OEM; and factory outlet stores. Since such establishments have more direct contact with the customer than does the factory, it makes sense to apply the quality management principles of ISO 9000 to them as well as to the factory.

RELATIONSHIP OF PROCESSES TO SERVICES

Everything we do is part of a **process.** In the manufacturing plant there are processes for making the product. Such processes should be documented. Organizations go to great lengths to make sure that their processes are completely understood and followed to the letter. This is necessary to ensure consistency regardless of which operator is on duty. In other words, it is done to ensure that the product will be the same as that produced when any other operator is at the controls. This is the only way to effectively assure the level of quality required and the unit-to-unit consistency that customers expect. Should an organization be any less concerned about the quality and consistency of its services? Does it take a different approach for services? The answer to both questions is an emphatic "no."

We have already made the point that services are an important segment of the economy, and that the customer dissatisfaction that results from substandard services can have a devastating impact on the service provider. Since services are provided through processes, as are products in the manufacturing sector, the same approach applies in both sectors. ISO 9001, in spite of the fact that some of the language is slanted toward the production of tangible products, applies to service organizations as well as to manufacturers. For a discussion of the use of processes in providing services, turn to Chapter 3.

QUALITY CHARACTERISTICS OF SERVICES

The word *quality,* when applied to a service, is no easier to define than when applied to a tangible product. Rather than getting caught up in the debate, we recommend a definition that includes the following simply stated criteria:

- Fitness for use
- Meeting or exceeding customer requirements
- Meeting or exceeding customer expectations

The first element of the definition, **fitness for use,** applies to the process of providing the service. For example, there are several processes for cleaning floors—sweeping, mopping, vacuuming, to name just a few. A hotel that prepares guest rooms by sweeping the carpeted floors is using a process that is unfit for the intended purpose. Carpeted floors require vacuuming. In order for a service to be of acceptable quality, the processes used in its delivery must be fit for use.

The second element of the definition, meeting or exceeding customer **requirements,** is critical because only the customer determines if the quality of service is acceptable. Unless the customer's requirements are met, his or her evaluation of service quality will be negative. Using the hotel analogy, suppose a customer is on a business trip, and intends to spend evenings working in his room. He will require a desk, adequate lighting at the desk, and perhaps a fax capability at the front desk. Should there be no desk in the room, if the illumination in the room is insufficient, or if a fax machine is not available, then the hotel service will not meet the customer's requirements, and the quality will be judged unsatisfactory. Hotels that cater to specific elements of the traveling public need to know what their customers require of them. The same is true of other service providers.

The third element of the definition, meeting or exceeding customer expectations, is closely related to customer requirements. However, while some accoutrements are *required,* others, though not required, may be *expected*—and greatly appreciated when provided. These are generally the things that set the best service providers apart from their counterparts who are merely adequate. "Extras" help service providers go beyond customer satisfaction to customer delight. Customers who receive outstanding service will be more than just satisfied, they will be delighted. The important thing is, they will be back. Customers who always come back have what all businesses should strive to create: customer loyalty.

The following is a list of service and service delivery characteristics.[4] They should be regarded as *quality characteristics*. These characteristics are measurable. Consequently, they can be used to monitor service quality. Quality characteristics are as follows:

- Facilities, capacity, number of personnel, and quantity of materials
- Waiting time, delivery time, and process times
- Hygiene, safety, reliability, security
- Responsiveness, accessibility, courtesy, comfort, aesthetics of environment, competence, dependability, accuracy, completeness, state of the art, credibility, and effective communication

An organization renting a seminar room from a conference center would clearly have specific requirements (e.g., size and capacity of the room, number of attendants to be employed, the audio-visual systems, etc.). All such requirements are easily specified and measured. These characteristics are typically related to the adequacy of the physical side of services.

The patient visiting a dental clinic may not be able to specify his or her requirements in so many words, but an unreasonably long wait (as determined by the patient) will result in customer dissatisfaction with the dental service. If the dentist is serving a half dozen patients at once, dividing his time among them assembly-line style, the dental work may take a long time to complete. This intermittent waiting may result in the dissatisfaction of all six patients. Once again, these quality characteristics are easily measurable.

An airline that is perceived to cut corners on maintenance or frequently cancels flights, or that has a poor on-time record, will soon feel the result of customer dissatisfaction with its service. Characteristics such as these are easily tracked and analyzed. They generally relate to the effectiveness of service processes and even to the issue of using the wrong processes.

Finally, the more esoteric quality characteristics (e.g., responsiveness, courtesy, comfort, aesthetics, dependability, accuracy, completeness, credibility, and communication) are just as important as other characteristics in the eyes of the customer. For the most part they are not difficult to track or measure. These are the characteristics which relate to *how well you execute your service processes*.

INSPECTING SERVICES

Final inspection in a factory is (theoretically) able to identify a nonconforming product that can then be returned for rework until it conforms to specification. However, a service is not, in this respect, like a manufactured product. In most cases, once a service is performed, that is it. Either it was performed satisfactorily or it was not. If not, it is too late to do anything about it except make amends with a dissatisfied customer. The point is that if an organization waits until the service is performed to apply control or monitoring, it will be unable to affect the outcome.

Does this mean that quality monitoring and control strategies cannot be applied to services? Certainly not. But it can usually be done only by controlling the process(es) that delivers the service (i.e., before the fact). This is done by designing processes that are as foolproof as possible, communicating with and training employees thoroughly, and making certain that employees understand and comply with procedures and policies.

Quality characteristics that are important to customers should be tracked, monitored, and analyzed to identify processes or individuals in need of improvement. Service delivery is typically played out in the presence of the customer. Consequently, it is especially important that the people doing it get it right. Although it is not mentioned in ISO 9000, service delivery is an area where a well-crafted reward and recognition program (for teams and individuals) can be effective at improving performance.

Customer feedback strategies are used routinely in the service sector. Even though they typically evaluate performance after the fact, and therefore cannot preempt poor performance, they can be an effective means of identifying needed improvements that will

ISO INFO
Customers who receive services from an organization have an innate sense of the quality of services they receive. Without even consciously thinking about it, they make judgments about such factors as the appearance of the facility, waiting time, attitudes of employees, apparent competence of the organization, courtesy, comfort, and aesthetics of the environment.

ISO ISSUE

What Would You Do?

Markee Powell is the new Manager of Service Quality for the Grandview Hotel, a luxury facility that caters to international business travelers. She wants to implement processes to ensure that the needs of customers are properly anticipated and fully met. However, she is unsure of what those processes should be. What advice would you give Powell?

affect future customers. ISO 9000 requires that service providers develop and use procedures to tap the invaluable information and insights of customers.[5]

Actual inspection of services by internal auditors may be appropriate for some service organizations. For example, hotels routinely inspect rooms after they have been prepared for the next night. This is analogous to final inspection in a factory. Until the new guest has checked in, there are still opportunities to correct any deficiencies found. At the least it provides performance data for the individual(s) concerned. Similar inspections are possible for many types of services.

Most services are candidates for audits to ensure that (a) the proper procedures are being used, and (b) those procedures are being followed to the letter. In the final analysis, claims that it is difficult to evaluate quality performance in the service sector are just rationalization. There are no real obstacles or unreasonable costs involved in assuring quality that are peculiar to the service industry.

DOCUMENTATION REQUIREMENTS FOR SERVICE PROVIDERS

In general, ISO 9000's documentation requirements apply equally to service providers and manufacturers. Chapters 4, 5, and 6 of this book apply to both. The difficulty may be—depending on the nature of the services—that you have less experience in documenting processes and procedures than do manufacturing firms. Service providers that are regulated by government agencies (e.g., transportation, utilities, broadcasting) or those that receive compensation through governmental agencies (e.g., medical and welfare service providers) are no strangers to mandatory documentation. Even in these cases, however, the emphasis has been on fiscal and safety issues, not on the processes and procedures that determine the quality of services. Under ISO 9000, the emphasis is where it should be: on quality-related processes, procedures, and policies.

WHICH CLAUSES APPLY TO SERVICE PROVIDERS?

ISO 9000:2000 has made it much easier to determine what applies to the service organization and what does not. Under the earlier version of the standard, there were eleven standards, including three actual registration standards. Now there are just three: ISO

9000, 9001, and 9004. ISO 9000:2000 exists to explain the quality management system fundamentals and the vocabulary. ISO 9001:2000 is the only standard to which an organization may register. It is the one with the requirements. ISO 9004:2000 is a companion guideline document for ISO 9001 and contains helpful information, especially for organizations that want to improve beyond the minimum requirements. All of these documents apply, ISO 9001 for requirements and registration, and the other two for information.

Under ISO 9000's 1994 version, three requirements standards—ISO 9001, 9002, and 9003—were aimed at three different kinds of organizations. ISO 9001:2000 replaces all three, but still allows some tailoring for the organization. The tailoring, though, is only allowed within **clause 7, Product Realization,** and then only by excluding subclauses which are not relevant to the organization's operations. Clause 1.2 states:

> All requirements of this international standard are generic and are intended to be applicable to all organizations, regardless of type, size and product [*service*] provided.
>
> Where any requirement(s) of this International Standard cannot be applied due to the nature of an organization and its product, this can be considered for exclusion.
>
> Where exclusions are made, claims of conformity to this International Standard are not acceptable unless these exclusions are limited to requirements within clause 7, and such exclusions do not affect the organization's ability, or responsibility, to provide product [*service*] that fulfills customer and applicable regulatory requirements.

Consider the example of a franchised service-providing organization. Its business is refinishing automobiles. Under the franchise rules, the equipment, finishing materials and supplies, and processes are all provided by the franchiser. The franchiser also provides training for the organization's employees, and requires the shop facilities to be laid out and arranged according to its model. The franchisee controls what is left. If this organization were to seek ISO 9000 registration, would it be appropriate to exclude any clauses? Let's examine the situation clause by clause.

- All requirements of ISO 9001 clauses 4 (Quality Management System), 5 (Management Responsibility), 6 (Resource Management), and 8 (Measurement, Analysis and Improvement) must be satisfied. (*Note:* Even though one might make the case that the franchiser provides training, therefore rendering clause 6.2.2, Competence, Awareness, and Training, beyond the organization's control, and therefore inapplicable, ISO will not allow it. The organization must address training and ensure that the training is provided as necessary.) There can be no exclusions except within clause 7.
- Clause 7.1, Planning of Product (Service) Realization. Applicable. The organization must address all the subclauses and should state, in reference to establishing processes, that the processes employed are provided by the franchiser.
- Clause 7.2, Customer Related Processes. Applicable. Requirements are under control of the organization.
- Clause 7.3, Design and Development. Possible exclusion. Since the franchiser designs the processes resulting in the services, the organization can rightly claim that it has no control over the requirements of clause 7.3 or one or more of its subclauses.

- Clause 7.4, Purchasing. Applicable. The organization does its own purchasing, even though a significant part of it is through the franchiser.
- Clause 7.5, Production and Service Provision. Applicable. This is the clause under which the organization provides its services. All subclauses apply.
- Clause 7.6, Control of Monitoring and Measuring Devices. Applicable. The organization is responsible for control, repair, and calibration of its monitoring and measuring devices.

No other clauses are candidates for exclusion. This is the situation that most small service organizations will find. Not a lot of excluding will be done.

CASE STUDY

ISO 9000 in Action

Jake Butler had to admit that service quality was a new concept to him. Fortunately, he had a couple of days to research the issue before his next meeting with AMI's executive management team. The issue of service quality came up because, in addition to producing its own products, the company also provides assembly services for other manufacturers.

What do the ISO standards say about service quality? What characteristics are important to AMI's service-oriented customers? How can those characteristics be measured? These are all questions Butler is researching in preparation for his upcoming meeting.

At this point, Butler is prepared to tell AMI's executive managers that ISO 9000:2000 also applies to services. He plans to prepare a summary of service applications for distribution at the upcoming meeting. This summary will give AMI's managers a good idea of the types of measurable characteristics that might be important to service customers. In Butler's opinion, the most important characteristics will turn out to be "waiting time," "delivery time," and "process time."

In fact, he plans to recommend that AMI establish a focus group with representatives from ten of the company's service customers to determine what characteristics are most important to them. He also plans to recommend a biannual survey of all service customers to get a broad base of feedback. Finally, he will explain why *all* quality service processes and procedures must be documented.

SUMMARY

1. Service is work performed for someone else. It may be in the form of providing something, supplying, installing, maintaining, or repairing. Examples of service categories are as follows: hospitality, communications, healthcare, maintenance, utilities, trading, finances, professional, administrative, technical, purchasing, and scientific. Throughout ISO 9000, wherever the term "*product*" occurs, it can also mean "*service*." Twenty-five percent of the ISO 9000 registered firms in North America are service firms.

2. If a service is work done for others, then the concept of the internal customer means that services are performed in every business of every type every day. Eliminating waste from services and infusing the services' quality is a high priority in the age of global competition. This is the mission of ISO 9000, and it is as applicable to the service sector as to manufacturing.
3. Some services are closely identified with products, while others are not. For example, the service provided in a restaurant and the corresponding meal are closely linked. In such cases, the quality of the service is as important as the quality of the product (the meal).
4. Organizations should be just as concerned about the quality of their services as they are about the quality of their products. Quality when applied to services means fitness for use, meeting or exceeding customer requirements, and meeting or exceeding customer expectations.
5. When a service is delivered, it is usually too late to correct imperfections. Consequently, it is important to control the processes that deliver the services. This involves designing processes that are as foolproof as possible, communicating with and training employees thoroughly, and making certain that policies and procedures are complied with.

KEY TERMS AND CONCEPTS

Administration
Clause 7
Communications
Conforming product
Design
Development
Financial
Fitness for use
Healthcare services
Hospitality services
Internal customer
Maintenance
Processes
Product realization
Production
Professional
Purchasing
Requirements
Scientific
Service
Technical
Trading
Utilities

REVIEW QUESTIONS

1. Define the term *service* from an ISO 9000 perspective.
2. Explain the reason why service organizations would pursue ISO certification.
3. List ten examples of service organizations in your community.
4. How does the concept of the internal customer support the imperative that service organizations seek ISO 9000 certification?

5. What is the relative importance of services and products in a restaurant? Is it different in an automobile dealership?
6. What is the relationship of services to processes?
7. Define the term *quality* as it relates to services.
8. List ten quality characteristics of services.
9. In most cases, once a service is performed it is too late to improve it. Consequently, it is important to control the processes by which services are delivered. Explain how this can be done.
10. Explain how ISO 9000 is applied to service organizations.

APPLICATION ACTIVITIES

1. Go to a fast food restaurant and observe the employees working. Make a list of all the processes you observe. Select one process and describe it as completely as possible in a step-by-step manner. How can this process be improved?
2. If you are pursuing your ISO studies as part of a college course, document the registration process using an annotated flowchart. How can your college's registration process be improved?

ENDNOTES

1. ISO 9001:2000, clause 1.2.
2. ISO 9001:2000, clause 0.2.
3. ISO 9004-2:1991, p. 1.
4. ISO 9004-2:1991, clause 4.1.
5. ISO 9001:2000, clauses 5.2, 7.2.1, and 8.2.1.

CHAPTER THREE

Decoding ISO 9000:2000

MAJOR TOPICS

- Component Parts of ISO 9000 and Their Relationships
- Language of ISO 9000
- Legal Considerations and Requirements

COMPONENT PARTS OF ISO 9000 AND THEIR RELATIONSHIPS

The 2000 release of ISO 9000 is vastly simplified in terms of numbers of documents and the often confusing numbering system used in the 1994 version. Where the ISO 9000:1994 used eight "guideline" documents to clarify and expand on the three "requirements" documents, there are now just three documents total. This is illustrated in Figure 3-1. The 2000 version of ISO 9000 is comprised of the following documents:

- ***ISO 9000:2000 Quality Management Systems—Fundamentals and Vocabulary.*** This document is the result of merging the old ISO 8402:1994 and some of the content of ISO 9000-1:1994. This "standard" is intended to provide the fundamental background information and to establish the meaning of words and phrases used in ISO 9000.
- ***ISO 9001:2000 Quality Management Systems—Requirements.*** This is the requirements standard to which ISO 9000 users must conform. The old 1994 standards, ISO 9001, 9002, and 9003 have been merged into this single document. Whereas under the 1994 version, ISO 9000 users had to decide on the basis of their activities which of the three standards to use for certification, under the 2000 version, there is no longer a choice. All certifications will be to ISO 9001:2000. This raises the issue of applicability. For example, organizations that engage only in the manufacturing of product (that is, they have no design or development role) could formerly have been certified under ISO 9002. Now that ISO 9002 no longer exists, is ISO 9000 still applicable to such firms? The answer is yes. ISO 9001:2000 is applicable to any kind of organization, regardless of its activities. Provisions have been made to exclude requirements that are not relevant to the organization. In

1987	1994	Purpose
ISO 9000:1987 None None None	ISO 9000-1:1994 ISO 9000-2:1993 ISO 9000-3:1991 ISO 9000-4:1993	Guidelines (G/L) for selection and use G/L for application of the standards G/L for appl. of ISO 9001 to software producers Guide to "dependability" program management
ISO 9001:1987 ISO 9002:1987 ISO 9003:1987	ISO 9001:1994 ISO 9002:1994 ISO 9003:1994	Standard—model for "full-house" producers Standard—model for production Standard—model for final inspection and test only
ISO 9004:1987 None None None	ISO 9004-1:1994 ISO 9004-2:1991 ISO 9004-3:1993 ISO 9004-4:1993	G/L for quality system elements and management G/L for services G/L for processed materials G/L for quality improvement

2000	Purpose
ISO 9000:2000	Fundamentals and Vocabulary
ISO 9001:2000	Requirements
ISO 9004:2000	Guidelines for Performance Improvements

Figure 3–1
Documents of ISO 9000

this case, the manufacturer that does no design work will be certified to ISO 9001, but with design-related requirements excluded.

- ***ISO 9004:2000 Quality Management Systems—Guidelines for Performance Improvements.*** This document represents a major revision of the old ISO 9004-1:1994, and is intended to assist organizations in establishing and improving their quality management systems. Although called a "standard," as all such ISO documents are, this one does not contain requirements for the user organizations. Rather, used with ISO 9001, it provides information and insights to help the organization conform to the standard, and to continually improve performance—in other words, to go beyond the minimum requirements.

ISO has also updated its guidance document for auditors with **ISO 19011—Guidelines for Quality and Environmental Auditing.** This new document is the result of merging the old ISO 10011, Parts 1, 2, and 3, with ISO 14010, 14011, and 14012. ISO

10011 was the former guideline for quality management system auditing. The 14010, 14011, and 14012 documents served the same purpose for auditing environmental management systems of ISO 14000 registered organizations. ISO has the long-range goal of blending the quality and environmental management systems, and ISO 19011 is one step in that direction. Many organizations registered under ISO 9000 will also be certified to ISO 14000, and the use of common audits makes sense for both the registrars and the auditees.

ISO 9001:2000 and ISO 9004:2000 have been developed to be what ISO calls a "consistent pair." Their structure and sequence have been made nearly identical to make them easy to use together. Their tables of contents parallel each other closely. One typically finds a difference only where necessary for the concepts of requirements and guidelines. For example, in both documents clause 5 is titled *Management Responsibility*. ISO 9001 clause 5.1 is "Management commitment." ISO 9004's clause 5.1 is "General guidance," explaining in detail how the requirement for management commitment is satisfied. Although structurally close, the consistent pair occasionally shows some inconsistency. For example, in both ISO 9001:2000 and ISO 9004:2000, clause 6 (Resource management) is generally consistent through clause 6.4. ISO 9004 adds clauses 6.5 through 6.8, which are not represented in ISO 9001. The reason seems to be that while there are no ISO 9001 requirements associated with the subjects of those clauses—Information, Suppliers and partnerships, Natural resources, and Financial resources—ISO 9004 is saying, and rightly so, that these are issues to be dealt with in any well run organization. So new clause numbers were added to ISO 9004 to accommodate them. It seems to the authors that with a bit more work, the consistency of clause numbers could actually have been accomplished while preserving ISO 9004's function of going beyond the requirements.

ISO 9000 users will find ISO 9001 and 9004 to be their core working documents. ISO 9000 and ISO 19011 will be used for reference. See Figure 3-2.

ISO states the purposes of the ISO 9001 and ISO 9004 documents as follows:[1]

> ISO 9001 specifies requirements for a quality management system that can be used for internal application by organizations, or for certification, or for contractual purposes. It focuses on the effectiveness of the quality management system in meeting customer requirements.
>
> ISO 9004 gives guidance on a wider range of objectives of a quality management system than does ISO 9001, particularly for the continual improvement of an organization's overall performance and efficiency, as well as its effectiveness. ISO 9004 is recommended as a guide for organizations whose top management wishes to move beyond the requirements of ISO 9001, in pursuit of continual improvement of performance. However, it is not intended for certification or for contractual purposes.

From this it is clear that while the structure and sequence of the two documents are the same, their scopes are not. ISO 9001 takes the user to compliance with the standard, but ISO 9004 can take the organization far beyond just compliance. This may prompt the reader to ask why an organization should want to go beyond compliance with ISO 9001. The answer comes from the philosophy of total quality management. In a word, it is "competitiveness." The more an organization's processes (and thereby its products) are improved, the more efficient and competitive the organization becomes. In today's global

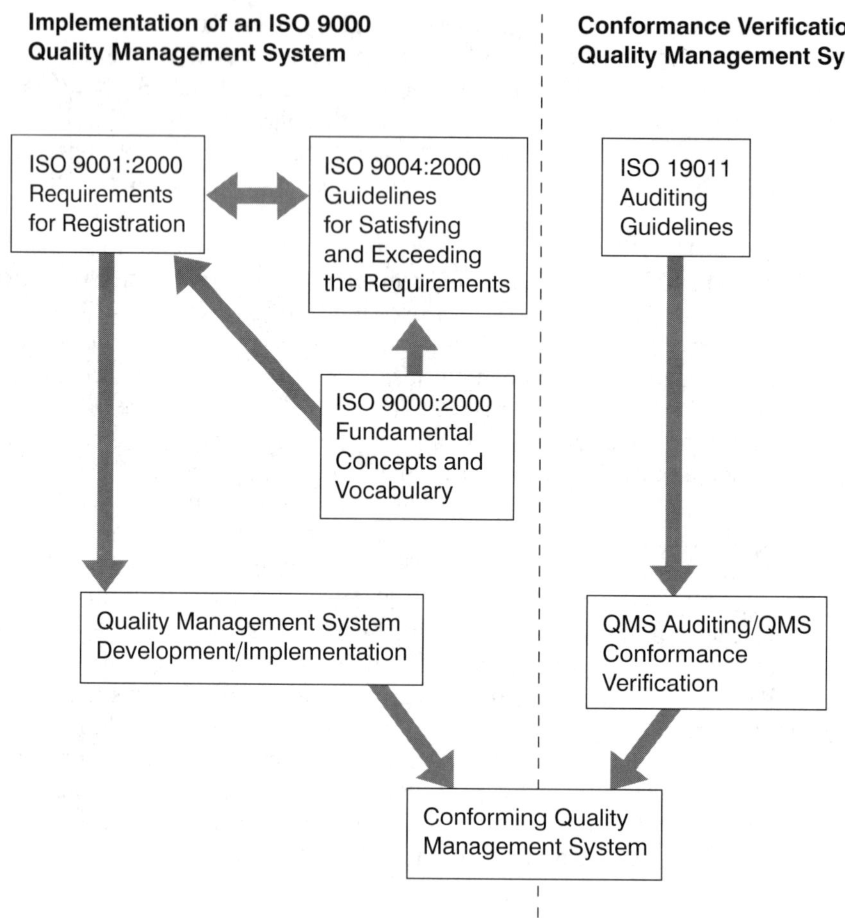

Figure 3–2
Relationships of ISO 9000 Standards from the Users Viewpoint

economy, being competitive is necessary for survival. Staying competitive requires continual improvement. ISO is to be commended both for simplifying the ISO 9000 standard, and for imbedding within it the philosophy that can move organizations toward total quality management. A lack of emphasis on continual improvement was a serious weakness of the 1994 standards.

LANGUAGE OF ISO 9000

The ISO 9000 standards use special words and phrases. The meanings of some are obvious, while the meanings of others may be obscure or may, in the language of ISO 9000, mean something different from what we might first think. In addition, there have been some significant changes in the language from the 1994 to the 2000 versions. To promote better understanding throughout, this section explains the meaning of key terms of ISO 9000:2000.

Accepted Order (ISO 9001, clause 7.2.2) An accepted order is a contract to provide the product as ordered. If the accepted order contains specifications, delivery conditions, and so forth, they are binding on the producer of the product.

Audit Criteria The set of policies, procedures, practices, and requirements related to ISO 9001 normally represent the audit criteria. This means that the auditors may examine the organization's conformance to the standard's requirements, and to the organization's own quality policy and relevant procedures and practices. It is customary for the registrar's lead auditor to establish the audit criteria with the organization prior to an audit.[2]

Audit Scope (ISO 9001, clause 8.2.2) ISO defines this term as the "extent and boundaries of an audit." A registration audit should be expected to cover the full extent and broadest boundaries of ISO 9000 influence in the organization. Surveillance audits are ordinarily more narrowly focused.

Audit Evidence This includes records, statements, and other information collected by the auditor that is relevant to the audit criteria.[3] A rule of auditing is that evidence should be verifiable through cross-checking. For example, one employee may make a statement that indicates top management is giving mere lip service to ISO 9000. However, unless other statements or evidence are forthcoming to corroborate that employee's claim, the statement will not be considered objective evidence[4] and will not be considered in the audit findings.

Audit Findings From the audit evidence the auditor(s) develop a set of findings. Audit findings can be positive or negative. They can point out the areas in which the organization is doing well, and also those in which it needs to put more attention or even make drastic changes.[5]

Audit Conclusions Audit conclusions are the final outcome of the audit. After weighing the audit findings, the auditor(s) will develop audit conclusions,[6] including those relative to the awarding or retention of registration.

Auditee This is the organization being audited.[7]

Certification Body (ISO 9001, clause 0.1) Certification bodies are commonly referred to as registrars. They are the entities that determine an organization's conformance to the standard, and issue certification.

Comply/Compliance (various clauses) Originally used in reference to the standard, or the organization's plans and procedures, as in "XYZ Company operates in compliance with its documented procedures and the requirements of ISO 9000." However, with the advent of ISO 14000, "comply" and "compliance" are increasingly being reserved for regulatory and legal requirements. The words "conform" and "conformance" are applied to the requirements of the standard and to the internal procedures and plans. Thus we might say that an organization *complies* with relevant regulatory and legal requirements, and *conforms* to ISO 9000 requirements.

Conform/Conformance If an organization satisfies ISO 9000 requirements, we say that it conforms, or is in conformance. *See* Comply/compliance.

Consistent Pair ISO refers to ISO 9001:2000 and ISO 9004:2000 as the consistent pair. The meaning here is that the requirements standard, ISO 9001, and the guidance (advice) standard, ISO 9004, have been designed to be used together. Structure and sequence are essentially identical.

Continual Improvement ISO's definition is "recurring activity to increase the ability to fulfill requirements."[8] Continual improvement is a cornerstone of Total Quality Management, and now becomes an explicit requirement of ISO 9000. (See ISO 9001 clauses 0.2, 4.1, 5.1, 5.3, 5.6, 8.1, 8.4, 8.5.1, and Figure 1.) The philosophical basis for continual improvement is that all human efforts, whether processes or products (including services), are imperfect and therefore can be improved. When improvements are made continually, even small incremental improvements, the processes and products become better, waste (representing cost) is decreased, the organization improves its competitive position, and customers benefit.

Corrective Action ISO offers the following definition for corrective action: "action taken to eliminate a detected nonconformity."[9] It is important to understand that the nonconformity can be associated with a process or its inputs as well as with its output product. In TQM, corrective action is meant to restore a process or product to the state that existed prior to the nonconformity, a condition that does not represent improvement. ISO, however, sees corrective action as part of the continual improvement process, and in ISO 9001, clause 8.5.2 requires corrective actions to *prevent recurrence* of the nonconformity. This would be seen as *preventive action* under TQM, and does represent improvement. (*See* Preventive action.)

Customer The *customer* may be the purchaser, user, or beneficiary of the *organization's* product or service. ISO defines *customer* as the "organization or person that receives a product." ISO 9000:2000 examples are consumer, client, end user, retailer, beneficiary and purchaser.[10] Client and beneficiary are usually associated with a service, such as legal or architectural services. A retailer is a customer of the organization from whom it buys its stock for resale. This view of the customer is accurate, but it does not go far enough. An objective of ISO 9000 is to assure the *fitness for purpose* of an organization's goods or services. In the final analysis, there can be only one agent who validates fitness for purpose. That agent cannot be ISO, or the registrar, or the auditors—it can be only the customer. So we would expand the definition to say that the customer is the purchaser, user, or beneficiary of your product, and the final voice for determining the product's quality, or fitness for purpose.

Note that in addition to these external customers there are also internal customers (i.e., within the organization). An internal customer is one who accepts work from a preceding process, and contributes more value adding work to the product. While the internal customer does not "purchase" the semifinished product or service from the pre-

ermine the quality or fitness for purpose resulting t must not be overlooked. At this point one might ch emphasis on the concept of the customer. It is ıs must be at the heart of any quality initiative, ıagement. But a focus *on external customers only* rovement. Just as organizations can improve their ıal customer's needs, they can also improve their s—by understanding the needs of internal cus- es for ISO 9000 registration, be sure to take advan- ell as external customers.

isfaction is achieved by fulfilling or going beyond tations. The common wisdom was once that sat- ıers. But we now understand that even a satisfied s product for any number of reasons, including s, price advantage, or simply for variety. To retain ond *satisfied,* all the way to *delighted*. A customer uch more likely to return to you for a subsequent ied. Customer satisfaction must be a key point of 00 organization.

ned as a set of processes that transforms require- ons of a product, process, or system.[12] For exam- red wristwatch. Classic wristwatches were spring n watch makers developed an automatic winding wind the spring. But after a couple of days off the te winding altogether, the battery powered watch ould run for several years, but then the battery had makers hear that they did not like to go through ry changes. At Seiko and Citizen, the consumer *nts* (needs), in effect, statements that "we want ttery replacement." Using this requirement, the es developed the plans for new electric powered eplacement. Light from any source falling on the an energy cell that powers the watch movement. the watch running for weeks or months. Seiko

O INFO

Will focusing on customers bring profit, or rs? The answer is the same, time and time ing plans and budgets around serving cus-

Brian L. Joiner

Kinetic watches transform motion into electrical energy. Neither represents perpetual motion, but both eliminate the need for battery replacement.

Development *See* Design and development. Development is often seen as a synonym for design, or as a phase of the design process.

Document *Webster's Collegiate Dictionary* provides a working definition of a document as "an original or official paper relied on as the basis, proof, or support of something." Documents form the basis for ISO 9000 registration. ISO 9000 requires that documented procedures be in place and that records be kept to verify that procedures are followed. Audits will focus on documentation, which may be maintained in any media. Documents include procedures, process flow diagrams, plans, drawings, contracts with specifications, all telling *what you are going to do,* and written records, logs, organization charts, internal audit reports, and the like, all telling *what you actually did.* For example, if in response to ISO 9001, clause 7.3.4, you document in your operating procedures that you will conduct formal design reviews at certain stages of the design process, that is *what you plan to do.* You must also document the reviews as written proof that the design reviews actually occurred (i.e., document *what you actually did*). In summary, under ISO 9000, documentation must cover:

- The organization's policies
- What you plan to do (plans)
- How and when you plan to do it (procedures)
- Proof that you complied with your own policies, plans, and procedures

Documented Procedure Any procedure that has been formally written on paper, or preserved in electronic or other media. ISO 9000 requires certain procedures to be documented. For others there is no explicit requirement to document. Where the term "documented procedure" appears in ISO 9001, it means that the procedure is established, documented, implemented, and maintained (used).[13]

Effectiveness ISO defines this as the "extent to which planned activities are realized and planned results achieved."[15] In other words, *effectiveness* is a measure of how nearly

ISO INFO

The essence of conforming to ISO 9000 is:[14]

- *Say what you do* — *(Document it)*
- *Do what you say* — *(Follow your plans and procedures)*
- *Record what you did* — *(Document the facts)*
- *Check on the results* — *(Analyze and record, i.e., document)*
- *Act on the difference* — *(Take corrective action, document it)*

the organization's activities match the aspirations of its plans for those activities. In cases where the results can be measured, the organization might achieve a defect rate of three parts-per-thousand. If that was their stated objective, then we could say that the activities that together produced the three parts-per-thousand rate was 100 percent effective. However, if the rate were six parts-per-thousand, the effectiveness rate would be 50 percent.

External Audit An audit performed by an entity other than the organization. In ISO 9000 this usually means an audit performed by the registrar, although audits performed by or for customers are not unusual.

Interested Party A person or group having an interest in the performance or success of the organization.[16] The interested party may be an employee, a stockholder, a neighbor, someone affected by the organization's operations, a customer, a supplier, or another.

Internal Audit An audit performed by members of the organization itself.

Legal Requirements These are requirements imposed by statute or governmental regulations. We know of no statutory or regulatory requirements governing the level of quality per se, but there are many such requirements concerning product safety. For example, automobile and truck tires must perform to safety specifications imposed upon the industry by the United States and other countries. In 2000, Bridgestone/Firestone found itself embroiled in lawsuits and congressional hearings concerning the safety of several models of its Firestone tires that allegedly did not meet the safety requirements. Many other industries are subject to similar laws and regulations.

Noncompliance In the case of ISO 9000, noncompliance has been used as a synonym for nonconformance, but is little used now except as regards legal and regulatory requirements. (*See* Comply/compliance.)

Nonconformity Failure to fulfill a requirement. In ISO 9000 it is typically used to indicate that a requirement of the standard or of the quality policy has not been met. (*See* Conform/conformance.)

Organization A group of people with an arrangement of responsibilities, authorities, and relationships, having the facilities and infrastructure necessary to carry out its intended functions. This may be a company, corporation, firm, institution, or other entity or parts thereof. It may be of any size, and may be public or private. In terms of ISO 9000, the *organization* is the registered entity or the entity seeking registration. With ISO 9000:2000, ISO shows the organization as part of the supply chain as follows:

Supplier → Organization → Customer

Note: *Organization* replaces the term *supplier,* and *supplier* replaces the term *subcontractor* used in earlier versions of ISO 9000.

Permissible Exclusions The 1994 version of ISO 9000 offered three QMS models under which organizations could register, depending upon their functions—ISO 9001, 9002, and 9003. For example, a firm that both designed and produced its products could register under ISO 9001:1994 since that standard had requirements for design and production. A firm that manufactured to designs produced elsewhere could register under ISO 9002:1994, since that standard had no design requirements. The 2000 version of ISO 9000, however, has only one QMS model—ISO 9001:2000. It contains requirements encompassing all possible relevant functions, as did the earlier ISO 9001. Under the 2000 version, organizations must register to ISO 9001. This raises an issue from the example above. What if an organization has no design capability, no staff for design, and no need or intention to develop one? To address that issue, ISO now allows the exclusion of certain requirements if "such exclusions do not affect the organization's ability or responsibility to provide product that fulfills customer and applicable regulatory requirements."[17] Exclusions may be made only from clause 7, Product realization. The intent here appears to be that if an organization has a function that can affect product quality, such as a design engineering department, that function must be included in the ISO 9000 registration. That was not the case before the 2000 version. That organization might have registered its manufacturing functions only, leaving the design function outside of ISO 9000.

Preventive Action ISO's definition of preventive action is "action taken to eliminate the cause of a potential nonconformity."[18] This fits with the TQM definition, but is narrower in the sense that it applies only to those situations where the particular nonconformity has never existed but is considered to be possible. ISO uses *corrective action* to cover the prevention of future nonconformities in cases where the same nonconformity has occurred at least once. (*See* Corrective action.)

Process *Process* is one of the most frequently used terms in the language of ISO 9000. Readers with backgrounds in Total Quality Management will have no difficulty understanding the concept. However, for others a discussion of the meaning of *process* may be helpful. Readers will find it helpful to remember that *all work is accomplished through processes*. Every task has its process. For example, let's say that you go into a restaurant and order a three-minute boiled egg. Preparing the egg is a process involving the steps shown in Figure 3-3. The three-minute egg requires ten distinct steps by the cook. These steps are as follows:

1. The cook goes to the refrigerator and takes out the egg.
2. The cook heats the water.
3. The cook waits for the water to boil.
4. The cook puts the egg in the boiling water.
5. The cook starts a three-minute timer.
6. The cook monitors the timer until the three minutes are up.
7. The cook removes the egg from the water.
8. The cook removes the shell from the egg.
9. The cook puts the egg in an egg cup.
10. The waiter serves the egg to the customer.

Figure 3–3
3-Minute Egg Process Flow Diagram

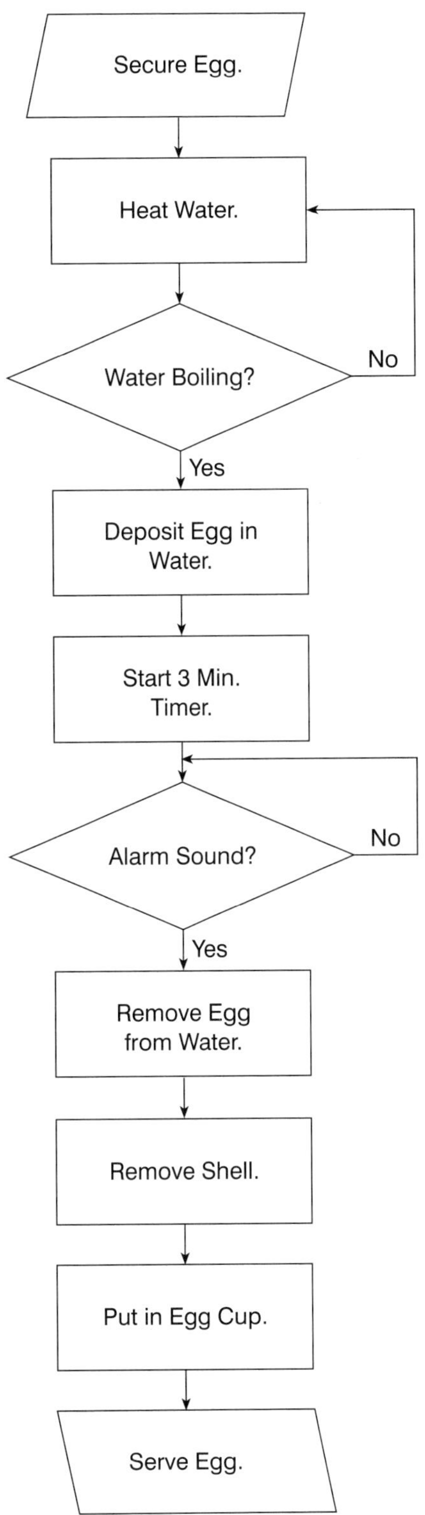

In summary, a process is defined as an activity or set of interrelated activities which transforms inputs into outputs. A *process* can, for example, transform raw materials into a finished product. Or in the case of a service, it can transform legal knowledge into a will. An important point to remember is that everything we do involves one or more processes.

Process Approach The process approach comes from TQM; it requires the organization to view its operations as a series of interrelated and interacting processes. The organization has to identify and understand its processes and consciously manage them and their interactions, paying particular attention to customer satisfaction. The process approach is a key element of continual improvement, since heightened attention at the process level will reveal continual opportunities for improvement.

Boiling an egg is an example of a simple process, as we saw in Figure 3-3. Virtually everything we do involves processes like this—whether we have ever thought of them in this way or not—and these processes can be flowcharted. When getting ready for bed at night, we go through a process that could be flowcharted. We don't consciously think of getting ready for bed as a process because it has become automatic. The same can be said for the way many people approach their jobs. ISO 9000:2000 now requires, as does Total Quality Management, that work be thought of in terms of processes so that the processes can be properly developed, implemented, and continually improved. (*See* ISO 9001, clause 0.2.)

This means that organizations seeking ISO 9000 registration must:[19]

- Recognize that their work is accomplished through many different interrelated, interacting *processes.*
- *Identify* the processes needed for the QMS (including processes for management activities, provision of resources and information, product realization, and measurement) and their application throughout the organization.
- Determine the sequence and interaction of these processes.
- Determine criteria and methods to ensure effective operation and control of these procedures.
- Ensure availability of needed resources and information for operating and monitoring these processes.

ISO INFO

Earlier versions of ISO 9000 were deficient in their approach to "processes directly affecting quality." There is in fact no process in the enterprise which does not, in the final analysis, affect quality. With the 2000 release of ISO 9000, this basic tenet of Total Quality Management has finally been embraced. See ISO 9001, clause 4.1a.

Goetsch and Davis

- Monitor, measure, and analyze these processes.
- Implement actions necessary for achievement of planned results and continual improvement of these processes.

Product The result of a process or set of interrelated processes. It may represent the final output of the organization (i.e., sold to external customers), or an intermediate output to be further acted upon by employees using subsequent internal processes (internal customers).

Product Realization This term is used to mean bringing the organization's product or service into reality. It is achieved through the various processes necessary to design, develop, produce, and test the product or service, working in conjunction with associated supporting processes throughout the organization. Previous versions of ISO 9000 have used the term *production system,* but it was apparently considered too narrow now that the standard is intended to encompass the entire organization.

Whether it results in tangible products such as a new car or a service such as a telephone connection, a linkage of many individual processes and people, technology, and procedures is the core of product realization. Earlier the process of preparing a boiled egg was described. In the larger context of *product realization* we would have to consider the other processes that come into play in the preparation of the three-minute egg. For example, when customers enter the restaurant they are seated and provided with a menu. This part of the process we call *taking the order* (Figure 3-4). After the customer is given time to consider the breakfast menu, the waiter secures her order, writes it down, and places it in the queue with other orders. The process of taking the order is *linked* to the chef's process for preparing the customer's order of one three-minute egg (Figure 3-5). The linkage occurs when the chef pulls the order from the order queue. He reads the order, sees that a three-minute egg is required, and goes through his process for preparing the egg. When the egg is ready, the chef places it on the ready-to-serve counter.

At this point there is a linkage to another process that we will call *closing the breakfast transaction* (Figure 3-6). The linkage is made when the waiter takes the egg from the ready-to-serve counter, and delivers it to the customer. At this point he will ask the customer if she would like something else. If the customer wants nothing else, he will prepare the bill

ISO INFO

Note that there is nothing in ISO 9001 requiring statistical process control (SPC). However, the use of SPC is clearly encouraged. ISO 9001, clause 8.1 and ISO 9004, clauses 8.1.2j and 8.4 mention "appropriate statistical techniques" for consideration for measurement, analysis, and improvement, and for supporting fact-based decisions. SPC may be a valuable process control technique for your organization, but it is up to you to decide.

Figure 3–4
Process Flow Diagram for Taking the Order

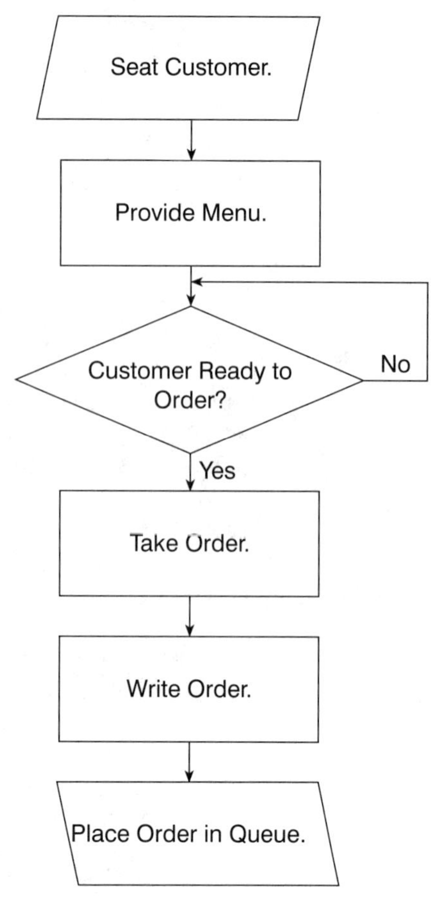

and present it to the customer. The customer will pay the bill—itself a process—closing out the transaction.

This rather simple illustration will hold for any linkage of individual processes that, together, result in the realization of a service or a product. The *linked processes,* along with the necessary *technology* (from pots and pans to computer controlled hydraulic presses), *people* (waiters and cooks to astronauts and mission controllers), and *procedures* (cookbooks to detailed instructions for assembling and testing jet engines) are *systems* of production. It is the *system of production,* necessary for product realization, that determines the quality of product or service, not just the people who work within the system. ISO 9000 recognizes this, and places great emphasis on documenting processes (for real understanding); on written, up-to-date work instructions (procedures); on control of processes and equipment; and on training of people.

Production System *See* Product realization.

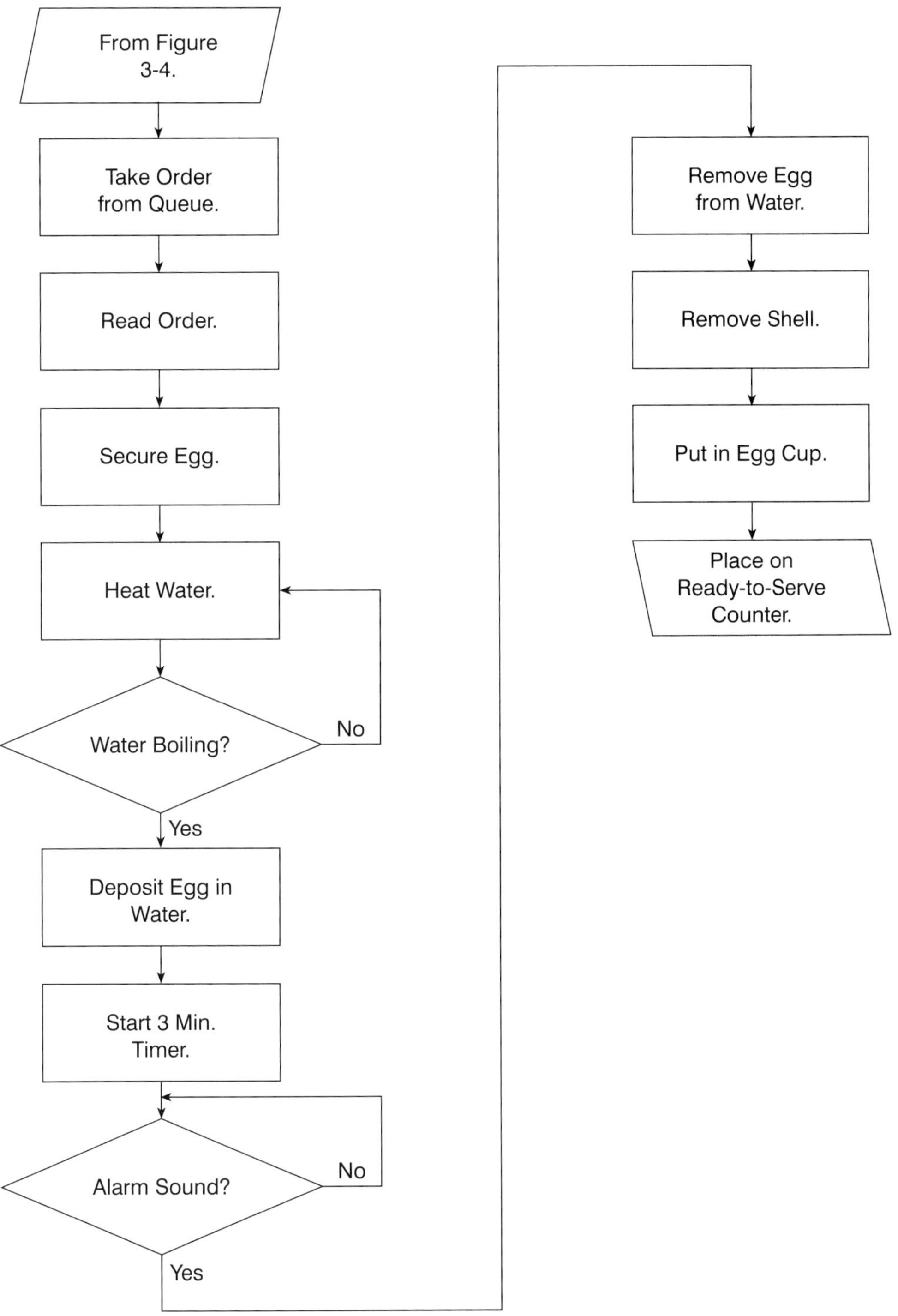

Figure 3–5
Preparing the Order

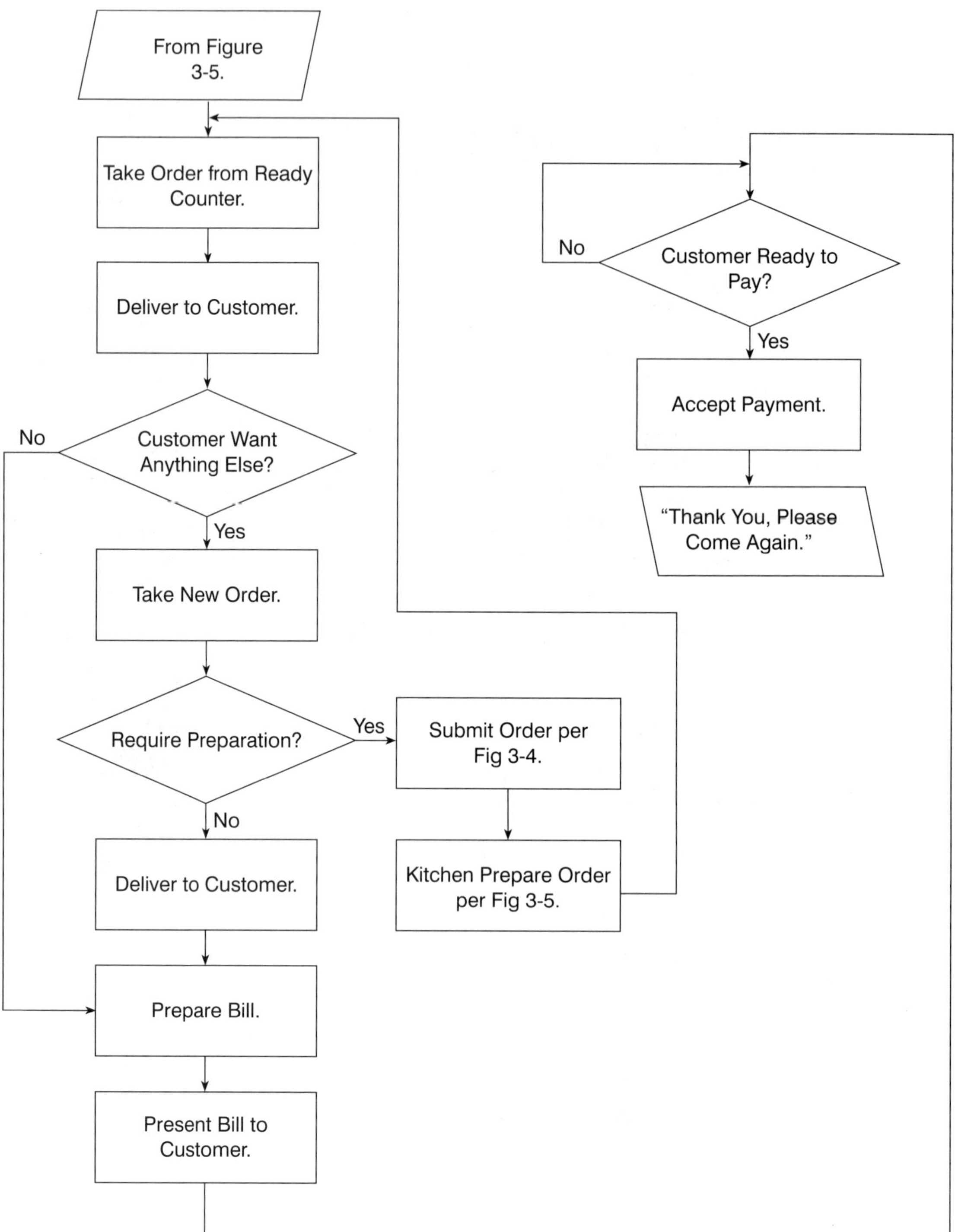

Figure 3–6
Closing the Breakfast Transaction

> **ISO INFO**
>
> *Dr. W. Edwards Deming, generally considered to be the father of the Total Quality movement, pointed out that production systems are controlled by management. Therefore, at least 85 percent of all organizational failures (including poor quality) are management failures. This is a good thought to keep in mind as you work toward ISO 9000 registration.*

Quality Manual ISO 9000 requires the organization to establish and continually use (maintain) a quality manual that includes:

- The scope of the organization's quality management system, including responses to ISO 9001 requirements and details and justification for any clause 7 exclusions.
- The documented procedures required for the QMS. (This may be by reference.)
- A description of the QMS processes and their interaction.

The quality manual should also include (or reference) the Quality Policy and Quality Objectives, and should provide cross-reference between the contents of the quality manual and external, referenced items.

Quality Objective A goal aimed for, related to quality. Quality objectives arise from the quality policy. They include broad strategic objectives that define "how the quality policy will be achieved" and narrower tactical objectives that define specifically "what must be done" to fulfill the strategy.

Quality Planning We can define this as a part of management related to quality that sets quality objectives, specifies and establishes the necessary operational processes, and provides the resources to fulfill the quality objectives. In other words, quality planning not only sets down what management wants to accomplish in terms of quality products or services, but also puts in place the processes and resources (people, technology, money, training and supporting infrastructure) to enable the accomplishment of the objectives.

Quality Policy Management's statement of the organization's overall intentions and direction with regard to quality. ISO 9000 requires top management to develop a documented quality policy and communicate it to all employees.

Regulatory Requirement A requirement imposed by a law or by an official agency having jurisdiction. In terms of quality, these may include product safety and performance standards.

Requirement This can be a need or an expectation. It may be stated, as an ISO 9000 requirement or a contract requirement, or it may be implied through common practice. It may also be obligatory, as a legal or regulatory requirement would be. Requirements may come from many various sources, including customers, governmental agencies, and other interested parties.

Standard *See* Chapter 1.

Supplier If the reader is familiar with the language of ISO 9000:1994, pay particular attention to this definition since it represents a significant change. In the standard's 1994 version the *supplier* was the entity seeking or holding registration. The requirements of the standard were phrased as, "The *supplier* shall establish and maintain . . ." With the 2000 release of the standard this odd perspective has been corrected. We now see the word *organization* where *supplier* had been, for example, "The organization shall establish and maintain . . ." In ISO 9000:2000 *supplier* is what the earlier version called a *subcontractor*. That is, a *supplier* is an organization or an individual that provides materials, goods, or products to the organization. This is the usual way of looking at a supplier. In addition, ISO correctly notes that suppliers may be external to our organization or, on a different level, may be internal—as the case of an internal customer receiving semi-completed work from and internal supplier (co-worker).

Top Management ISO defines top management as the person or group of people who direct and control an organization from the highest level within the organization.[20] This means the senior manager, whether the title is President, CEO, or something else, and the executive staff of vice presidents, directors, or similar titles. *Top management* is used repeatedly in the standard, making it clear that the action required or suggested must be taken by managers at the top of the organization. This recognizes that in ISO 9000, as in TQM, initiatives that change the way things are done in an organization can be done only from the top down.

LEGAL CONSIDERATIONS AND REQUIREMENTS

Unlike ISO 14000, which exists in a climate of environmental laws and regulations set down by the governments of the communities in which organizations operate, there are no corresponding laws and regulations dictating quality levels for products or services. However, ISO 9000 users must be aware that there is an ever-growing collection of governmental regulations concerning safety, many of which may relate directly to an organization's processes and products or services. These regulations are concerned with two general categories of safety: safety of the worker within the organization, and safety of the user (consumer).

The Occupational Safety and Health Administration (OSHA) in the United States is the regulating agency for on-the-job safety. To take a well-known example, OSHA's regulations require private and public employers to provide alarms on workplace vehicles to alert people to the fact that the vehicle is backing up, and that they may be in harm's way. Another requires employees to be notified of any toxic substances they use or may be exposed to in their work processes. OSHA requirements cover a vast array of regulations over the work environment, such as ventilation, lighting, and so on. This is not something restricted to the United States. Many other countries have their equivalent of OSHA.

User/consumer safety requirements in the United States and many other nations are generally statutory. These may cover virtually any product or service. We have already used the automobile/truck tire example. Many examples are found in the transportation industry, which is subject to a complex web of legal requirements concerning the equipment (airplanes, trucks, ships, trains) and the people who operate it (qualification, work schedule, physical condition). Few industries escape similar requirements. Even toy makers are subject to these laws. For example, they must design and build toys that are not going to harm children—even if the toy is misused. There are endless examples of laws designed to protect the users of products and services.

These workplace safety and user/consumer safety requirements may not be directly "quality" related, but this does not mean they are irrelevant as far as ISO 9000 is concerned. If any process used by the organization in designing or manufacturing the product or executing the service is subject to one or more of these laws or regulations, then it is a procedural issue and should be treated as such under ISO 9000. In other words, the action required by the law or regulation should be built into the applicable operational procedures and work instructions. At that point it becomes an ISO 9000 issue. Once the required action becomes a part of an ISO 9000 related procedure, consistent adherence must be verified by the auditors.

As a final point on these legal and regulatory requirements, it is the responsibility of the organization to be aware of those that apply to its operations and products and services. An organization should not expect the registrar to advise it of any such requirements. To do so, the registrar would be taking on a responsibility that opens the risk of culpability—something registrars are not willing to do, and rightly so.

CASE STUDY

ISO 9000 in Action

During his presentation to AMI's executive management team it became obvious to Jake Butler that the company's CEO, Arthur Polk, wanted to take a limited approach to ISO 9000 registration. He was still thinking in terms of the old 1994 version of the standard that allowed organizations to select ISO 9001, 9002, or 9003. He made it clear to Butler and the other participants that he wanted AMI to take the narrowest possible approach to certification. In his words, "We don't have time to get the entire company tied up with this thing. I don't want us to involve any department that doesn't have to be involved." What worried Butler even more was Polk's constant references to the "marketing benefits" of ISO 9000 certification.

Undeterred but cautious, Butler began to make his case that the ISO 9000 certification process and the standard had changed. "Unlike in the old days of ISO 9000 registration, we don't have the option of limiting our efforts to just manufacturing or just inspection and test," countered Butler. "We must involve every department that affects the quality of our performance as an organization." Butler had decided to overlook the CEO's constant references to the marketing benefits of ISO 9000 certification. He knew that the benefits Arthur Polk found so inviting were real, and that they would come with certification. Therefore, he decided to argue for the most comprehensive approach to seeking registration, knowing that the benefits of registration would take care of themselves.

After Butler finished his initial presentation, the executive team seemed to form itself into two factions; one that favored a comprehensive approach to certification and one that favored a limited approach. Arthur Polk remained the leader of the limited faction. Fortunately, Polk was not the kind of CEO who forced his views on the company's other executives. Consequently, a spirited discussion took place with each faction pressing its argument with equal conviction.

After more than an hour of discussing the matter, Polk's faction seemed to be gaining an edge when Butler, who had remained silent during most of the debate, raised his hand to speak. "Maybe we are viewing the entire issue from the wrong perspective," said Butler. "We have been thinking only of securing ISO 9000 certification. What if we step back and change our perspective to one of becoming a Total Quality company. If we do this, ISO 9000 registration becomes not an end unto itself, but a means to an even better end. Being ISO 9000 certified is important, but it will make us just one more of thousands of companies worldwide that have the same certification. In my view this just keeps us in the ballgame. We want to do more than just play the game, don't we? We want to win it, or at least that is what I think. If we do want to win on a global scale, then continual improvement of every process in every department of this company will be necessary. ISO 9000 registration, if approached in the broadest possible way, can get our company started down the path to world-class performance."

When he had stated his opinion, Butler sat down and thought, "Well, that probably ends my career with AMI." But to his surprise, Arthur Polk said, "That's a good point. Instead of using ISO 9000 registration to market harder, maybe we should use it to perform better. Does anybody have serious objections to taking a comprehensive approach to ISO 9000 registration?" When no objections were raised, Polk instructed to Butler to "get started."

SUMMARY

1. The number of documents in the "ISO 9000 family" has been reduced from eleven in the 1994 version to just three in ISO 9000:2000. These three are as follows: ISO 9000:2000, Fundamentals and Vocabulary; ISO 9001:2000, Requirements; and ISO 9004:2000, Guidelines for Performance Improvements.
2. The language of ISO 9000 has changed somewhat from the 1994 version of the standard. There are many special terms in the language of ISO 9000 that must be understood by users of the standard. However, the most important changes in the language are in the use of the following terms: *compliance* and *supplier. Compliance* is increasingly used to indicate adherence to legal and regulatory requirements. The term *conformance* is usually used when speaking of adherence to ISO standards. In the 2000 version of the standard the term *supplier* has been replaced by *organization* when speaking of the entity seeking ISO 9000 registration. The term *supplier,* when used in an ISO 9000:2000 context means *subcontractor* (an entity that provides materials, goods, or products to the organization).
3. Although there are no legal or regulatory requirements dictating the level or quality expected of products and services, there is an ever-growing body of governmental regulations concerning employee safety and product (end user) safety. These regulations can have a direct bearing on product and service quality. When governmental regulations relating to safety or any other aspect of product or service quality affect a process, it can become a procedural issue and must be treated as such when pursuing ISO 9000 certification.

KEY TERMS AND CONCEPTS

Accepted order
Audit conclusions
Audit criteria
Audit evidence
Audit findings
Audit scope
Auditee
Certification body
Comply/compliance
Conform/conformance
Consistent pair
Continual improvement
Corrective action
Customer
Customer satisfaction
Design and development
Development
Document
Documented procedure
Effectiveness
External audit
Interested party
Internal audit
ISO 9000:2000 Quality Management Systems—Fundamentals and Vocabulary
ISO 9001:2000 Quality Management Systems—Requirements
ISO 9004:2000 Quality Management Systems—Guidelines for Performance Improvements
ISO 19011—Guidelines for Quality and Environmental Auditing
Legal requirements
Noncompliance
Nonconformity
Organization
Permissible exclusions
Preventive action
Process
Process approach
Product
Product realization
Production system
Quality manual
Quality objective
Quality planning
Quality policy
Regulatory requirement
Requirement
Standard
Supplier
Top management

REVIEW QUESTIONS

1. List and explain the component parts of ISO 9000:2000 and their relationships to each other.
2. Explain the difference between the terms *compliance* and *conformance.*
3. What is a "corrective action" in the language of ISO 9000?
4. What does ISO mean by the term *consistent pair?*
5. Define the term *customer* as it is used in the language of ISO 9000.

6. Explain the "supplier–organization–customer" chain in terms of the language of ISO 9000.
7. What is a "product" in the language of ISO 9000?
8. Explain the "organization/supplier" and "supplier/subcontractor" changes in the language of ISO 9000.
9. Explain how governmental regulations can become procedural issues during the ISO 9000 registration process.

APPLICATION ACTIVITIES

1. Assume that yours is a build-to-print firm that does no design. Identify the criteria in the ISO 9001:2000 standard that could be excluded from the registration process.
2. Go to the web site of the Occupational Safety and Health Administration (OSHA). Review several of OSHA's safety regulations with a view to determining how they might become procedural issues in the ISO 9000 certification process. The web address is: *http://www.osha-slc.gov*

ENDNOTES

1. ISO 9004:2000, clause 0.3.
2. ISO 9000:2000, clause 3.9.3.
3. Ibid, clause 3.9.5.
4. Ibid, clause 3.8.1.
5. Ibid, clause 3.9.5.
6. Ibid, clause 3.9.6.
7. Ibid, clause 3.9.8.
8. Ibid, clause 3.2.13.
9. Ibid, clause 3.6.5.
10. Ibid, clause 3.3.5.
11. Brian L. Joiner, *Fourth Generation Management: The New Business Consciousness* (New York: McGraw-Hill, 1994) p. 90.
12. ISO 9000:2000, clause 3.4.4.
13. ISO 9001:2000, clause 4.2.1, note 1.
14. Robert M. Cumbers, Robert Cumbers Associates, Meath, Ireland, 1996.
15. ISO 9000:2000, clause 3.2.14.
16. Ibid, clause 3.3.7.
17. ISO 9001:2000, clause 1.2.
18. ISO 9000:2000, clause 3.6.4.
19. Adapted from ISO 9001:2000, clause 4.1.
20. ISO 9000:2000, clause 3.2.7.

CHAPTER FOUR

Requirements of ISO 9000

MAJOR TOPICS

- Philosophical Approach of the 2000 Release of ISO 9000
- New Requirements in ISO 9000
- ISO 9001 Structure
- Requirements of ISO 9001:2000
  - Quality Management System—Requirements of Clause 4
  - Management Responsibility—Requirements of Clause 5
  - Resource Management—Requirements of Clause 6
  - Product Realization (Production)—Requirements of Clause 7
  - Measurement, Analysis, and Improvement—Requirements of Clause 8

PHILOSOPHICAL APPROACH OF THE 2000 RELEASE OF ISO 9000

Fundamental philosophical changes have been introduced into ISO 9000 with the 2000 release. ISO 9000 is now closely aligned with Total Quality Management as indicated by the adoption of the eight quality management principles described in Chapter 1:

1. Customer focus
2. Leadership
3. Involvement of people (employees)
4. Process approach
5. System approach to management
6. Continual improvement
7. Factual approach to decision making
8. Mutually beneficial supplier relationships

The standard is now designed around a "process approach" to management. (The reader may want to refer to the definitions and discussion of *process* and *process approach* in Chapter 3.) ISO has stated:

> For organizations to function, they have to define and manage numerous inter-linked processes. Often the output from one process will directly form the input into the next process. The systematic identification and management of the various processes employed within an organization, and particularly the interactions between such processes, may be referred to as the "process approach" to management.
>
> The revised quality management system standards [2000 release] are based on just such a process approach, in line with the guiding quality management principles.[1]

NEW REQUIREMENTS IN ISO 9000

For users of ISO 9000:1994, the following is provided as a synopsis of the new requirements introduced with the 2000 version of the standard.[2] These new requirements fall into nine distinct categories.

- Continual improvement
- Increased emphasis on the role of top management
- Consideration of legal and regulatory requirements
- Establishment of measurable objectives for all relevant functions and levels
- Monitoring of customer satisfaction/dissatisfaction as a measure of system performance
- Increased attention to resource availability
- Determination of training effectiveness
- Measurements extended to system, processes, and product
- Analysis of collected data on the performance of the QMS

Each of these will be explained in the Requirement sections of this chapter.

ISO 9001 STRUCTURE

Readers who are familiar with the 20 requirement elements of the 1994 version of the standard will appreciate the more logical structure and sequence of this version. There are eight single-digit clauses in ISO 9001:2000. The first three are:

Clause 1—Scope

Clause 2—Normative Reference

Clause 3—Terms and Definitions

These three clauses simply set the stage for the requirements that are contained in the next five single-digit clauses:

Clause 4—Quality Management System (QMS)

Clause 5—Management Responsibility

Clause 6—Resource Management

Clause 7—Product Realization

Clause 8—Measurement, Analysis and Improvement

Figure 4-1 is based on the model of a process-based **quality management system** depicted in ISO 9000, 9001, and 9004. It illustrates the process linkages presented in the final five clauses. Note that the quality management system (clause 4) encompasses the four clauses 5 through 8. These are the operating elements that make the QMS work. It also illustrates that requirements are drawn at least partly from customers and other interested parties (such as regulatory agencies), and that customer satisfaction feedback is an essential element of the QMS. It is important to remember that at many process levels, customer feedback will be from internal customers. These are the users of the previous process's output, and only they can determine if that process output meets their requirements and preferences. See the discussion of *process approach* in Chapter 3.

REQUIREMENTS OF ISO 9001:2000

The actual requirements, that is, the provisions of the standard with which the organization must conform, are contained in clauses 4 through 8. In this book we treat each of these major clauses separately in five sections devoted to the requirements of clauses 4 through 8 respectively. Figure 4-2 illustrates the book's presentation scheme for ISO 9001 requirements.

Quality Management System—Requirements of Clause 4

Clause 4 is concerned with the establishment of a quality management system within the organization. The subclauses of clause 4 define the general requirements for the organization (4.1), for documentation (4.2.1), for a quality manual (4.2.2), and for control of documents (4.2.3) and records (4.2.4).

ISO 9001:2000, Clause 4.1: General Requirements

The organization shall establish, document, implement and maintain a quality management system and continually improve its effectiveness in accordance with the requirements of this International Standard.

The organization shall . . . [*continued in 4.1 a through f*]

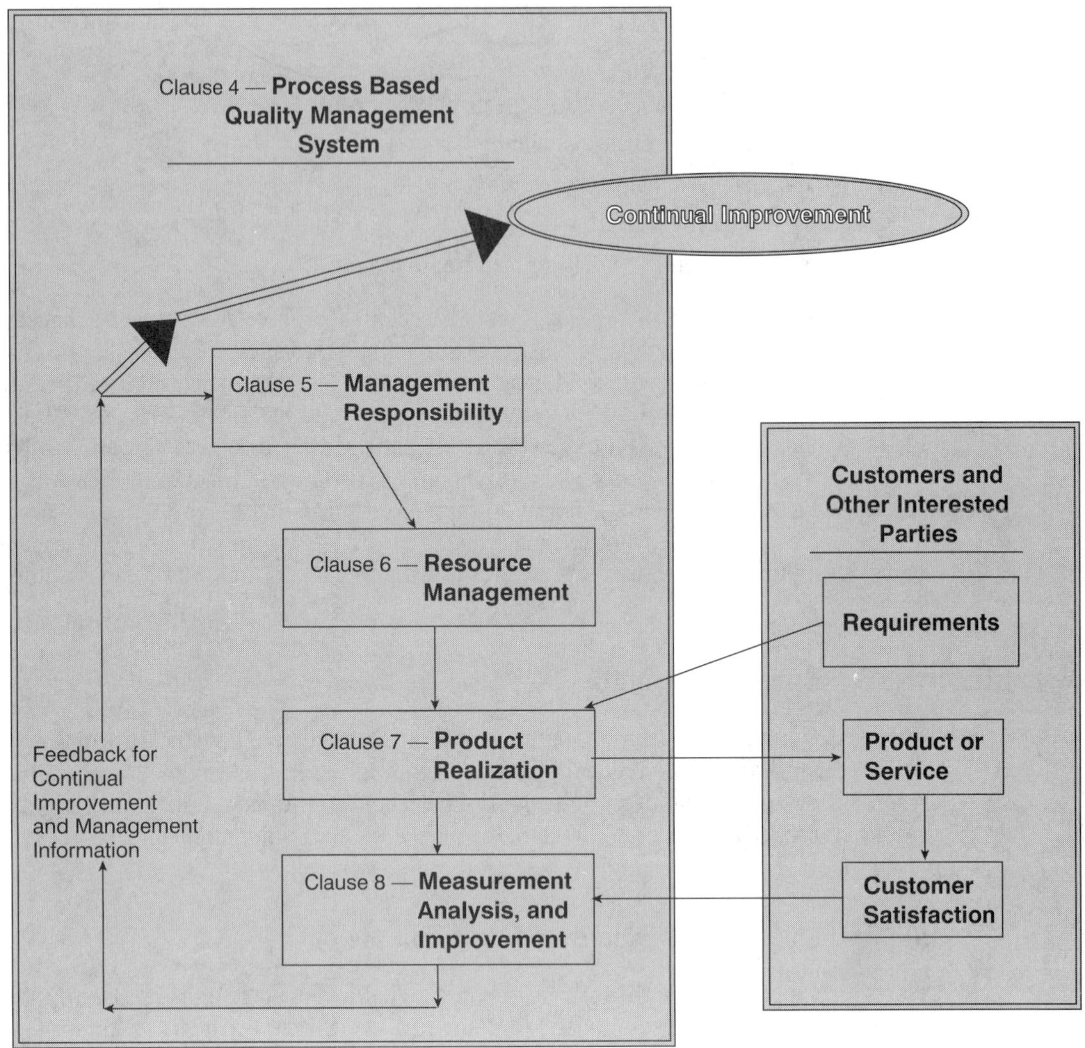

Figure 4–1
Model—A Process-Based Quality Management System (Showing process linkages represented by clauses 4 through 8, ISO 9001.)

Requirement—Clause 4.1

This is the first of many instances of the use of *shall* in ISO 9001. Wherever the word appears, it signifies a requirement. In this case the clause requires a registered organization, or one seeking registration, to develop for itself, document, implement, constantly use and keep up-to-date (maintain), and continually improve a QMS as specified by the whole of clause 4, following the model outlined in the whole of ISO 9001. In addition, the organization must do the things listed in the subclauses a through f that follow.

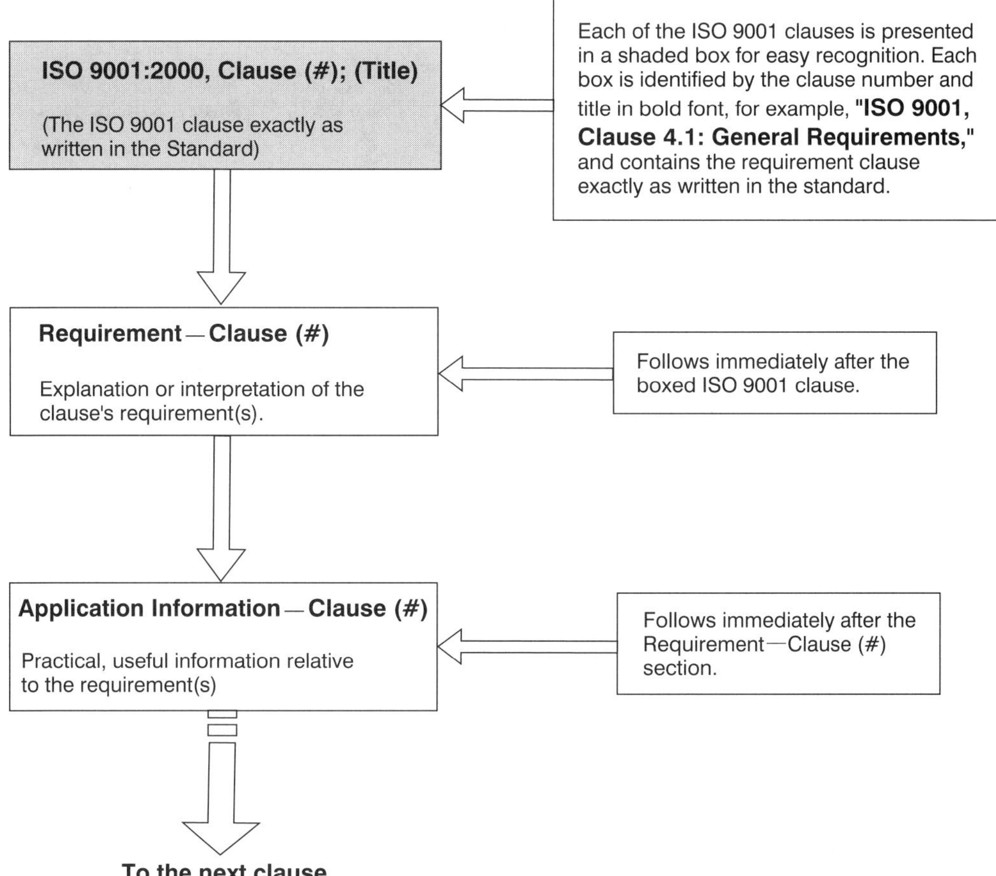

Figure 4–2
Presentation Scheme for ISO 9001 Requirements

Application Information—Clause 4.1

Requirement 1 Clause 4.1 in its statement "The organization shall *establish* . . ." means that the organization must determine what it needs in a QMS that conforms to the standard, and must design the QMS accordingly. To do that the organization must take the actions specified in subclauses a through f, and the remaining clauses under clause 4.

Requirement 2 The word *shall* also applies to documenting the QMS. That requires the organization to document (on paper or other suitable media) those elements of the QMS that are specified in succeeding clauses.

Requirement 3 Similarly, the organization must implement and maintain its QMS. This means that once the organization has determined what it needs in a conforming QMS, the QMS must be implemented and activated throughout the relevant activities,

ISO INFO

Leading and operating an organization successfully requires managing it in a systematic and visible manner. Success should result from implementing and maintaining a management system that is designed to continually improve the effectiveness and efficiency of the organization's performance by considering the needs of interested parties.

Top management should establish a customer-oriented organization

a) by defining systems and processes that can be clearly understood, managed, and improved in effectiveness as well as efficiency, and
b) by insuring effective and efficient operation and control of processes and the measures and data used to determine satisfactory performance of the organization.[3]

and constantly used and kept up-to-date. It is necessary to keep the QMS current with revisions to processes, new products or services, revised procedures, and so on.

Requirement 4 The organization must also continually improve all aspects of the QMS in order to improve its effectiveness. This means that every element of the QMS is a potential subject for improvement, not just once, but on a continuing basis. For example, emphasis on continual improvement may lead to a process change that reduces the chance for operator error. When such a change is implemented and operator errors are reduced, is that the end of it? Not in a continual improvement world. The process is still not perfect, so further improvement is possible. The continual improvement initiative will place continued emphasis on developing improvements for all processes, procedures, policies, products, tooling, and so forth. (For more information on continual improvement, see Chapter 9.)

To accomplish the general requirements of this clause, it will be necessary for the organization to take the steps indicated in subclauses a through f. Please note that the next boxed item from Clause 4.1 appears after subclause f in ISO 9001, but applies to all of the subclauses. Since we are treating the subclauses individually, we have taken the liberty to move the boxed item ahead of the subclauses so that it can more easily be associated with them. We suggest that you refer to this box after reading each subclause.

ISO 9001:2000, Clause 4.1: General Requirements, Continued

These processes shall be managed by the organization in accordance with the requirements of this International Standard.

ISO 9001:2000, Clause 4.1: General Requirements, Continued

Where an organization chooses to outsource any process that affects product conformity with requirements, the organization shall ensure control over such processes. Control of such outsourced processes shall be identified within the quality management system.

NOTE Processes needed for the quality management system referred to above should include processes for management activities, provision of resources, product realization and measurement.

ISO 9001:2000, Clause 4.1a: General Requirements, Continued

a) [*The organization shall*] identify the processes needed for the quality management system and their application throughout the organization (see 1.2),

Requirement—Clause 4.1a

The organization must (note the clause's use of the word *shall*) identify the processes used in the production of its products and services, and any other processes relating to or interacting with them. These will be the processes that will have to be addressed in the quality management system. Once identified, the organization must develop an understanding of the application of the processes throughout the organization. The note in the boxed element of clause 4.1 above cautions the organization to include processes *for management activities, provision of resources, product realization and measurement.*

Application Information—Clause 4.1a

Any system of production, whether for manufacturing products, processing data, or providing a service, is made up of a series of processes. Before the organization can develop a quality management system, it has to identify those processes and understand how they are operated by the various elements of the organization, or even by outside organizations if any processes that could affect product conformity are outsourced (subcontracted). That is what clause 4.1a is asking for. It is important that all processes that can have any effect on the quality or fitness-for-use of the product or service be included. In a typical organization it will be found that several departments will be involved. The processes of each of the departments can have direct or indirect effect on product or service. For example, the engineering department's design processes will affect the ease or difficulty of execution of processes in the production department. The purchasing department's processes can have impacts on production and the quality of product or service. Production processes directly affect efficiency and quality. Quality assurance processes clearly affect quality. Even support departments such as human resources and accounting can have impacts, usually through morale, training, and availability of resources. Top management should not be overlooked, because policies established at the top always affect the processes at every level of the organization.

Requirement To satisfy the requirement of clause 4.1a, the organization should:

- Identify the processes (including relevant outsourced processes) that result in the development and production of products and services.
- Identify the processes necessary to support the development and production processes.
- Determine the organizational activities responsible for operating the processes and providing input to and taking output from them.
- Develop a solid understanding of the processes, their inputs and outputs. This is best done by flow-charting the processes.

It is important to understand that at this point the organization seeking ISO 9000 registration is not asked to develop any new processes, but simply to identify and understand the processes that are already in place and operating. This may seem to be an odd requirement since the organization is in business to produce its product and therefore must, one might think, already have complete and intimate knowledge of its processes and how they work together. The fact is, however, that unless such a deliberate effort is taken, the typical organization will not have an adequate knowledge of its processes, how they *really* work, and who is involved in each. Although a large number of people will have knowledge of bits and pieces of the processes, a complete knowledge of how the entire system of processes works is not usually to be found. This is true of employees at all levels, even to the top management level. Gaining that knowledge is essential for the success of a process-based quality management system, as required by ISO 9000 or by Total Quality Management.

ISO 9001:2000, Clause 4.1b: General Requirements, Continued

b) [*The organization shall*] determine the sequence and interaction of these processes,

Requirement—Clause 4.1b

Having determined the processes to be addressed in the quality management system in response to clause 4.1a, the organization must now determine the chronological sequence of those processes, and how they interact with each other.

Application Information—Clause 4.1b

Requirement The organization must determine the sequence and interactions of the processes identified under clause 4.1a. The best way to do that is by flow-charting the processes. If that is done, developing the chronological sequence and interactions will be almost automatic, because the flow charts will provide graphic information on how the output of one process forms the input for another, what resources or inputs are required from other processes, and so forth.

Once detailed process flow diagrams are available, it is quite simple to develop a network of the processes using a higher level (less process detail) overall flowchart. To illustrate, refer to three of the process flow diagrams in Chapter 3, Figures 3-4, 3-5, and 3-6. We can represent each of these detailed flow diagrams by a single box in a higher level diagram. We can then use these boxes to generate a high-level flow diagram that clearly portrays the process sequence and interrelationships. This is illustrated in Figure 4-3. This simple example will hold for any set of process flow diagrams that, taken together, represent a production system. High-level flow diagrams can become a lot more involved when interrelationships, such as approval loops or supporting processes, are introduced.

ISO 9001:2000, Clause 4.1c: General Requirements, Continued

c) [*The organization shall*] determine criteria and methods needed to ensure that both the operation and control of these processes are effective,

Requirement—Clause 4.1c

Clause 4.1c requires the organization to establish criteria and methods to ensure that the processes are always operated properly, and that they are effectively controlled.

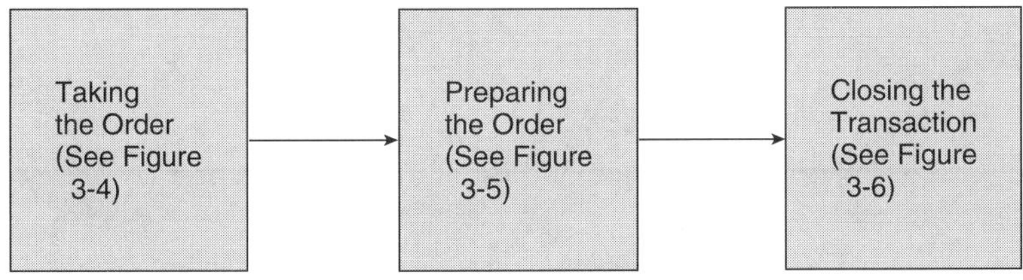

Note: These three boxes contain the 31 discrete steps contained in the process flow diagrams of Figures 3-4, 3-5, and 3-6.

Figure 4–3
High-Level Flow Diagram Illustrating Process Sequence

Application Information—Clause 4.1c

In response to clauses 4.1a and b, the organization has identified the processes that must be addressed by its quality management system and has studied their use throughout the organization and their sequence and interrelationships. Now the organization must do two things:

Requirement 1 Establish the criteria and methods under which the processes are to be operated.

Requirement 2 Establish controls to ensure that the processes are always operated according to the established criteria and methods.

Criteria and methods usually refers to specifications and procedures. Procedures will normally provide the work instructions (methods) for process operation and input and output requirements. For processes intended to operate with specified parameters such as temperature, time, tolerance, speed, and so on, these will be included in the process specifications, and should be included in the procedures. Interestingly, ISO 9001:2000 does not explicitly require that such procedures be documented. ISO now leaves it to the organization to determine which procedures must be documented to "ensure effective planning, operation and control of its processes" (see clause 4.2.1d). In very small organizations with the simplest of processes, documentation of procedures may not be necessary. However, for the vast majority of organizations, ensuring effective operation and control of processes over time is next to impossible without documented procedures. Under ISO 9001:2000 if the organization cannot ensure effective operation and control of a process in the absence of documented procedures, then the procedures must be documented. This will be a continuing issue throughout the standard's clauses.

ISO 9001:2000, Clause 4.1d: General Requirements, Continued

d) [*The organization shall*] ensure the availability of resources and information necessary to support the operation and monitoring of these processes,

Requirements—Clause 4.1d

To support the operation and monitoring of its processes, the organization must:

1. Ensure the availability of needed resources.
2. Ensure the availability of needed information.

Application Information—Clause 4.1d

Having determined under clauses 4.1a, b, and c the processes used by the organization, the sequence and interaction of the processes, and the criteria and methods to ensure effective operation and control of the processes, the organization—or more correctly, top management—must provide the necessary resources and information to support the use of the processes. In other words, it is not enough to simply say these are our processes, this is the sequence of operation, and here's how the processes interact among themselves and the various organizational departments; to make sure the processes do what we want them to do; and to specify these are the criteria for operating and controlling them. All of that is necessary, but it will not get the job done. Top management has to provide the means by which work will be accomplished. These are in the form of two broad categories—physical resources and information resources.

Requirement 1 The organization must ensure that needed physical resources are available. Physical resources include:

- Competent (proficient) people to actually operate and control the processes
- Tools, fixtures, machines, and test equipment required by the processes
- Materials used in the processes
- Work environment suitable for the processes, to include considerations of shelter, cleanliness, lighting, environmental control, noise level, work space, and appropriate power

Requirement 2 The organization must ensure the availability of needed information resources that may include:

- A quality policy that sets the general tone for the operation of all processes
- Process specifications and operating parameters
- Product specifications as appropriate to the processes
- Drawings, sketches
- Procedures and work instructions
- Test instructions

Under this subclause we should substitute *top management* for *the organization*. Only top management can authorize the expenditures of money and time necessary to provide the obligatory resources and information.

For the most part the physical resources—the manpower, machines, materials, and environment—are obvious requirements. Clearly, a process that requires a stamping machine must have one. The material from which the machine forms its product (say, sheet steel) must be available, or there can be no process operation. The stamping machine requires a trained, proficient operator, or it cannot be operated. And the machine and operator must have a suitable space in which to work. Failure to provide any one of these elements will prevent process operation, or at least impact effectiveness of the process.

The second category of things top management must provide is not as clear cut. For example, one might question a process operator's need for the quality policy. While the operator, as much as any employee, should be familiar with the quality policy, we are not suggesting that the company needs to provide a copy at each work station. We are suggesting, however, that the quality policy should be reflected in every document and instruction that the operator uses. So the quality policy is required at a higher level to ensure that documents and instructions reflect the intent of top management in regard to quality management.

Process specifications and operating parameters are required by the operator in order to properly set up and run the process and maintain its control. The same is true of operating procedures and work instructions. While the clause requires that they must be provided, there is a question about whether they must be documented. Is it sufficient to tell the operator orally how the process is to be operated and controlled? That would be sufficient for only the very simplest of processes. ISO says that written documents must be used if they are necessary "to ensure the effective planning, operation and control of its processes."[4] This can be translated to mean that if the absence of such a documented procedure or instruction could impact the effectiveness of planning, operation, or control, then it must be documented. With the 2000 release of ISO 9000, ISO has tried to back off the explicit requirement for so much documentation, leaving the decision to document or not with the organization. However, the registrars will be basing their judgment of conformance on how effective they believe the organization can be in operating and controlling its processes without written documentation. For most organizations, we believe that all process specifications, procedures, and work instructions should be documented.

If tests or inspections are to be conducted, perhaps as part of the process's control system, then test instructions and inspection procedures must be provided. Whether or not they must be documented instructions and procedures is determined by the same criteria given above. Product specifications and drawings and sketches may also be necessary to satisfy test and inspection requirements.

ISO 9001:2000, Clause 4.1e: General Requirements, Continued

e) [*The organization shall*] monitor, measure and analyze these processes . . .

Requirements—Clause 4.1e

This clause requires the organization to

1. Monitor its processes
2. Measure the processes against the process and/or product specifications
3. Analyze the data that is collected

Application Information—Clause 4.1e

This clause requires the organization to do three things relative to the operation of its processes—it must monitor, measure, and analyze.

Requirement 1 Under ISO 9000, a process is not simply turned on and left to run without further attention. The process must be monitored to ensure that it continues to operate within its intended specifications and parameters. There are many ways to do this, and the organization has to determine which is appropriate for its needs. It may be that continuous monitoring of gauges is required, or that frequent checks are made to ensure that critical parameters stay within specified ranges. In some cases parameters of process or product are used for statistical process control (SPC).

Requirement 2 Monitoring techniques require some sort of measurement or sampling to ascertain that the process is operating within its specified tolerance range. This may be nothing more than making sure the gauges "stay in the green." It may involve sophisticated techniques like statistical process control (SPC) that will indicate that the process is drifting out of adjustment or that some special cause, that is, a cause outside of the normal variation of the process, is affecting the process. The organization must choose, and then implement appropriate measurements on its processes.

Requirement 3 The data collected in the monitoring phase must be analyzed. The output of that analysis is expected to indicate steps to be taken to maintain the process within its normal operating parameters. In the event that measurement revealed the process to be operating outside its parametric range, then data analysis should indicate what must be done to restore the process to normal operation and to prevent a recurrence. The latter becomes a potential for improvement of the process, and is important for the next clause.

ISO 9001:2000, Clause 4.1f: General Requirements, Continued

f) [*The organization shall*] implement actions necessary to achieve planned results and continual improvement of these processes.

Requirements—Clause 4.1f

This clause requires the organization to take whatever actions are necessary

1. To achieve planned results from the processes
2. To bring about continual improvement of the processes

Application Information—Clause 4.1f

The preceding clause (4.1e) required the organization to monitor, measure, and analyze the processes. On the basis of the data collected, clause 4.1f requires the organization to take two further steps.

ISO INFO

A major difference between ISO 9000:2000 and ISO 9000:1994 is the new approach for requiring documented procedures. *Except for six explicit requirements for documented procedures, the 2000 release leaves it to the organization to determine the need to document its procedures. The general criteria for documenting procedures is whether they are "needed by the organization to ensure the effective planning, operation and control of its processes.*[5]

Requirement 1 The organization must take the actions necessary to achieve planned results from the process. The data collected and analyzed under 4.1e determines the action required to assure that the processes are operating as intended.

Requirement 2 The organization must implement actions necessary to achieve continual improvement of the processes. The analysis from clause 4.1e represents one channel of input for the continual improvement program.

ISO:9001:2000, Clause 4.2: Documentation Requirements

This clause's requirements are all contained in its subclauses. Chapter 6 of this book is a discussion of ISO 9000 documentation as required by ISO 9001:2000. It may be helpful to see Chapter 6 for a perspective of QMS documentation.

ISO's philosophical approach to documentation is considerably different in the 2000 release than it was in the 1994 version, where the explicit requirements for documentation were much more numerous. ISO now leaves it up to the organization to determine the need for most documentation. In ISO's own words,

> Management should define the documentation including the relevant records needed to establish, implement and maintain the quality management system and to support an effective and efficient operation of the organization's processes.
>
> The nature and extent of the documentation should satisfy the contractual, statutory and regulatory requirements, and the needs and expectations of customers and other interested parties and should be appropriate to the organization.[6]

As in clause 4.1, we are moving the clause 4.2 notes to this location, ahead of the subclauses.

> NOTE 1 Where the term "documented procedure" appears within this International Standard, this means that the procedure is established, documented, implemented and maintained.
>
> NOTE 2 The extent of the quality management system documentation can differ from one organization to another due to
>
> a) the size of organization and type of activities,
> b) the complexity of processes and their interactions, and
> c) the competence of personnel.
>
> NOTE 3 The documentation can be in any form or type of medium.

In order to establish a common understanding of these notes, the following explanations are provided.

- Note 1 uses the phrase "established, documented, implemented and maintained." That means that a procedure shall have been developed (established) by the organization, formally written (documented) so as to be available to any relevant employee when needed, put into practice (implemented) and used continually, and kept current (maintained) as conditions, technology, and products/services evolve and as improvements are introduced.
- Note 2 simply recognizes that there is no "one-size-fits-all" quality management system. Larger organizations with many varied products and services will likely require more documented procedures. Organizations with complex, interacting processes will require more documented procedures than those operating a few simple processes. The authors have reservations about Note 2 c, which suggests that organizations having more competent personnel may require fewer documented procedures. In the world of twenty-first century business, about the only constant is change. Products change from day to day, processes change, tooling and equipment change, and—not the least of the issue—personnel change. The organization that has long-standing, well understood processes, operated by experienced, trained, long-term employees—and as a result doesn't feel the need for documenting procedures—cannot avoid the inevitable. The organization has no guarantee that key employees will remain in place for the next five years, or even five weeks. Better offers can siphon experienced employees away. Retirement, ill health or injury, or even death also take employees. Furthermore, few employees want to remain in the same job forever. They want to advance to better positions, and by doing so, create experience voids where they had been. The organization's best interests are served by documenting all key procedures while the experience is there to do it. Once that is accomplished, the process can be continued much more effectively when the experienced operator is no longer available.

- Note 3 is important. **Documented procedures** do not necessarily have to be printed on paper. They may be on electronic media, viewable on a screen. They may be on microfiche or any other media or combination of media. Any of these constitute *documented* procedures. The important thing to remember is that whatever medium is selected by the organization for its documentation, it must be readily available to those who need it.

Documentation Requirements, Continued

ISO 9001:2000, Clause 4.2.1: General

The quality management system documentation shall include [*the elements required by the subclauses 4.2.1a through e.*]

Documentation Requirements, Continued

ISO 9001:2000, Clause 4.2.1a: General

[*The quality management system documentation shall include*] documented statements of a quality policy and quality objectives,

Requirements—Clause 4.2.1a

This clause requires the documenting of two elements of the organization's quality management system:

1. The quality policy
2. The quality objectives

Application Information—Clause 4.2.1a

The reader might well ask, "What quality policy and what quality objectives?" Chronologically speaking, the structure of ISO 9001 is a bit convoluted. The problem is, we have not seen a requirement for a quality policy or for quality objectives at this point. Well, don't be concerned. They will be required by clauses 5.1b and 5.1c respectively. In addition, Clause 5.3 is devoted entirely to requirements for the quality policy, and 5.4.1 for quality objectives. Quality objectives are also required by clause 7.1a. However, none of those clauses require the policy or objectives to be documented, since the requirement is found in this clause, 4.2.1a.

Requirement 1 The organization must document its quality policy. That means that the quality policy has to be more than "just the way we operate" as held in the collective head of management. It has to be a written document that can be accessed by anyone needing it. This requirement is not just for large organizations. It applies to every organization, large or small.

Requirement 2 The organization must document its quality objectives. A quality objective has a short life—at least it should. It is developed, achieved, and becomes a record. This means that documented quality objectives are not something static in the QMS documentation. A continuing effort will be required to add new objectives, and to remove those that have been achieved. For its QMS documentation the organization has two options for objectives: (a) place the actual objectives in the QMS documentation, or (b) substitute a list of current quality objectives there, with reference to their actual location. Either way, the file of objectives or list of active objectives must be kept up to date.

Documentation Requirements, Continued

ISO 9001:2000, Clause 4.2.1b: General

[*The quality management system documentation shall include*] a quality manual,

Requirements—Clause 4.2.1b

This clause requires that the organization's quality manual be written down (documented).

Application Information—Clause 4.2.1b

Requirement The quality manual must be a written document. The requirement for a quality manual is found in clause 4.2.2, although that clause does not address documenting the manual. It is difficult to envision a manual that is not documented, but the requirement to do so is given in this clause. All elements of the quality plan must be written down (documented).

ISO INFO

Although there is a general chronological flow or sequence in each of the five major requirements clauses (4 through 8), this structure does not support a consistent chronological sequence from one major clause to another, nor is it intended to. In other words, the organization cannot meet the requirements of clause 4, then proceed to clause 5, and so on. For working with ISO 9001's requirement clauses, the organization must consider the whole of the standard simultaneously.

Documentation Requirements, Continued

ISO 9001:2000, Clause 4.2.1c: General

[*The quality management system documentation shall include*] documented procedures required by this International Standard,

Requirements—Clause 4.2.1c

Any documented procedures that are required by the standard must become part of the quality management system documentation.

Application Information—Clause 4.2.1c

Requirement The requirement of 4.2.1c is that all documented procedures that are either required explicitly by the standard, or implicitly by the requirement for effective operation and control of processes, must become part of the quality management system documentation. This is to cause them to be subject to the standard's documentation control requirements. So doing also provides the auditors a means of comparing actual operations with the plans.

Scattered throughout the standard are explicit requirements for documented procedures. For example, there are explicit requirements for documented procedures in clause 4.2.3 (for control of documents), clause 4.2.4 (for control of records), clause 8.2.2 (for audits), clause 8.3 (for dealing with nonconforming product), clause 8.5.2 (for corrective action), and 8.5.3 (for preventive action). Every organization registered to ISO 9000 must have written procedures to satisfy these clauses, and they must be a part of the quality management system documentation. Are these the only documented procedures the organization must have? Probably not. If you refer back to the Application Information for clause 4.1c you will recall that the organization must (shall) determine the criteria and methods needed to ensure effective operation and control of processes. Organizations customarily use procedures and specifications for this purpose, and in many instances these specifications and procedures are documented. The rule of thumb should be, if the organization cannot guarantee that the processes will always be effectively operated and controlled without such documents, then they must be documented. Said another way, if the absence of a written procedure could lead to an impact on the product's quality or its fitness for use, then documented procedures should be used.

Whatever the organization's approach, it is the organization's responsibility to decide whether a documented procedure is necessary. However, if a registrar finds that the absence of a procedure is having a deleterious impact, then the organization must be found nonconforming.

Documentation Requirements, Continued

ISO 9001:2000, Clause 4.2.1d: General

[*The quality management system documentation shall include*] documents needed by the organization to ensure the effective planning, operation and control of its processes,

Requirements—Clause 4.2.1d

Any documents that the organization needs in order to ensure that it can effectively plan, operate, and control its processes must become part of the quality management system documentation.

Application Information—Clause 4.2.1d

Requirement This requirement hinges on the meaning of "planning, operation and control" of processes. Of the three operative words, *planning* may be the most obscure. ISO provides the following definitions:

- **Quality planning**—part of quality management focused on setting quality objectives and specifying necessary operational processes and related resources to fulfill the quality objectives.[7]
- **Quality plan**—document specifying which procedures and associated resources shall be applied by whom and when to a specific project, product, process or contract.[8]

The first set of documents covered by this clause is that which is needed (1) to enable the organization to develop quality objectives and determine and specify the processes and resources to be used to meet them, and (2) to develop the necessary employee assignments and schedules for its products, projects, or contracts. This may include customer/interested party information, product specifications, process development information, contractual requirements, employee data, resource availability, statutory and regulatory requirements, and so forth. Documents that fall into this category must be made part of the quality management system documentation. If the organization determines that the procedures by which this planning is accomplished should be documented, then the procedures must also be included in the quality management system documentation.

Similarly, as we discussed under clause 4.2.1c, any documents required for the effective operation or control of processes must also be part of the quality management system documentation. If these processes procedures are covered by clause 4.2.1c, what kind of documents are we addressing in this clause? The list may include engineering specifications for the product or process, vendor manuals, specifications, operating instructions, operator qualification requirements, and statutory and regulatory requirements.

Documentation Requirements, Continued

ISO 9001:2000, Clause 4.2.1e: General

[*The quality management system documentation shall include*] records required by this International Standard (see 4.2.4).

Requirements—Clause 4.2.1e

This clause requires that the records required by the various clauses of the standard become part of the quality management system's documentation.

Application Information—Clause 4.2.1e

Requirement ISO 9000:2000, clause 3.7.6 defines record as a "document stating results achieved or providing evidence of activities performed." In other words, ISO 9000 records are the proof that something was achieved or some activity was performed. Clearly records are valuable to the organization from the historical sense (how did we do this before?), and from the standpoint of learning from experience, paths to follow and paths to avoid. Records are invaluable to auditors and other interested parties, providing objective evidence of conformance (or nonconformance). There are twenty-one explicit requirements in ISO 9001 for various kinds of records.

Clause	Requiring records of
5.6.1	Management reviews
6.2.2e	Education, training, skills, and experience
7.1d	Realization processes and product—meeting requirements
7.2.2	Results of product requirement reviews and actions
7.3.2	Product requirement inputs
7.3.4	Results of design and development reviews and actions
7.3.5	Results of design and development verification and actions
7.3.6	Results of design and development validation and actions
7.3.7	Design and development changes
7.3.7	Results of review of changes and actions
7.4.1	Results of supplier evaluations and actions
7.5.2d	Results of process validation (where output cannot be verified)
7.5.3	Product traceability (when required by customer or other)
7.6a	Basis of calibration of monitoring and measuring devices
7.6	Validity of previous measuring results (when the equipment is found not to conform)

7.6	Results of calibration and verification of measuring devices
8.2.2	Results of internal audits
8.2.4	Evidence of conformity and product release
8.3	Nonconforming product and actions
8.5.2e	Results of corrective actions
8.5.3d	Results of preventive actions

Each of these requirements has an explicit link to clause 4.2.4, Control of Records. Such records are considered part of the quality management system's documentation. See clause 4.2.4.

Documentation Requirements, Continued

ISO 9001:2000, Clause 4.2.2: Quality Manual

The organization shall establish and maintain a quality manual that includes

a) the scope of the quality management system, including details of and justification for any exclusions (see 1.2)
b) the documented procedures established for the quality management system, or reference to them, and
c) a description of the interaction between the processes of the quality management system.

Requirements—Clause 4.2.2

Clause 4.2.2 explicitly requires the organization to develop and use a quality manual. Through subclauses a, b, and c, it specifies three distinct topics to be documented in the quality manual.

a) A description of the quality management system's scope, including details of and justification for any exclusions of clause 7 requirements
b) By inclusion or reference, all documented quality management system procedures
c) Descriptions of interaction between the quality management system processes

Application Information—Clause 4.2.2

Requirement Clause 4.2.1 stated the requirement that the quality management system documentation include a quality manual. Clause 4.2.2 requires the organization to "establish and maintain" the quality manual. This means that the organization is responsible for developing the quality manual, using it, and keeping it up to date. In addition, clause 4.2.2 establishes the quality manual's core content. In practice, it will contain more than these three elements.

The quality manual is the document specifying how the organization is managed with regard to quality. In other words, it is the document that guides the direction and control of the quality management system. It is not provided by ISO. ISO specifies the generic requirements, but the organization must respond to those requirements in a manner appropriate for the organization and its activities. Top management is responsible for the development of the quality manual. That does not mean that is has to be written by top management—part or all of it may be delegated—but top management must endorse it, enforce it, and make certain that it is kept up to date in the face of continuing change and continual improvement. ISO 9000 documentation is typically comprised of four levels. See Figure 4-4. The four levels are:

1. Policy
2. Procedures
3. Practices
4. Proof

Of these four levels, at least the first two are part of the quality manual. At the first level the quality policy (clause 5.3) represents management's statement of the organization's intentions and principles related to its quality performance, and is the document from which the quality management system flows. Beyond the quality policy statement, there should be policies to address each of the requirements of the standard. These are individual policies describing how the organization intends to conform to each of the ISO 9001 clauses. For example, clause 4.2.2 requires a quality manual, stating, *The organization shall estab-*

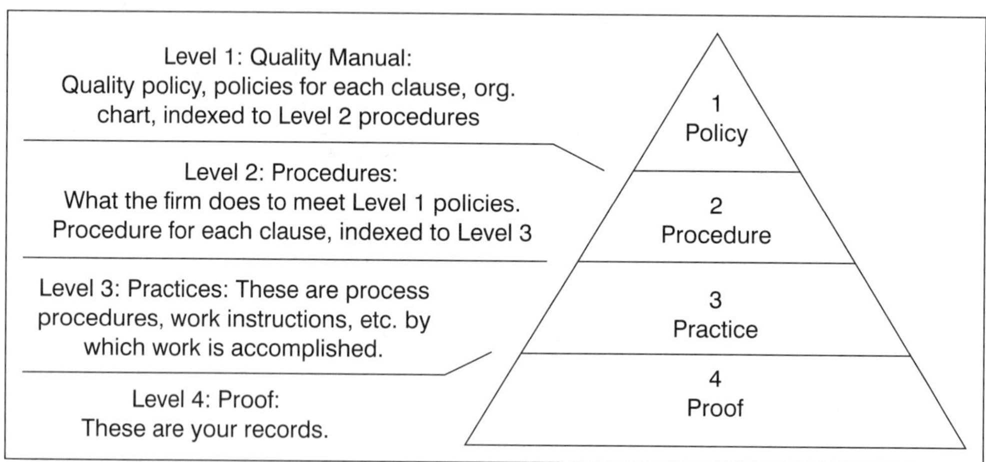

Figure 4–4
The ISO 9000 Documentation Pyramid

lish and maintain a quality manual that includes . . . The organization has to respond to this and all other requirements, and one way to do it is to rephrase the ISO 9001 clause by changing *the organization shall* to (*organization name*) *will,* as follows: *Expo Technology will* establish and maintain a quality manual that includes . . . Text may be included as the organization feels necessary for clarification. This statement and accompanying text becomes the organization's policy relative to clause 4.2.2.

In response to clause 4.2.2a, through the policies and other documents such as organization charts and process flow diagrams, the quality manual establishes the scope of the quality management system. It defines the processes that are involved, and the responsibility and authority delegated to employees for following them, and thus, the scope of applicability of the Product Realization clauses (clause 7).

Clause 4.2.2a also requires the organization to explain why any section 7 clause is excluded. For example, a contract manufacturing company that only produces products for other companies may not have a design capability. Excluding clause 7.3, Design and Development Planning, is appropriate, but a detailed explanation and justification must be included in the quality manual. (Note: This company would have been certified to ISO 9002 under the 1994 version of the standard.)

Clause 4.2.2b requires that all documented procedures either be a part of the quality manual or be referenced by the manual. This applies to procedures that are required by the various clauses to be documented, and to those without which nonconformance may be possible (determined by the organization). Some organizations may have too many documented procedures to include them in the quality manual. It is permissible to locate the procedures elsewhere, but their titles, subject, location, and so forth must be referenced in the quality manual.

Clause 4.2.2c requires the quality manual to have descriptions of the interaction between the quality management system's processes. That includes all processes that can have an impact on quality performance—few processes fall outside those bounds. This information may be conveyed by high-level process flow diagrams (see Figure 4-3) that illustrate how processes take outputs from one and provide inputs to another, and how some processes require support of other processes, and so on. Organization charts and job descriptions also help by providing information of responsibility and authority of employees. The authors, and we believe most registrars, prefer this information in chart form rather than in narrative form, which quickly becomes very complex and difficult to comprehend.

Documentation Requirements, Continued

ISO 9001:2000, Clause 4.2.3: Control of Documents

Documents required by the quality management system shall be controlled. Records are a special type of document and shall be controlled according to the requirements given in 4.2.4.

Documentation Requirements, Continued

ISO 9001:2000, Clause 4.2.3: Control of Documents

A documented procedure shall be established to define the controls needed

a) to approve documents for adequacy prior to issue,
b) to review and update as necessary and re-approve documents,
c) to ensure that changes and the current revision status of documents are identified,
d) to ensure that relevant versions of applicable documents are available at points of use,
e) to ensure that documents remain legible and readily identifiable,
f) to ensure that documents of external origin are identified and their distribution controlled, and
g) to prevent the unintended use of obsolete documents, and to apply suitable identification to them if they are retained for any purpose.

Requirements—Clause 4.2.3

1. This clause requires that all non-record documents required by the quality management system be controlled.
2. This clause further requires the organization to establish (and consistently use) a documented procedure (or procedures) to define the controls that are needed for the control of non-record documentation. (Control of records is the subject of clause 4.2.4.) The procedure(s) must address the seven points listed in subclauses a through g:
   a) Approval of documents prior to issue
   b) Review, updating as necessary, and reapproval of documents
   c) Identifying change and current revision status of documents
   d) Making relevant versions of applicable documents available where needed
   e) Ensuring document legibility and ready identification
   f) Ensuring that documents of external origin are identified, and their distribution appropriately controlled
   g) Preventing unintended use of obsolete documents, but if retained for any purpose, identifying them as obsolete

Application Information—Clause 4.2.3

Requirement 1 The first requirement of this clause is that the organization must (*shall*) control the documents (excluding records, which are the subject of clause 4.2.4) required by the quality management system.

Requirement 2 To satisfy requirement 1, the organization must have a document control system. Any system as complex as a document control system must be operated in accordance with written procedures, and that is the second requirement of the clause. The organization must (*shall*) have a *documented* procedure (or procedures) establishing and defining the documentation controls. This is ISO 9001:2000's first explicit requirement for a *documented procedure.*

Many organizations have a tendency to complicate document control systems. This does not best serve the organization, nor is it what ISO 9000 wants. The best document control system is the simplest one that meets the needs of the organization while satisfying the requirements of ISO 9000. The best systems allow for easy, rapid—but controlled—access; ease and timeliness of necessary modification and updating; and ready availability of (only) current versions at point of use. The worst systems are slow and ponderous in responding to access or modification and updating needs, and are so complex and rigid that having documentation at points of use where it is needed is difficult. This leads to all sorts of "workarounds" and subversion of the document control system by employees who are not well served by the system. Workarounds typically subvert the very things that the control system is supposed to provide, such as having legitimate control or ensuring that only current versions are in use. ISO does not ask for a complex document control system.

If the organization already has a document control system in place, it should be tested against the requirements of clause 4.2.3. If it satisfies the requirements, continue to use it. If it does not, then modify it to conform. If an organization has no document control system, it must develop one. A good approach is to benchmark a document control system that is successfully used by another organization. In any event, clause 4.2.3 must be satisfied.

Clause 4.2.3 has seven subclauses (a through g) specifying the mandatory elements of the documentation control procedure. The documentation concerned here represents the first three "P"s of the ISO 9000 document hierarchy—policies, procedures, and practices. *Practices* include work instructions, process flow diagrams, drawings, job descriptions, and others. In addition, any documents of external origin—such as legal and regulatory requirements, vendor manuals and instructions, and contractual paper—required by the quality management system are also subject to this clause.

Clause 4.2.3a requires that the document control procedure define how internally generated documents (procedures, work instructions, drawings, etc.) are to be approved prior to issue. The reason for this is clear. The quality management system must prevent the use of unauthorized documentation. When new process instructions, for example, are developed, the proper authority must approve them for adequacy and intent prior to their issue for use. The document control procedure must spell out the process for obtaining this approval. The procedure should include the designation of individuals responsible (usually by position title, not name, to avoid revisions when employees move about) for originating (or modifying) specific categories of documentation and for authorizing them. For example, the ISO 9000 management representative might be designated as the person responsible for procedures concerned with record retention. He or she would be the person to develop, or have developed, procedures spelling out the retention periods of various kinds of quality management system records. Usually a second person of senior status would also be designated to approve certain categories of procedures. For example, the final signature approval authority for procedures dealing with records and improvement might be the vice president of operations. The approval authority for procedures and work instructions dealing with manufacturing processes could be the vice president of manufacturing. For more detailed information and for formats of procedures, refer to Chapter 6.

Clause 4.2.3b requires the documentation control procedure to establish and define controls to ensure that documentation is reviewed and updated as necessary, and then reapproved prior to release. The issue here is that procedures, drawings, work instructions and processes have to change as business requirements and improvement initiatives dictate. ISO wants to make certain that as processes and requirements change, relevant documentation reflects the changes. When it is necessary to modify a document, it must be reapproved prior to its issue for use. The process for doing that is similar to that discussed in the previous paragraph.

Clause 4.2.3c requires the document control procedure to provide a process to ensure that changes to documents and their current revision status are identified on the documents. This requires two things. First, any changed text or drawing section must be highlighted or in some way identified as different from the previous version. In text, this is often done by placing a vertical bar in the margin beside the new entry. There is no preferred technique. Second, the modified document must be identified as a new revision. This is usually designated in a revision block on the document. Typically the revision block entry will advance to the next letter with each revision. For example, a procedure at the Revision B level would be marked Revision C after another change. See Chapter 6.

Clause 4.2.3d requires that the document control procedure contain a process for ensuring two things. First, relevant documents must be available at the point of use, that is, where they are needed. Second, all controlled documents at the points of use must be of the appropriate revision level.

Clause 4.2.3g is closely related to 4.2.3d. It requires the document control procedure to have processes for two more things. The first is to prevent the unintended use of obsolete documents, and the second is to identify as obsolete (or, at least not current), any superseded documents that the organization retains for any reason. Between subclauses 4.2.3d and 4.2.3g, ISO intends that relevant revision level documents, and only those, are available at the point of use, and that documents can be easily determined to be current or obsolete. Something that must be covered in the procedure to satisfy 4.2.3g is the recall of obsolete documents. It is not enough to issue only current versions. When a document is made obsolete by a new version, the old version must be removed from the points of use.

Clause 4.2.3e requires the document control process to provide the means of ensuring two things:

1. That documents *remain* legible
2. That documents are readily identifiable

ISO's use of the word *remain* implies a fear that a document which was once legible might become illegible. This is a valid storage-process concern, and has to be addressed by the procedure. However, in our view, the bigger concern is initial legibility. Before any document is approved, both its legibility and its understandability should be confirmed. If the users of the document do not understand its meaning, it might as well be illegible. As for the requirement that documents be readily identifiable, procedures and work instructions should carry both titles and purpose, and be cross-referenced to a requiring clause, process, procedure, work instruction, legislation, or regulation as appropriate. Drawings must have titles and identifying numbers. The point here is that some precise means of identifying all documents is required. Refer to Chapter 6.

Clause 4.2.3f requires the document control procedure to ensure that all documents of external origin required by the quality management system be

- Identified
- Subject to controlled distribution.

Documents of external origin may include contractual documents, operating and maintenance instructions and manuals for equipment used by the organization, legal and regulatory documents dealing with product safety, and others. ISO intends that they be clearly identified for what they are, and distributed to the activities that need them or are subject to them.

Documentation Requirements, Continued

ISO 9001:2000, Clause 4.2.4: Control of Records

Records shall be established and maintained to provide evidence of conformity to requirements and of the effective operation of the quality management system. Records shall remain legible, readily identifiable and retrievable. A documented procedure shall be established to define the controls needed for the identification, storage, protection, retrieval, retention time and disposition of records.

Requirements—Clause 4.2.4

1. This clause establishes the requirement for a records system to be established and used by the organization. The records involved are those that demonstrate conformity of the quality management system and the effectiveness of its operation.
2. This clause requires the organization to establish and continually use a documented procedure defining the processes needed for the control of such records. The procedure must address:
   - Identification
   - Storage

- Protection
- Retrieval
- Retention time
- Disposition of records

As long as the records are retained, they must remain legible, be readily identifiable, and be retrievable.

Application Information—Clause 4.2.4

Requirement 1 This clause requires the organization to establish and maintain records related to the quality management system through an organized, stuctured, orderly, controlled process. Maintaining records means that they must (*shall*)

1. Remain legible
2. Be readily identifiable
3. Be retrievable

The requirement for *legibility* seems almost condescending, but often record files are undecipherable and obscure. Any record that cannot be read or understood by employees or auditors is useless and will not be considered valid, objective evidence. Be sure that handwriting and signatures signature are readable.

Records must be readily *identifiable.* Make sure that before any record is made part of the file it can be easily identified, even if it means attaching a title page. An important component of identification of records is a date. For example, in the case of records of repetitive events, the titles of the record will be the same, and only the date will identify a specific record. ISO does not mention dates in this clause, but for identification it is an implicit requirement, and dating records should not be overlooked.

The record control process must ensure that every record is easily *retrievable.* ISO 9000 does not require that all records be stored in the same way (electronic media or paper, for example), or that they be retained at a single location. Whatever storage media and locations are selected by the organization, the means for ready retrieval are required.

Requirement 2 The second documented procedure explicitly required by ISO 9001:2000 is the procedure called for in this clause. This documented procedure by which control of the quality records process is maintained must address the following:

1. **Identification**—set a system for identification of all QMS related records.
2. **Storage**—determine storage media to be used, location(s), and responsibility.
3. **Protection**—establish the means of safeguarding records from the elements, decay, loss, and unauthorized access.
4. **Retrieval**—establish a process for retrieval of records, responsibility for records retrieval, and access authorization.

5. **Retention time**—establish appropriate periods of retention for the various categories of records.
6. **Disposition of obsolete records**—establish a procedure for the disposition of records when they are no longer needed.

Refer to Chapter 6 for additional information on document and records control.

Management Responsibility—Requirements of Clause 5

Clause 5 is concerned with **management responsibility** in connection with the quality management system. Its requirements are contained in six two-digit clauses:

5.1 Management Commitment
5.2 Customer Focus
5.3 Quality Policy
5.4 Planning
5.5 Responsibility, Authority and Communication
5.6 Management Review

The reader is encouraged to review Figure 4-1. So far we have been involved with Clause 4's basic requirements for what we might call the shell of the quality management system. Clause 5 presents the first of the four elements comprising the operation of the process-based quality management system.

The importance of active, visible leadership and support by management cannot be overemphasized when an organization is attempting to achieve the kind of major cultural change required by ISO 9000:2000 or Total Quality Management. Not only must the top level manager lead the change, but the other high-level managers, the top manager's staff, must also take leadership roles. Neither from personal experience, nor from the recorded experiences of dozens of companies of which we are aware is there a single success story of changing a corporate culture without the total commitment of top management. If the message from the top is not crystal clear, and if the person at the top is not seen as being totally involved, that will be all the encouragement some will need to toss monkey wrenches into the gears.

ISO INFO
Leadership, commitment and the active involvement of the top management are essential for developing and maintaining an effective and efficient quality management system . . .

Management Responsibility

ISO 9001:2000, Clause 5.1: Management Commitment

Top management shall provide evidence of its commitment to the development and implementation of the quality management system and continually improving its effectiveness by

a) communicating to the organization the importance of meeting customer as well as statutory and regulatory requirements,

b) establishing the quality policy,

c) ensuring that quality objectives are established,

d) conducting management reviews, and

e) ensuring the availability of resources.

Requirements—Clause 5.1

Clause 5.1 establishes two requirements for top management.

1. Top management must be committed to the development, installation, and continual improvement of the quality management system.
2. Top management must provide evidence of its commitment by doing the things listed in the five subclauses a through e, and in the succeeding clauses. These are:
   a) Top management must demonstrate its commitment by communicating to employees the importance of meeting customer requirements and statutory and regulatory requirements. (See clause 5.2.)
   b) Top management must demonstrate its commitment by establishing the quality policy. (See clause 5.3.)
   c) Top management must demonstrate its commitment by ensuring that quality objectives are established. (See clause 5.4.)
   d) Top management must demonstrate its commitment by conducting management reviews of the quality management system. (See clause 5.6)
   e) Top management must demonstrate its commitment by providing the resources necessary for the effective operation of the quality management system. (See Clause 6.)

Application Information—Clause 5.1

Requirement 1 Implementation of a conforming quality management system and registration to ISO 9000 require the total commitment of the organization's top management. The first step in our *Fifteen Steps to Registration* (Chapter 9) is *Secure Commitment from the Top.* Commitment to achieve and maintain certification from the highest level of management is an essential prerequisite to the starting of the ISO 9000 journey because:

- Only top management can authorize the necessary significant investment of resources including money and employee time.
- Only top management can overcome the inevitable differences of opinion—or outright resistance—over the development and implementation of a conforming quality management system.

Requirement 2 An expression of top management commitment is not enough. Top management must demonstrate its commitment on a continual basis. The entire organization must be convinced that the commitment exists, that it is real, and that it will be sustained. Employees take their cue from management by discerning from their actions what is important to management and what is not. Only when employees see that top management is committed to the quality management system will they commit themselves. Top management demonstrates its commitment by being actively and visibly involved in the development and implementation of the quality management system and its continual improvement. Evidence of its commitment is provided by top management's active role in developing a customer focus philosophy, establishing and exemplifying the quality policy, and ensuring that quality objectives are established, tracked, and achieved; by its leadership in conducting management reviews of the QMS; and by its habitually ensuring the availability of needed resources. Like Total Quality Management, ISO 9000 cannot be achieved or maintained without top management's full, active support.

Related to these requirements, ISO 9000 sees top management's role within the quality management system as having the following nine functions:[10]

a) establishing and maintaining (using) the organization's quality policy and quality objectives;
b) promoting the quality policy and objectives to increase employee awareness, motivation and involvement;
c) ensuring that the entire organization is focused on customer requirements;
d) ensuring that the processes necessary for achieving customer requirements and quality objectives are in place;

ISO INFO

Through leadership and actions, top management can create an environment where people are fully involved and in which a quality management system can operate effectively. The quality management principles can be used by top management as the basis of its role. . . .

ISO 9000:2000, Clause 2.6

e) ensuring that an effective, efficient quality management system is established, implemented and maintained (used) to achieve the organization's quality objectives;
f) ensuring that necessary resources are available;
g) reviewing the quality management system regularly and periodically;
h) providing leadership and decision-making regarding the quality policy and quality objectives;
i) making decisions for quality management system improvement actions.

Each of these functions is necessary not only for getting the job done, but also for demonstrating top management's continuing commitment.

Management Responsibility, Continued

ISO 9001:2000, Clause 5.2: Customer Focus

Top management shall ensure that customer requirements are determined and are met with the aim of enhancing customer satisfaction (see 7.2.1 and 8.2.1)

Requirements—Clause 5.2

Clause 5.2 requires top management to do two things to enhance customer satisfaction:

1. Ensure that customer requirements are determined (related to clause 7.2.1).
2. Ensure that those customer requirements are met (related to clause 8.2.1).

Application Information—Clause 5.2

Requirement 1 There is much more to *customer focus* than sending out surveys. Customer focus means learning what customers want and applying that to develop new products and to improve existing products. It means not just listening to customers, but seeking their active participation and involvement. It means using every method possible to determine how effectively the organization has met customer requirements. It means examining everything you do *through customers' eyes,* that is, with a customer's perspective.

The 2000 version of ISO 9000 has borrowed this major tenet of Total Quality Management and placed it squarely at the center of the requirements assigned to top management. We think it was entirely appropriate for TC 176 to do so. The question is, what can the organization do to enhance customer satisfaction? The answer, of course, is deliver what the customer wants in terms of a product (or service): reliability, longevity, cost, ease of use, and features, among many other characteristics. It takes the efforts of the

ISO INFO

This International Standard promotes the adoption of a process approach when developing, implementing and improving the effectiveness of a quality management system, to enhance customer satisfaction by meeting customer requirements.[11]

entire organization to do that, but the first requirement of clause 5.2 makes top management directly responsible for ensuring that it happens. (Clause 7.2.1 addresses the processes for determining customer requirements.)

Requirement 2 The second requirement of clause 5.2 is that top management must ensure that customer requirements are met. Once again, many people within the organization will have to be involved in fulfilling customer requirements, and determining the degree of success, but top management is responsible for making it happen. Although there may still be some senior managers who disagree, this concept is entirely correct. Who else can cause customer focus to be integrated into the organization's culture? The answer is, no one. With the requirement to ensure that customer requirements are determined and fulfilled, top management must infuse the organization with the concept of customer focus.

Management Responsibility, Continued

ISO 9001:2000, Clause 5.3: Quality Policy

Top management shall ensure that the quality policy

a) is appropriate to the purpose of the organization.
b) includes a commitment to comply with requirements and continually improve the effectiveness of the quality management system,
c) provides a framework for establishing and reviewing quality objectives,
d) is communicated and understood within the organization, and
e) is reviewed for continuing suitability.

Requirements—Clause 5.3

Notice that although clause 5.3 is titled "Quality Policy," it does not require the development or establishment of the quality policy. That was done by clause 5.1, Management Commitment, subclause b. (One has to go back to clause 4.2.1a to find the requirement that the quality policy be documented.) Remember that clauses 5.1 and 5.3 both fall under "Management Responsibility," clause 5. That places the responsibility for developing (or guiding the development of) the quality policy on top management. It is at this juncture that clause 5.3's six specific requirements come into play—that the quality policy

a) Be appropriate to the purpose (vision and mission) of the organization
b) Include a commitment to comply with requirements
c) Include a commitment to continually improve QMS effectiveness
d) Provide a framework for establishing and reviewing quality objectives
e) Be communicated to all employees, and be understood by them
f) Be periodically reviewed to ensure that it remains suitable over time

Application Information—Clause 5.3

Requirement 1 Articulating the organization's **quality policy** is the first step in developing the quality management system because it sets the stage for everything that is done under ISO 9000. It is important to have the quality policy defined and harmonized with the aims and policies of the organization. (See clause 5.3a) It must flow from, and support, the corporate vision and guiding principles, assuming these exist. Like the vision, the quality policy acts as a beacon for guiding activity within the organization. The quality policy should be considered coequal with the vision, guiding principles, or other overall policies of the organization.[12]

Requirement 2 It is a firm requirement that the quality policy include a top management commitment to *comply with requirements*. "Requirements" include those imposed by the standard, as well as those self-imposed in the policies, procedures, work instructions, and other provisions of the QMS. They also include all applicable laws and regulations.

Requirement 3 The clause also requires a top management commitment to the continual improvement of the QMS. Before putting that commitment into the quality policy, top management must understand what it means. That is, improvement of the QMS will have to be an ongoing, continual project. A continual improvement commitment, followed through, has the potential for improving not just the QMS, but also the product and the organization's competitiveness and bottom line. But once the commitment is made, the organization must consistently demonstrate QMS improvement over time. Failure to do so may oblige the registrar to terminate ISO 9000 registration. On the other

ISO INFO

Only top management can "establish" the quality policy, because only top management has the leadership responsibility to do it. Note that this does not mean that top management must necessarily write the policy, but that the intentions of top management must be articulated by subject. The quality policy must be seen as top management's intentions for the organization.

hand, if management cannot make the commitment, then there will be no registration in the first place. This is a major departure from the 1994 version of the standard.

Requirement 4 This clause requires the quality policy to provide a *framework* for establishing and reviewing quality objectives. What is meant here is that the quality policy, when put into action, promotes the development of ideas for quality objectives throughout the activities and levels of the organization, and that management enthusiastically transforms them into formal quality objectives, assigns responsibility and authority to achieve them, and maintains an active interest in their progress and results. How does this happen? Several things are necessary within the organization including:

- All employees must feel free to voice their ideas or to suggest that a certain process is a problem and needs improvement.
- The organizational culture must support the continual improvement concept.
- There must be open channels of communication, and they must be omnidirectional.
- Management, including top management, must be actively involved in the QMS and must openly and honestly promote quality in all aspects of the organization and its products.

It is obvious that we are talking about fundamental organizational characteristics, employee involvement, devotion to continual improvement, effective communication, and an involved, facilitating management. None of these organizational characteristics are native to the traditional organization, but spring from Total Quality Management, and now from ISO 9000. The quality policy comes into play as the philosophical roadmap of the organization that, through its written statements, causes the organization to develop these organizational characteristics. An appropriate quality policy, therefore, is the genesis of the organizational characteristics that are the real framework for developing and reviewing quality objectives and everything else the QMS does.

Requirement 5 Once the quality policy is crafted it is not to be placed on the shelf to be dusted off in time for the next auditor visit. It is a working document that must infuse every activity by every employee having contact with the QMS. That cannot happen if the employees are not completely familiar with it. Employees become familiar with the quality policy, like so many other things, through communication. Top management is required to communicate the quality policy within the organization—and to ensure that it is understood. In a practical sense, that requires more than just an initial communication to explain its details and to underscore its importance. It also requires continual reinforcement, and perhaps even salesmanship, through every communication medium available. One of the strongest reinforcers of the employees' understanding will come from observing top management conducting their activities in accordance with the quality policy. This is something to which management will always have to devote attention.

Requirement 6 The final requirement of this clause is that top management implement an ongoing process to review the quality policy for continued suitability. It has been said that the only constant in today's business world is change. Every organization seems involved in change—if not internally, then certainly in its business environment. When changes take place, they can have all kinds of secondary effects. This requirement guards against the quality policy becoming obsolete and ineffective as a result of changes in or around the organization. The authors suggest that it is appropriate to incorporate this review of the quality policy into the management reviews required by clause 5.6.

Requirement to Document Although clause 5.3 does not mention it, the quality policy must be documented. The explicit requirement is found in clause 4.2.1a: *The quality management system documentation shall include documented statements of a quality policy . . .*

Things to Consider ISO 9004:2000, clause 5.3 offers a list of considerations when establishing the quality policy:

- The level and type of future improvement needed for the organization to be successful
- The expected or desired degree of customer satisfaction
- The development of people in the organization
- The needs and expectations of other interested parties
- The resources needed to go beyond ISO 9001 requirements
- The potential contributions of suppliers and partners

These may or may not be particularly useful taken at face value, but considering them might result in a well thought out quality policy.

Quality Policy Criteria ISO 9004:2000, clause 5.3 also offers the following criteria for the quality policy:

- Must be consistent with top management's vision and strategy for the organization's future
- Must permit quality objectives to be understood and pursued throughout the organization
- Must demonstrate top management's commitment to quality and the provision of adequate resources for achievement of objectives
- Should aid in promoting a commitment to quality throughout the organization, with clear leadership by top management
- Should include continual improvement as related to satisfaction of the needs and expectations of customers and other interested parties
- Should be effectively formulated (well written) and efficiently communicated

Pointers for Developing a Quality Policy

- A good ISO 9000 quality policy is short, while representing a lofty ideal. It is generally free of details, but it must contain two commitments (clause 5.3b):
    1. A commitment to comply with requirements stemming from ISO 9000, legal and regulatory authorities, and the organization's own QMS
    2. A commitment to the continual improvement of the QMS' effectiveness

- The quality policy reflects top management's intentions for the organization's quality reputation, desired level of quality, highest level quality objective, and commitment to continual improvement; it states the approach used to deploy, communicate, manage, and achieve quality objectives (clause 5.3c and 5.3d); and it stipulates the role of employees in that pursuit. The quality policy should also provide for periodic and special reviews to ensure that it remains viable and effective over time.
- The responsibility for a quality policy rests with the highest level of management, but this does not mean that the CEO should draft the policy in isolation. It is better for the senior manager and those directly reporting to that manager to develop the quality policy as a team. This will result not only in a better document, having input and discussion from all the senior managers, but will promote buy-in by all departments.

A sample quality policy statement is provided in the next chapter, in Figure 5.2. For additional information, see the section Conforming Quality Policy in that chapter.

Management Responsibility, Continued

ISO 9001:2000, Clause 5.4: Planning

The requirements of clause 5.4 are contained in the subclauses 5.4.1, Quality Objectives, and 5.4.2, Quality Management System Planning. Why is this section called *Planning?* The ISO 9000 quality management system is intended to coordinate the activities that direct and control (i.e., manage) an organization with regard to quality.[13] Directing and controlling (managing) with regard to quality would generally require the establishment of:

- A quality policy (to set the organization's intent with regard to quality)
- Quality objectives (goals for achievement)
- Quality planning (the strategy and tactics for achieving the objectives)
- Quality control (to ensure that quality requirements are met)
- Quality assurance (providing confidence that quality requirements will be met)
- Quality improvement (to improve the ability to meet quality requirements)

Quality planning focuses on at least the following three elements:[14]

1. Setting quality objectives (What do we hope to accomplish?)
2. Specifying the necessary processes (How are we going to do it?)
3. Specifying the resources necessary to operate and support the processes (What resources will be required?)

For quality planning to work most effectively, it must be linked to and supportive of the organization's strategic planning. Strategic planning attempts to define the strategies to be employed over a one to five year period to guide the organization toward its long range vision. The strategic plan will cover all aspects of the business—a broader outlook than ISO 9000. The quality plan is simply a mini strategic plan focused on the elements of the quality management system. Assuming the organization has a strategic plan, the quality plan must contribute to its achievement. If it does not, then either the quality plan or the strategic plan may need revision. If one does not support the other there is something seriously wrong, and the organization needs to take another look. For example, suppose that a manufacturer's strategic plan called for investing in a machine to automate a certain process. If the quality plan has a quality objective to benchmark the existing unautomated process, then it is not in harmony with the strategic plan. One or the other must be changed. Under no circumstances should resources be expended to benchmark a process that is already planned for a major change.

Planning, Continued

ISO 9001:2000, Clause 5.4.1: Quality Objectives

Top management shall ensure that quality objectives, including those needed to meet requirements for product [see 7.1a], are established at relevant functions and levels within the organization. The quality objectives shall be measurable and consistent with the quality policy.

Requirements—Clause 5.4.1

Clause 5.4.1 establishes mandatory requirements for three things:

1. That top management ensure the establishment of **quality objectives** at relevant organizational functions (departments) and levels. These are to include any objectives needed to meet product requirements.
2. Quality objectives must be measurable.
3. Quality objectives must be consistent with the quality policy.

In addition, there is a link to clause 4.2.1 that requires quality objectives to be documented.

Application Information—Clause 5.4.1

What are these quality objectives? You will recall from Chapter 3 that a quality objective is *a goal aimed for, related to quality*. Quality objectives are based on the organization's quality policy, meaning that they are in harmony with or support the quality policy. Quality objectives must also support the organization's long-range vision or strategy. They are assigned to relevant functions and levels within the organization.

Requirement 1 The organization develops its own quality policy, and based on that it establishes quality objectives toward which it strives. These objectives may be broad and strategic in nature, in that they define *how* the quality policy will be achieved. Typically the strategic objectives give rise to narrower tactical quality objectives that must be accomplished in order to achieve the strategic-level objective. Strategic-level quality objectives may be assigned to multiple functions, but, preferably, with a single responsible leader. The tactical-level quality objectives are ordinarily assigned to a single function, or even a single individual. To illustrate, suppose an organization's quality policy endorses, as it must, the continual improvement of all processes. Several processes in the organization's production department are known to be inferior to the industry's best practices, so a quality objective is developed with the goal of making those processes competitive within eighteen months. That is a broad, strategic, quality objective that could have originated with top management, or in the production department itself with approval from top management. The objective of making the deficient processes competitive offers no clue as to what needs to be done or what new performance level is expected of each of the processes. To that end, more specific, tactical quality objectives for each of the processes can be developed. These lower level tactical objectives detail what must be achieved in each of the processes in order to accomplish the broad strategic objective. One such objective could be assigned to manufacturing engineering, giving that function the lead responsibility for benchmarking a particular process, to determine world-class performance parameters, and to apply the benchmark results to bring the processes up to world-class standards.

Notice that clause 5.4.1 gives top management the responsibility for the establishment of quality objectives. This does not mean that top management must craft all quality objectives, but that top management must cause the development and deployment of quality objectives to happen. Top management's interest in quality objectives has to be clear to all employees. It must be seen as a priority. Although some of the organization's quality objectives may be developed by top management, that is not the thrust of this requirement. Rather, it is that top management must ensure that, regardless of who actually develops the objectives, quality objectives are established and acted upon.

Requirement 2 Clause 5.4.1 requires that quality objectives be measurable. This can be difficult. Generally speaking the lower level tactical quality objectives lend themselves well to measurability in terms of change, performance parameters, and date. That is not always so with the broader strategic quality objectives. They may not even carry an anticipated date for accomplishment. The point here is that there should be some kind of evaluation to determine whether the accomplishment of an objective yielded the desired result, the degree of improvement achieved, or even whether one could say the objective was completed. In those instances where metric or other concrete evaluation schemes are not applicable, it may be necessary to involve a secondary measurement, such as customer satisfaction. A very useful technique for discovering appropriate measurements for objectives, and the changes that flow from them, is to ask what is important to the customer (internal or external), and develop a measurement from that.

Requirement 3 Whatever their origin, all quality objectives should be reviewed by top management to assure their consistency with the quality policy and the strategic plan.

ISO INFO

It is important that measurements support the overall objective. It is easy to develop measures that seem logical, but work counterproductively to the objective. Brian Joiner tells of an airline that, in the interest of improving customer satisfaction, established "on-time departure" as a quality measure. Few of us would argue with that—unless, like Joiner, we happened to arrive at the gate one minute late, just as the attendants closed the door and began moving the jetway. Facing a four-hour wait for the next flight, Joiner pleaded with the attendants to reopen the door for him. But management had defined "departure" as door-closed and jetway-moved. *With the attendants and flight crew being graded for on-time departures, no amount of cajoling could convince them to open the door, even though the airplane sat at the gate for more than an hour, burning its image into Joiner's memory. As a result of a misguided measurement that was implemented to make customers happier, that airline permanently lost one of its very best customers.*[15]

In deploying quality objectives, it is critical that they be assigned to the organizational function that can deal with the objective, and at an organizational level commensurate with the task.

Quality objectives must be documented. The link to clause 4.2.1a provides an explicit requirement for doing so. Documenting objectives is best done through a standard objective format, an example of which is illustrated in Figures 4-5 and 4-6.

Planning, Continued

ISO 9001:2000, Clause 5.4.2: Quality Management System Planning

Top management shall ensure that

a) the planning of the quality management system is carried out in order to meet the requirements given in 4.1, as well as the quality objectives, and
b) the integrity of the quality management system is maintained when changes to the quality management system are planned and implemented.

Requirements—Clause 5.4.2

This clause has two requirements for top management.

1. Top management must take responsibility for quality planning, and focus it on defining the processes needed to effectively and efficiently meet the requirements of clause 4.1 and its quality objectives.

Quality Objective Action Plan

1. **Objective:** ______________________________
   ____

2. **Accomplishing Objective Will:** ______________________________
   ____

3. **Measurement:** ______________________________

4. **Plan for Accomplishing:** ______________________________
   ____
   ____
   ____

5. **Objective Leader:** ______________________________

6. **Tasks** | **Dept.** | **Assigned to**

Tasks	Dept.	Assigned to
______	______	______
______	______	______
______	______	______
______	______	______

7. **Schedule**

8. **Objective Management Review**

   Format: ______________________________
   Required Content: ______________________________
   Review Schedule: ______________________________

Approved by: ____________________ Date: ____________________

Use additional sheets as required.

Figure 4–5
Quality Objective Action Plan

Instructions for Quality Objective Action Plan form:

- Entry 1 is the objective's name or title.
- Entry 2 documents the purpose of the objective and confirms that it supports the quality policy (and the organization's vision or strategic plan). This satisfies clause 5.4.1 and 4.2.1a.
- Entry 3 defines the measurement that determines success.
- Entry 4 provides a brief plan for accomplishing the objective.
- Entry 5 designates the leader of the effort to achieve the objective.
- Entry 6 lists specific tasks, and the departments and individuals responsible.
- Entry 7 is a schedule of milestone events leading to achievement of the objective.
- Entry 8 is for establishing the format of progress reviews with management, content of the reviews, and a schedule for the reviews.

If more space is needed for any entry, additional pages may be attached to the form.

Quality Objective Action Plan

1. **Objective:** *Upgrade PCB cleaning process performance and efficiency*

2. **Accomplishing Objective Will:** *Result in cleaner PCB product, lower cost, and eliminate a hazardous waste problem. These support the Quality Policy and Vision.*

3. **Measurement:** *(1) Higher first pass acceptance, (2) cleaning cost reduced > 25%.*

4. **Plan for Accomplishing:** *Develop and implement a CFC-free PCB cleaning process over the nine month period ending next July 1.*

5. **Objective Leader:** *Carlos Reyes, Manager, Manufacturing Engineering*

6. **Tasks**

Tasks	Dept.	Assigned to
Assess available cleaning agents	*Mfg. Engineering*	*Chris Dudley*
Select agent & any new equip. req'd.	*Mfg. Engineering*	*Dudley*
"	*Manufacturing*	*Steve Owens*
Develop & implement the new process	*Mfg. Eng./Mfg.*	*Owens & Dudley*

7. **Schedule**

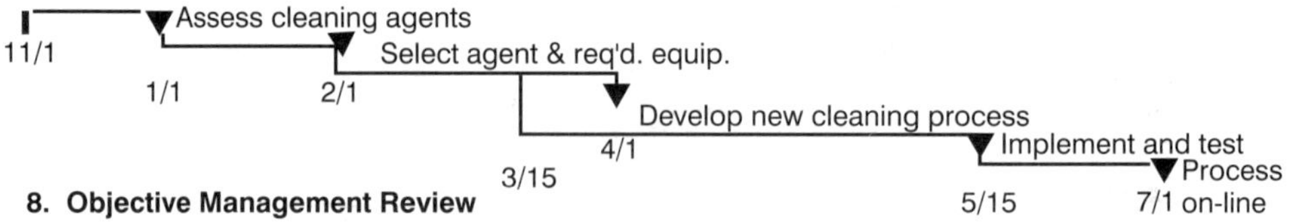

8. **Objective Management Review**

Format: *Progress and problems presentations*
Required Content: *Achievements/misses against milestones, issues*
Review Schedule: *Monthly (with program reviews)*

Approved by: *Gerald Welniak VP Manufacturing* Date: *10/24/2001*

Use additional sheets as required.

Figure 4–6
Completed Quality Objective Action Plan

2. Whenever changes to the QMS are being considered, part of the planning process must be directed at ensuring the continued integrity of the QMS.

Application Information—Clause 5.4.2

Requirement 1 Clause 4.1 lists the general requirements for a quality management system. Under that clause the organization is to identify the processes needed by the QMS, their sequence and interactions, and the criteria and methods for ensuring their effectiveness. In addition, the organization is to ensure the availability of resources and information necessary to support the processes, to monitor and measure the processes, and to implement actions to achieve planned results and continual improvement. Now we have clause 5.4.2 referring us back to clause 4.1. As convoluted as this may seem, the purpose is reasonable. Clause 4.1 told us that "the organization shall" do the things enumerated above, that is, develop and implement a QMS. Clause 5.4.2 comes along and makes it emphatically clear that it is top management's responsibility to get it done. Remember, everything in clause 5 deals with management responsibility. Between clause 4.1 and 5.4.2 the requirement is to plan and implement an effective QMS, with top management responsible for doing so. This is all part of ISO 9000:2000's thrust into total quality management. Top management must be actively and visibly involved in all the activities of the QMS, from planning to review.

Requirement 2 The second requirement of clause 5.4.2 is also a top management responsibility. That is to ensure that when changes to the QMS are planned and then implemented, the integrity of the QMS is not adversely affected. Under ISO 9000:2000 continual improvement will drive changes to the QMS to improve performance. Obviously, the organization is not knowingly going to implement a change that will harm the QMS. Unfortunately, change for all the right reasons may sometimes produce unanticipated consequences. That usually happens when the people sponsoring the change have become so focused on eliminating one problem that they are unable to see that the change will create another, possibly for another group of people. It can also happen when an objective is set without testing it to be certain that the objective supports the organization's vision, strategy, and policies. These situations, and others, that can lead to undermining the integrity of the QMS are not uncommon. The requirement of subclause b is simply that top management must make sure it does not happen. Why top management? First, top management always carries the burden of leadership. Second, top management is in the best position to have the "big picture" view of the operation, and therefore, should be best able to spot these kinds of problems. Below the level of top management, the perspective can easily become department oriented rather than organization oriented. Sometimes departmental agendas that promote changes benefiting one department even though causing problems for another are intentional, although Total Quality Management and ISO 9000 should minimize that. Sometimes it is accidental—the result of just not understanding the potential consequences in another department. Top management, by virtue of its broader view of the organization and its processes, and its authority to deal with such issues, is the only entity that can ensure that nothing will be done to harm the integrity of the QMS.

What this really means is that as suggestions and proposals for changes come forward, top management must lead the questioning and probing throughout the organization that will reveal any unintended consequences before the changes are implemented.

Management Responsibility, Continued

ISO 9001:2000, Clause 5.5: Responsibility, Authority and Communication

The requirements of this clause are found in three subclauses, 5.5.1, 5.5.2, and 5.5.3. The first deals with responsibility and authority, and the third with internal communication. So what is the other subclause for? Apparently this was the place that had the best fit for the management representative clause. It does make some sense, because the management rep clause itself is concerned with responsibility and authority of that individual.

Responsibility, Authority and Communication, Continued

ISO 9001:2000, Clause 5.5.1: Responsibility and Authority

Top management shall ensure that responsibilities and authorities are defined and communicated within the organization.

Requirements—Clause 5.5.1

This clause carries two requirements.

1. Top management must *define* organizational responsibilities and authority.
2. Once they are defined, responsibility and authority assignments must be *communicated* within the organization.

Application Information—Clause 5.5.1

Requirement 1 ISO's intent with the requirement to *define* organizational responsibilities and authorities is that the roles, tasks, obligations, and authority to act of everyone involved with the QMS be clearly defined. Except in very small organizations, there will be many people involved in one way or another with the QMS. Only when all these employees understand the tasks, obligations, and authorities delegated can the QMS function effectively. ISO wisely used the words *responsibilities* and *authorities*. It did so to prevent the all too common situation wherein management delegates the responsibility for a task to an employee without assigning him or her the authority necessary to complete it. Under ISO 9000, when management assigns responsibility, the requisite authority must also be given. No hierarchical level is implied in the defining of responsibility and authority. An effective QMS in most organizations will require the effort of employees at all levels. Thus roles, responsibility, and authority must be defined for all levels of employees who participate in or interact with the QMS.

Although there is no explicit requirement to document responsibility and authority assignments, it is difficult to conceive how the information can be communicated and utilized except through a document. Most registrars will require that it be documented. What we are talking about are organization charts, job descriptions, or any other format that will be easily interpreted by all employees. This documentation should be included in the quality manual, especially for the middle and top levels. Lower levels may be covered in procedure documents.

Requirement 2 In addition to defining the responsibilities and authorities, the information must be communicated within the organization. Of course the employee assigned the role for which the responsibility and authority is extended must be informed, but other employees with whom the employee might interact in executing the responsibilities must also know that this person has the responsibility and authority to execute the task.

Responsibility, Authority and Communication, Continued

ISO 9001:2000, Clause 5.5.2: Management Representative

Top management shall appoint a member of management who, irrespective of other responsibilities, shall have responsibility and authority that includes

a) ensuring that processes needed for the quality management system are established, implemented and maintained,

b) reporting to top management on the performance of the quality management system and any need for improvement, and

c) ensuring the promotion of awareness of customer requirements throughout the organization.

NOTE The responsibility of a management representative can include liaison with external parties on matters relating to the quality management system.

Requirement—Clause 5.5.2

Clause 5.5.2 places two requirements on top management:

1. Create a position of management representative, whose duties must include those enumerated in subclauses a, b, and c.
2. Appoint a member of management to that position.

Application Information—Clause 5.5.2

Requirement 1 ISO's approach to implementing and operating the QMS is to have a person from the management ranks designated to *manage, monitor, evaluate and coordinate* the quality management system.[16] For organizations not already registered to ISO 9000 this will be a new position, and must therefore be created. Since the QMS functions extend across several organizational departments, including engineering, purchasing, manufacturing, quality assurance, and so on, the implication is that the management representative will have responsibility and authority across departmental boundaries, something normally associated with the top executive. If such a management representative is implanted in any organization structure that we can think of, many of the employees will have two bosses (a situation to be forcefully avoided). (Refer to Figures 4-7 and 4-8.) The only way to prevent that and meet the letter of ISO 9004's clause 5.5.2 is for the top manager himself to take on the functions of the management representative. Top-manager-as-management-representative has much to recommend it, and was in fact our recommendation in our 1998 book.[17]

In the 1994 version of the standard, the management representative could be seen as a *staff position* to senior management. That person could monitor, evaluate, and coordinate, but rather than manage the QMS directly, would have to work through the top manager or the various department heads. If one goes strictly by ISO 9001:2000, clause 5.5.2 (which is essentially identical to its 1994 counterpart), ignoring its ISO 9004:2000 companion clause, the *staff position* management representative can still be justified. That is what we are going to do here, and what we suggest organizations do. (See Figure 4.9)

The management representative position must have responsibility and authority for the following:

- Ensuring that QMS processes are identified (established)
- Ensuring that QMS processes are implemented
- Ensuring that QMS processes are maintained
- Reporting to top management on QMS performance
- Reporting to top management on needs for QMS improvement
- Ensuring the promotion of customer requirements awareness throughout the organization

Though not a requirement, the management representative is usually the QMS point-of-contact with the registrar and other external parties.

The management representative should be a staff position, reporting to the top manager or to the top management steering committee if such exists in the organization. (See Figure 4-9). This structure will fulfill ISO 9001's requirements and will facilitate the functions listed above without creating the organizational nightmare illustrated in Figure 4-8.

Management representative may or may not be a full-time position. It may be set up as an additional duty position with a manager splitting time between the management representative function and his or her normal duties. The size and complexity of the organization will usually determine whether the position needs to be full time.

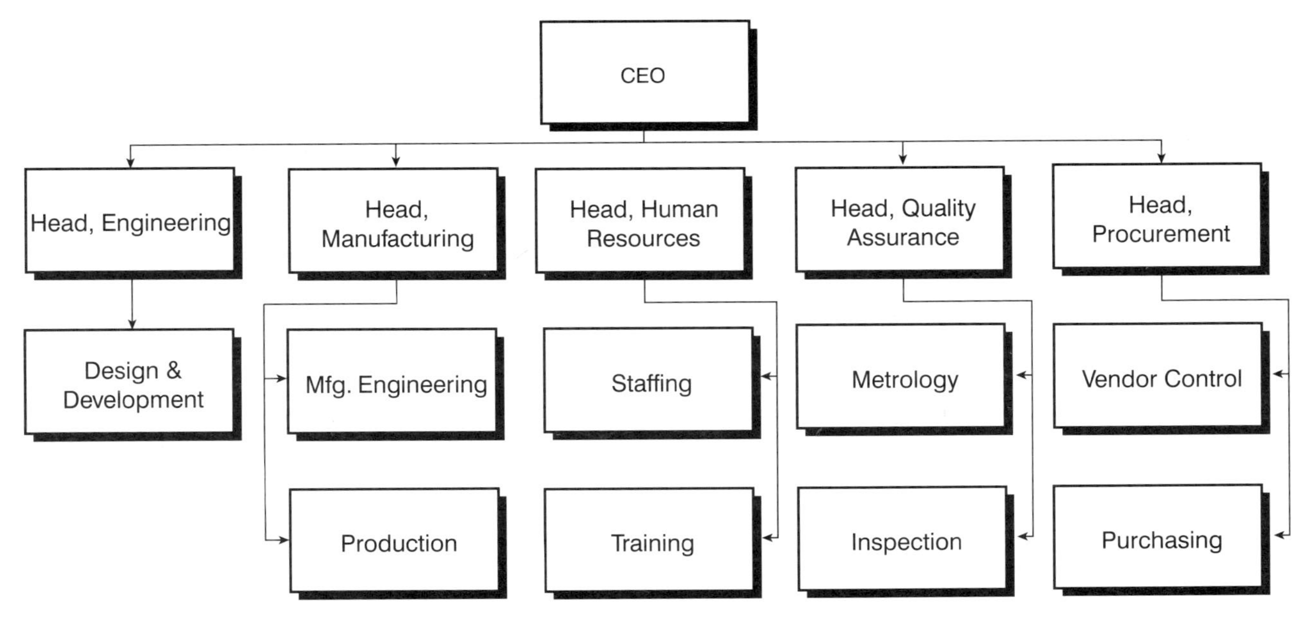

Figure 4–7
An Abbreviated Organization Chart (Traditional)

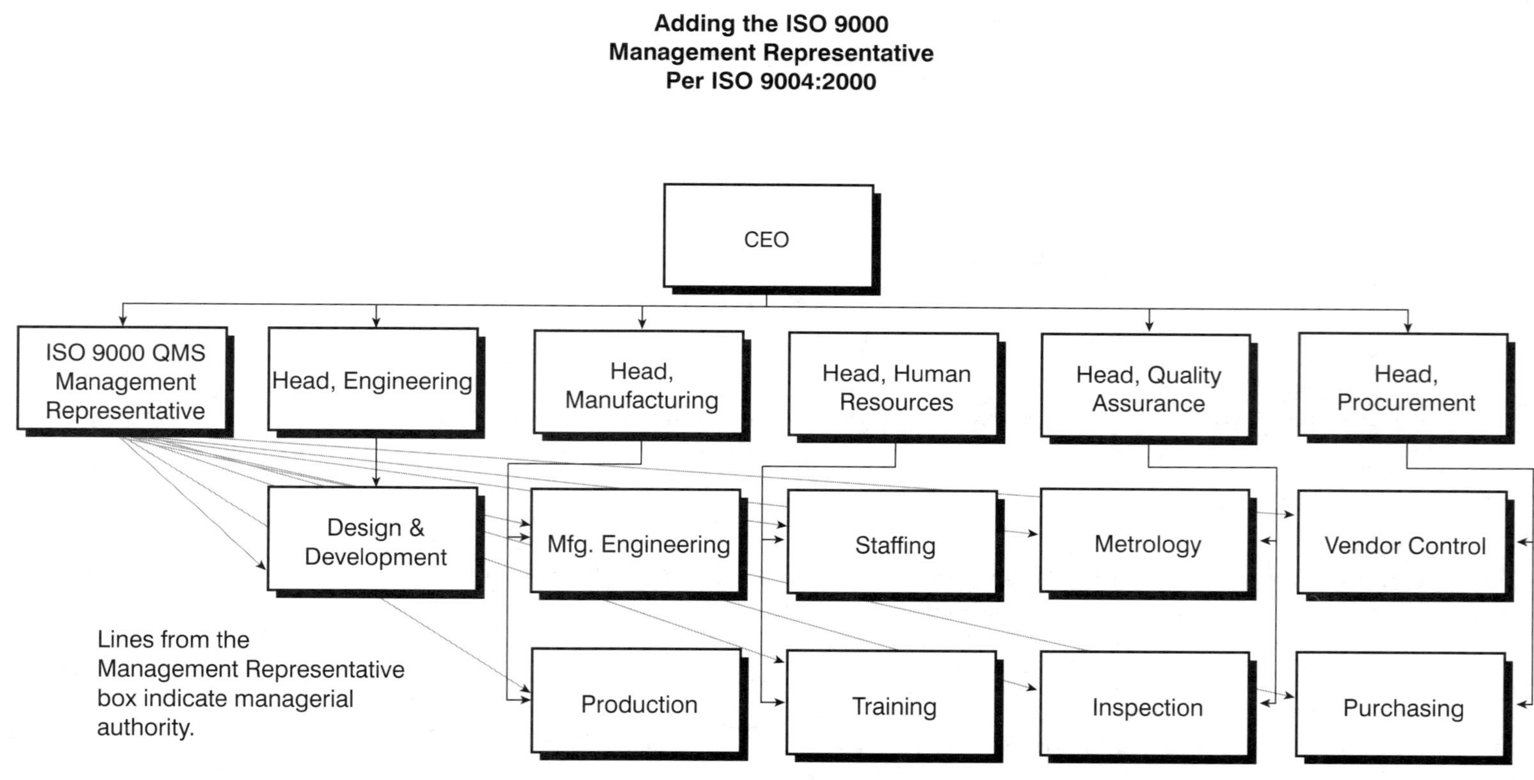

Figure 4–8
Traditional Organization Chart With Management Rep Per ISO 9004

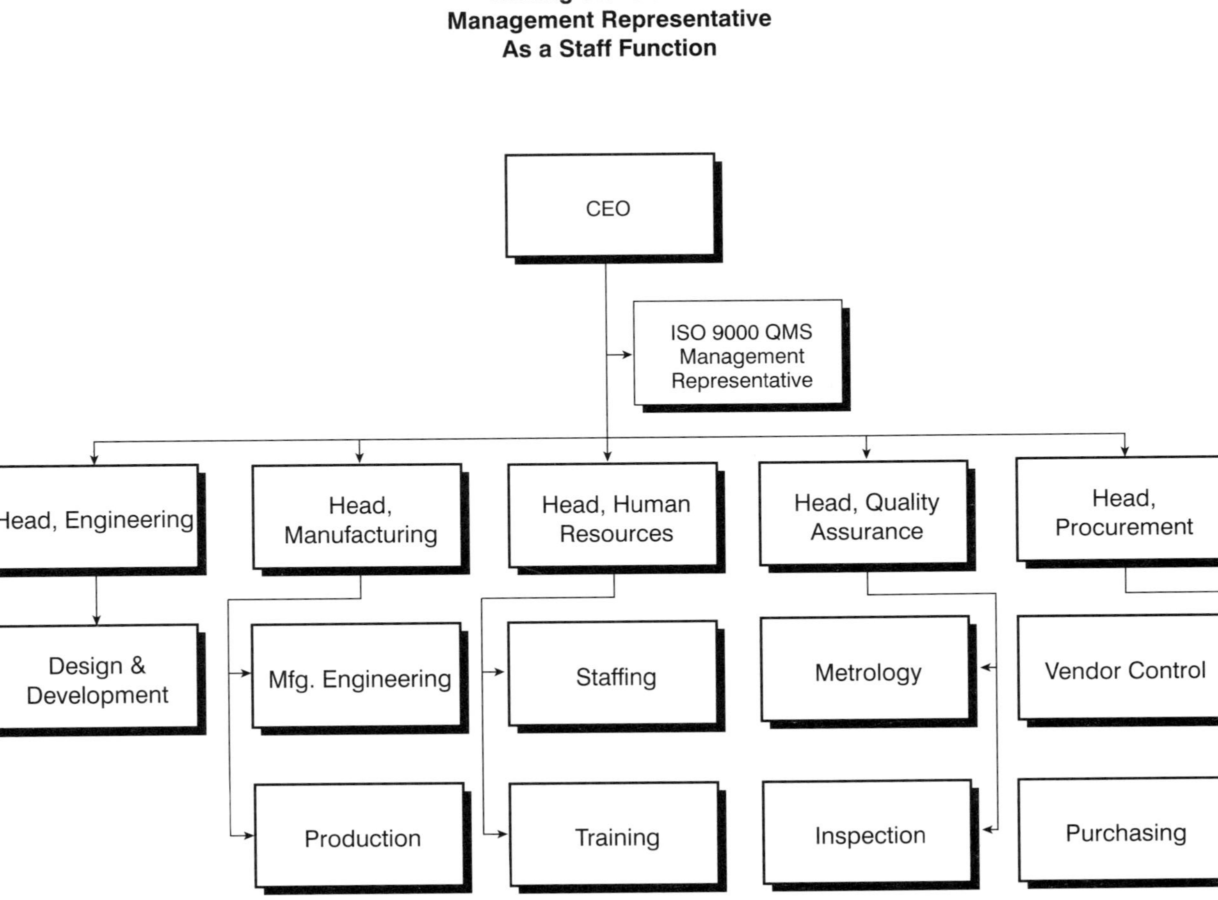

Figure 4–9
Traditional Organization Chart With Management Representative a Staff Position

Requirement 2 Appointing the management representative is the second requirement of clause 5.5.2. The clause requires the organization to appoint a member of management to the position. Most organizations appoint directly from their management team, and usually from the quality assurance or manufacturing disciplines. Although the clause is easily read that way, we do not believe it is ISO's intent to prohibit filling the position with a qualified outside candidate (i.e., not currently an employee of the organization) should the organization wish to do so.

Caution Organizations should be sensitive to the potential for conflict of interest in the management representative position. It must be guarded against or the integrity of the quality management system may be compromised.

Responsibility, Authority and Communication, Continued

ISO 9001:2000, Clause 5.5.3: Internal Communication

Top management shall ensure that appropriate communication processes are established within the organization and that communication takes place regarding the effectiveness of the quality management system.

Requirements—Clause 5.5.3

This clause places two requirements on the organization's top management.

1. Top management must ensure that appropriate internal communication processes are in place.
2. Top management must ensure that these communication processes are used to the benefit of the QMS.

Application Information—Clause 5.5.3

Requirement 1 It is ISO's intent that the organization's management should define and implement the communication processes necessary to effectively and efficiently communicate QMS related subjects, including the quality policy, process and product requirements, quality objectives, and accomplishments. These communication processes are more than the physical means of communicating, such as telephones, audiovisual aids and e-mail. They also include activities through which communication takes place. They include:

- Management led communication meetings. These may be mass meetings of the entire organization or of smaller groups. They may be formal or informal in nature. They may be held in an auditorium or on the work floor.

  One of the most effective processes of this type we have observed is the manufacturing floor "Ladder Meeting." Once a month the manufacturing head of a large upstate New York plant held a meeting with all manufacturing employees strictly for the purpose of providing information, promoting the quality agenda, answering their questions, and making several awards. This was done in the work area with everyone standing. As the first meeting got underway, he was having difficulty seeing all the employees, and many could not see him. Someone suggested he stand on a nearby stepladder. He did, and the meeting took the name of his improvised stage. He could have made better arrangements for subsequent meetings, but he believed the ladder gave ownership of the meetings to the employees and added an identifying character. It must have worked, because of all the communication schemes in place, the Ladder Meeting was considered by the employees to be their favorite.

- Meetings held for the purpose of briefing teams, reviewing the QMS or quality objective progress, recognizing achievements, and other events. As teams are formed to achieve assigned quality objectives or for continual improvement projects, management has the opportunity to communicate "up close and personally." It is the perfect setting to reinforce the QMS philosophy while providing the teams with needed information. Similarly, meetings for reviewing the QMS, progress on quality objectives, and recognition of achievement offer a unique opportunity to promote the QMS.
- Employee surveys can provide management with information they might not otherwise receive.
- Employee involvement, if taken seriously by management, will enable upward communication that is nonexistent in traditionally managed (i.e., not TQM, not ISO 9000) organizations. When employees believe that management is interested in what they think, they come forward with information concerning improving processes, products, and efficiency that management could obtain nowhere else. The key word here is *believe*. Employees have to trust and believe that management sincerely wants employee participation and involvement, wants their ideas, and will act on them. In other words, employee involvement cannot be window dressing—something management talks about, but is not backed up by actions.

Requirement 2 There are endless ways to communicate. Each organization must define those that best fit the organization, and then it must employ them. ISO is not talking about a barrage of communication when the QMS is first implemented. This has to be a continuing, persistent, focused effort. Employees will always need information from management, and management will always need information from employees. In addition to these vertically oriented communications, it is just as important to make the most of horizontal communication—manager-to-manager, and employee-to-employee. There is no such thing as over-communication in the workplace. Lack of communication, according to data from hundreds of employee interviews in North America, rates as management's number one failure.

Management Responsibility, Continued

ISO 9001:2000, Clause 5.6: Management Review

The requirements of clause 5.6 are contained in three subclauses. Subclause 5.6.1 is concerned with the implementation of the **management review** process, 5.6.2 is concerned with the inputs to the review, and 5.6.3 with the review's outputs.

Management Review, Continued

ISO 9001:2000, Clause 5.6.1: General

Top management shall review the organization's quality management system, at planned intervals, to ensure its continuing suitability, adequacy and effectiveness. This review shall include assessing opportunities for improvement and the need for changes to the quality management system, including the quality policy and quality objectives.

Records from management reviews shall be maintained (see 4.2.4).

Requirements—Clause 5.6.1

This clause has four requirements for the organization.

1. Top management must conduct periodic reviews of the QMS to ensure that it remains suitable, adequate, and effective.
2. The reviews must include an assessment of continual improvement opportunities.
3. The reviews must include an assessment of the need to change the QMS or its constituent parts, including the quality policy and quality objectives.
4. The organization must maintain records of the management reviews in a manner compatible with clause 4.2.4

Application Information—Clause 5.6.1

The intent of clause 5.6.1 is that top management engage itself in the operation of the quality management system's continual improvement process. This fits with the model of the process-based QMS of Figure 4-1, with the management reviews providing feedback data from all the QMS elements as primary inputs to drive improvements. It also fits with the *check* function of the **Plan-Do-Check-Act (PDCA) Cycle** that is now offered by ISO as a methodology that can be applied to processes for continual improvement.[18] Figure 4-10 illustrates the PDCA Cycle.

Management is to evaluate the feedback data and make corrections or improvements where indicated. This is the vehicle that will enable ISO 9000 organizations to continually improve performance.

Requirement 1 As changes occur within the organization and the environment in which it operates, ISO wants management to ensure that the QMS as established:

- Continues to be suitable to the organization's needs
- Does what the organization wants it to do
- Is effective in producing the desired results and performance

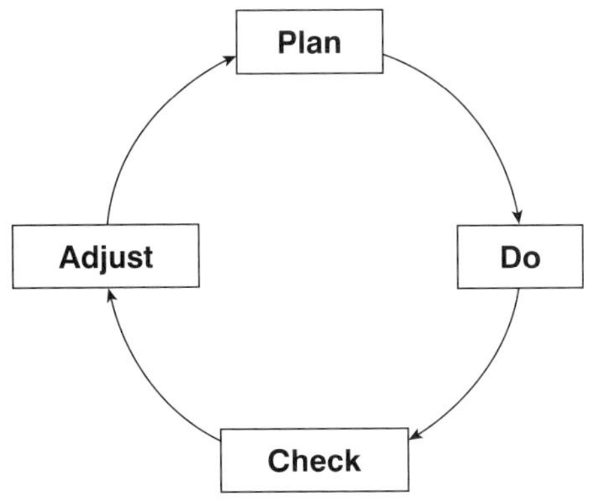

The **PDCA Cycle** operates like this:

1. **Plan** the improvement.
2. **Do** (or implement) the planned improvement.
3. **Check** the results to determine whether the anticipated improvement occurred.
4. **Adjust** the improvement based on the **check** results, and go through the cycle again. Or, if satisfactory, make the improvement permanent and monitor.
5. Continue the cycle, refining the improvement or developing new improvements.

Note: The authors routinely substitute *Adjust* for *Act* since it more accurately describes the action being taken in a continual improvement loop.

Figure 4–10
The PDCA Cycle

To accomplish this, top management must implement a program of management reviews of the QMS. These reviews must be conducted at **planned intervals,** meaning on a regularly scheduled basis. ISO allows management to determine the appropriate review interval. A monthly review should be considered; intervals longer than a month tend to result in the data becoming stale and interest in the issues waning.

ISO INFO

Continual improvement is one of the essential elements of Total Quality Management, and is based on the ***Shewhart Cycle,*** *also called the* ***Deming Cycle*** *or the Plan-Do-Check-Act (PDCA) Cycle.*[19] *ISO has incorporated the PDCA Cycle as an engine for continual improvement in the ISO 14000 Environmental Management System, and with the advent of the 2000 release, ISO 9000. ISO 14000 went so far as to use PDCA as the basis of its EMS model. The relationship is also there in the ISO 9000 QMS model*[20]*, but is far less perceptible. Nevertheless, ISO's TC 176 is to be commended for bringing the benefits of PDCA to ISO 9000.*

Requirement 2 At every management review, top management must assess continual improvement opportunities. Those providing input to the reviews will often put forward improvement recommendations. Improvement opportunities may also become apparent as a result of inspiration engendered by the discussion and exchange of information that takes place in the review meetings. Whatever the source, management should look at every opportunity for improvement, and proceed with those that are in the best interests of the organization and its customers and compatible with available resources.

Requirement 3 At every management review, top management must assess the need for changes to the QMS and its elements, including quality policy, procedures, processes, and objectives. This reflects the reality that sometimes it takes more than continual improvement to achieve what the organization wants. For example, if a production process is found to be incapable of meeting its goals or too inefficient for further use, then the solution is found in *change* rather than improvement. If the quality policy is no longer appropriate for the organization's business, then it must be changed. It goes without saying that the quality objectives will continually be changing. As existing objectives are achieved, new objectives will replace them. Requirement 3 is intended to ensure that top management stays abreast of the current situation and changes the QMS and its elements as necessary.

Requirement 4 This clause requires that records be kept of all management reviews, and that they be maintained and controlled in accordance with the requirements of clause 4.2.4. The records will be useful to the organization and to auditors, internal and external.

Management Review, Continued

ISO 9001:2000, Clause 5.6.2: Review Input

The input to management review shall include information on

a) results of audits,
b) customer feedback,
c) process performance and product conformity,
d) status of preventive and corrective actions,
e) follow-up actions from previous management reviews,
f) changes that could affect the quality management system, and
g) recommendations for improvement.

Requirements—Clause 5.6.2

This clause spells out the mandatory input items to be included in the management reviews.

Application Information—Clause 5.6.2

Requirement Top management is required to appraise the seven categories of input information specified in clause 5.6.2 at each management review to satisfy the requirement of this clause. The seven input categories have data from

a) Audits, internal and external
b) Customer feedback, internal and external
c) Process performance and product conformity measurement
d) Preventive and corrective actions underway and completed during the review period
e) Follow-up reports on action items continuing from previous reviews
f) Information concerning changes that could affect the QMS
g) Recommendations for improvement

The best way to handle this is by developing and implementing a documented management review procedure that describes the review process in detail, specifies the presentation format, establishes the schedule, defines attendance requirements, and assigns responsibility for the various inputs. There is no ISO 9001 requirement for a documented management review procedure, but having one will make it much easier to consistently conduct the reviews in accordance with ISO requirements.

Management Review, Continued

ISO 9001:2000, Clause 5.6.3: Review Output

The output from the management review shall include any decisions and actions related to

a) improvement of the effectiveness of the quality management system and its processes,
b) improvement of product related to customer requirements, and
c) resource needs.

Requirements—Clause 5.6.3

This clause specifies the mandatory output to be obtained from the management reviews.

Application Information—Clause 5.6.3

Requirement 1 Having reviewed the input data, management must make decisions and take actions as appropriate. Specifically, the decisions made and actions taken must concern:

a) Improving the QMS and its processes, procedures, policies, and so on
b) Improving the product or service, bringing it up to or beyond customer requirements and expectations
c) Resources of all categories needed by the QMS

Requirement 2 In addition to the decisions made and actions taken, management must also record these decisions and actions for the QMS records, and maintain them in accordance with clause 4.2.4.

Resource Management—Requirements of Clause 6

Clause 6 is concerned with the essential resources of the quality management system. Its requirements are found in four two-digit clauses:

6.1 Provision of Resources
6.2 Human Resources
6.3 Infrastructure
6.4 Work Environment

The reader is once again encouraged to review Figure 4-1. In this chapter so far we have discussed Clause 4's basic requirements for the QMS shell, as well as the management requirements found in Clause 5, the first of the four elements comprising the operation of the process-based QMS. Clause 6 represents the second of those elements.

The thrust of Clause 6 is that the quality management system, like any other operating system, will require a variety of resources to enable it to function. The resources that are needed include people, information, infrastructure, and capital. Management must provide all of these resources in the necessary types and numbers.

Resource Management, Continued

ISO 9001:2000, Clause 6.1: Provision of Resources

The organization shall determine and provide the resources needed

a) to implement and maintain the quality management system and continually improve its effectiveness, and
b) to enhance customer satisfaction by meeting customer requirements.

Requirements—Clause 6.1

This clause levies a requirement on the organization's management to do two things:

1. Management must determine what resources will be required in order to
   - Implement a functional QMS
   - Operate and maintain it
   - Continually improve its effectiveness
   - Enhance customer satisfaction by meeting customer requirements
2. Management must provide those resources.

Application Information—Clause 6.1

Requirements for the generic *types* of resources are found in the three clauses (6.2 through 6.4) that follow, but this clause is concerned with management's requirement to define the organization's resource needs and to make them available.

Requirement 1 As the organization's quality management system is developed, management must determine the specific resources required by the QMS and what quantities and excellence levels will be necessary for implementation and operation. Management must further define the resources needed in order to continually improve QMS effectiveness, as well as consistently meet customer requirements. There are two messages here. The first is, it is the organization's responsibility to define the resources needed for its QMS implementation and long-term operation and maintenance. The second is that just installing and operating a quality management system is not good enough. To conform to ISO 9000, the QMS must be continually improved, and products must meet customer requirements with the aim of enhancing customer satisfaction. These efforts require additional resources, and management must determine what they are.

Requirement 2 The organization's management must make all of these resources available in a timely manner.

Resource Management, Continued

ISO 9001:2000, Clause 6.2: Human Resources

The requirements of clause 6.2 are contained in the subclauses 6.2.1, General, and 6.2.2, Competence, Awareness and Training. This section deals with the people who must operate the processes associated with the QMS.

> **Human Resources, Continued**
>
> **ISO 9001:2000, Clause 6.2.1: General**
>
> Personnel performing work affecting product quality shall be competent on the basis of appropriate education, training, skills and experience.

Requirement—Clause 6.2.1

This clause requires, in a straightforward manner, that every employee whose work can affect the quality of the product or service must be competent to perform that work. Competence is to be judged on the basis of education, training, skills, and experience.

Application Information—Clause 6.2.1

Requirement Any job position that can have an impact on product quality must be manned by an employee competent to do that job. Employee records and job requirement documents will be required to substantiate conformance.

> **Human Resources, Continued**
>
> **ISO 9001:2000, Clause 6.2.2: Competence, Awareness, and Training**
>
> The organization shall
>
> a) determine the necessary competence of personnel performing work affecting product quality,
> b) provide training or take other actions to satisfy these needs,
> c) evaluate the effectiveness of the actions taken,
> d) ensure that its personnel are aware of the relevance and importance of their activities and how they contribute to the achievement of the quality objectives, and
> e) maintain appropriate records of education, training, skills and experience (see 4.2.4).

Requirements—Clause 6.2.2

The organization is required to do five things:

a) It must determine the level of competence necessary for employees whose work could affect quality of the product.
b) It must provide training, or in some other manner ensure that employees are competent.

c) It must evaluate the effectiveness of training provided or other actions taken.
d) It must ensure that employees are aware of the relevance and importance of their activities and efforts, and how they contribute to the achievement of the organization's goals and objectives.
e) It must maintain records of employee qualifications, including education, training, skills, and experience in accordance with clause 4.2.4

Application Information—Clause 6.2.2

If subclause 6.2.2a is taken in isolation, it appears that the organization is only required to *determine* the essential competence for employees working in activities that affect product quality. That is part of it, but then, referring back to clause 6.2.1, once the organization has established the competence requirements, it must put competent employees into those positions. So the requirements for the organization are actually the following.

Requirement 1 ISO requires that the organization determine the necessary education level, specialized training, skills, and experience required of the employees assigned to the various activities that may affect product quality. This usually means that documented job descriptions or job qualification specifications must be developed if they do not already exist.

Requirement 2 The organization must assign competent employees to these activities. To ensure conformance, job qualification requirements must be compared against the employees' résumés or other records. Many organizations get along without formal job descriptions or job qualification documentation. It may be difficult for them to convince registrars that the organization has met the competence requirement.

Requirement 3 The organization must provide training or make other suitable arrangements to bring employee competence up to the required levels. If the organization has the in-house ability for such training, that is usually the best approach. Otherwise, the necessary training can be contracted to consultants, colleges, or other training specialists. A third way to satisfy the competence requirement is by hiring new employees already possessing the necessary competence. This should be done only when training is not a viable alternative.

Requirement 4 Whatever actions the organization takes to achieve the necessary competence level, the actions must be evaluated for effectiveness. If training has been provided, have the results of the training met the organization's expectations? The implication is that the organization should learn from its experiences, and make improvements to the training program for the next time around. Similarly, if competency has been imported from outside the organization, are the results satisfactory? How can better results be achieved next time? Actually, the bottom line is that if the training or other actions have not produced the competence level required for the job position, conformance with the clause has not been achieved, and further actions are necessary.

Requirement 5 The organization must ensure that employees understand how their activities fit into the overall picture of producing a product or providing a service. ISO refers to that as the *relevance* of their activities. The organization must ensure that employees understand that their activities are important, and that they contribute to the organization's objectives, quality and otherwise. This may seem like something that automatically goes with the job, but unfortunately, millions of people in the workforce do not have a clue as to the relevance or importance of their work, or whether it is contributing to an important objective or not.

To illustrate this point, several years ago we were involved in an electronics plant's transition from a traditional manufacturing system to the *lean production,* or *just-in-time,* system. In the existing traditional setup, various cables were manufactured in one part of the plant, circuit boards in another, and so on. These subassemblies were shipped to another plant where they were put together, forming complex multi-rack test systems that were used for testing and troubleshooting electronic systems aboard military aircraft. Even though the company tried diligently to inform the employees, few, if any, in the first plant knew what they were making, what it was for, or how it was ultimately to be used. The change over to lean production brought all of the production elements together in a compact work group that fabricated all of the subassemblies, tested them, assembled them into the finished test system product, and tested that. For the first time in their careers, employees could see and appreciate the result of their labors, and could understand how critically important their work was. In employee surveys we learned that they considered this to be a major motivational force, and a source of enormous personal satisfaction. The point is that employees work at their best when they know that what they are doing is significant, and is important to the organization and to its customers.

Requirement 6 The organization must maintain records of all relevant training provided, and educational, training, skills, and experience records of employees working in activities that can impact quality. These are the records from which the registrar will determine whether or not the organization conforms to the requirements of clause 6.2. Records must be maintained in accordance with clause 4.2.4.

Resource Management, Continued

ISO 9001:2000, Clause 6.3: Infrastructure

The organization shall determine, provide and maintain the infrastructure needed to achieve conformity to product requirements. Infrastructure includes, as applicable

a) buildings, workspace and associated utilities,
b) process equipment (both hardware and software), and
c) supporting services (such as transport or communication).

Requirements—Clause 6.3

Clause 6.3 requires the organization to do three things.

1. Determine the infrastructure needed for its activities in terms of
   - Buildings, workspace, and utilities
   - Process equipment in hardware and software categories
   - Supporting services
2. Provide these infrastructure resources.
3. Maintain them.

Application Information—Clause 6.3

For most organizations seeking ISO 9000 registration, the majority of these infrastructure resources will already be in place. However, the QMS will invariably put some demands on the operation that require different or additional resources. Management should focus on requirements for infrastructure resources beyond those that already exist.

Requirement 1 The organization's management must determine what existing resources must be replaced or upgraded for the QMS, and what new resources will be required. It would be unusual to require new buildings, although the existing ones may need to be upgraded. It may be necessary to arrange workspaces differently to support the requirements of Clauses 7 and 8. New equipment and software may be required to support any new processes and the measurement and monitoring required. The organization's communications systems may require modernization. Whatever they may be, the organization must determine its infrastructure needs.

Requirement 2 The organization must make these infrastructure resources available in a timely fashion. When monitoring and measuring commences, it must be supported by the necessary equipment and software. The same is true for new office or factory layouts and all the other resources that will be needed.

Requirement 3 Something that is critically important to any organization that wants to produce quality products or services is maintenance. Unfortunately the North American

ISO INFO

Total Productive Maintenance is a Japanese maintenance system that maintains all systems and equipment continuously and promptly all of the time. In a rushed workplace, it is not uncommon to slack off on machine and system maintenance. This is unfortunate because poorly maintained machines cannot achieve the quality and productivity to be competitive. Manufacturers who have embraced the TPM philosophy have achieved productivity increases of 60 percent, with breakdowns reduced to 1 percent or less compared to their previous experience.[21]

record for factory infrastructure maintenance is lamentable. We tend to install expensive machines and run them until they break down. We should be applying preventive maintenance techniques such as **Total Productive Maintenance (TPM)**, to keep the machines in tip-top, better-than-new shape throughout their useful lives. Although not mentioned in this clause, we should also be applying continual improvement to our machines and tooling. This is a requirement under clause 5.1.

Under clause 6.3, ISO requires that the total infrastructure be continually maintained. That means machines and equipment, buildings, utilities, support services, and everything else that is part of the organization's infrastructure.

Resource Management, Continued

ISO 9001:2000, Clause 6.4: Work Environment

The organization shall determine and manage the work environment needed to achieve conformity to product requirements.

Requirement—Clause 6.4

The requirement of clause 6.4 is for the organization to determine the kind of work environment required to enable the achievement of conformity to product requirements, and then to provide that kind of environment (manage it) on a continuous basis.

Application Information—Clause 6.4

To satisfy this requirement the organization has to look at the totality of the work environment, which includes:[22]

- Creative work methods and opportunities for greater involvement to realize the potential of people in the organization
- Safety rules and guidance, including the use of protective equipment
- Ergonomics
- Workplace location
- Social interaction
- Facilities for people in the organization
- Heat, humidity, light, and airflow
- Hygiene, cleanliness, noise, vibration, and pollution

Requirement The organization must determine the kind of work environment it requires to conform with ISO 9000 and product requirements, and to enable the QMS to function effectively. Then it must manage—that is, control—the environment accordingly.

Product Realization (Production)—Requirements of Clause 7

Clause 7 is concerned with what ISO calls **product realization.** Product realization is a term that is used to cover the entire production cycle, starting with the product concept, and concluding with the finished, tested product. The requirements of Clause 7 are found in six two-digit clauses:

7.1 Planning for Product Realization
7.2 Customer Related Processes
7.3 Design and Development
7.4 Purchasing
7.5 Production and Service Provision
7.6 Control of Monitoring and Measuring Devices

A review of Figure 4-1 will show that we have arrived at the third of the four blocks of the process-based quality management system. This is the section of ISO 9001 that places the controls on all actions that deal with designing the product, purchasing the input materials needed for production, manufacturing the product (or delivering the service), and ensuring that monitoring of the processes is valid. This is the only clause within which organizations may obtain exclusions, and then only when a requirement is not applicable to the organization's activities.

The supposition here is that a new product is to be conceived. At the starting point the organization does not know what the product is or what processes will be required to design, develop, manufacture, and test it. The clauses of this section take the organization through all these phases, and establish requirements that the organization must satisfy.

Product Realization, Continued

ISO 9001:2000, Clause 7.1: Planning of Product Realization

The organization shall plan and develop the processes needed for product realization. Planning of product realization shall be consistent with the requirements of the other processes of the quality management system (see 4.1).

In planning product realization, the organization shall determine the following, as appropriate:

a) quality objectives and requirements for the product;
b) the need to establish processes, documents, and provide resources specific to the product;
c) required verification, validation, monitoring, inspection and test activities specific to the product and the criteria for product acceptance;
d) records needed to provide evidence that the realization processes and resulting product meet requirements (see 4.2.4).

Product Realization, Continued

ISO 9001:2000, Clause 7.1: Planning of Product Realization

The output of this planning shall be in a form suitable for the organization's method of operations.

NOTE 1 A document specifying the processes of the quality management system (including the product realization processes) and the resources to be applied to a specific product, project or contract, can be referred to as a quality plan.

NOTE 2 The organization may also apply the requirements given in 7.3 to the development of product realization processes.

Requirements—Clause 7.1

The total requirement set of clause 7.1 is as follows:

1. Plan and develop the needed product realization processes.
2. The planning and development must be consistent with clause 4.1, and with the requirements of other QMS processes.
3. As a part of product realization planning the organization must determine (as appropriate):
   a) Quality objectives for the product or its processes, and requirements for the product
   b) The need for product-specific new processes, documentation, and resources
   c) Requirements for verification, validation, monitoring, inspection, and test activities specific to the product; and product acceptance criteria
   d) What records will be required as evidence that the processes and product meet requirements
4. The organization must put the output of its planning into a form that is suitable for its method of operations.

Application Information—Clause 7.1

The primary requirement of clause 7.1 is for the organization to *plan and develop* the processes that will be needed for product realization. That is followed by a series of secondary requirements that must be carried out as well.

Requirement 1 The organization must plan and develop all the processes that will be required by the new product. Planning must be consistent with clause 4.1, which deals with a similar planning phase for the QMS itself. Clause 7.1's subclauses go into detail for what is to be covered in the required planning and development activities.

Requirement 2 Subclause a can be confusing. It reads, " . . . the organization shall determine . . . *quality objectives and requirements for the product."* It is not clear what TC 176 had in mind here. One can interpret the entire phrase as requiring quality objectives and product requirements. Another interpretation is that the adjective *quality* is meant to apply to both *objectives* and *requirements.* If that is the case, then the requirement is for quality objectives and quality requirements, that is, the level of product quality required. We like this interpretation because *product requirements* are covered in clause 7.2. and it would not make sense to have requirements for the same thing in two places. Subclause a, then, requires the organization to determine quality objectives and quality requirements for the new product. Quality objectives may relate to product characteristics or to the processes related to the product. Quality requirements may apply to any of the product's specifications, characteristics, or attributes.

Requirement 3 Subclause b requires the organization to determine (as appropriate)

- The need to establish new processes
- The need for documentation specific to the new product
- Any product-specific resources that will be needed

Sometimes new products may be designed, developed, manufactured, and tested using the same processes, documentation, and resources already in place for other products. Often that is not the case, however, and the requirement of this clause is simply to determine what additional resources, documentation, and processes will be needed.

Requirement 4 One requirement of subclause c is similar to b. In this case the organization must determine any additional processes and activities required for the verification, validation, monitoring, inspection, and test of the new product (and its product-specific processes).

Requirement 5 Subclause c also requires the organization to determine acceptance criteria for the product, that is, determine the basis or test criteria upon which the product is accepted or rejected. If the product is designed to a set of performance specifications, then it may only be necessary to verify that the product meets the specifications. It is not always so cut and dried, but whatever the acceptance criteria turns out to be, it should be documented for consistency.

Requirement 6 Subclause d requires the organization to determine the records that will be needed to provide evidence that the processes and the product meet requirements. There is a direct link from this clause to clause 4.2.4, Control of Records. Records required may include monitoring procedures and data for processes associated with the product, and product and process verification, validation, monitoring, inspection, and test data.

Requirement 7 The final requirement of clause 7.1 is that the result of all the required planning—that is, the planning output—must be put into a form that is suitable to the organization's methods of operations. Remember, the planning output will be used to establish and operate and monitor the needed new processes. Some organizations may decide that no documentation is required. (We doubt that they would convince the registrar, however.) For any but the simplest of products and processes, the planning output should be documented. Why ISO does not say that directly is a mystery. The result of the planning may be years of production. For the sake of consistency, which is what ISO 9000 is about, all the product requirements, processes, procedures, specifications, tests, and objectives should be documented. Note 1 to clause 7.1 provides a clue to ISO's preference, stating that if the organization should document the planning output, it may be called a quality plan.

Note 2 is not a requirement. It states that the organization *may* apply the requirements of clause 7.3 to the development of product realization processes. Where the development planning has been treated lightly in clause 7.1, it is much more thoroughly detailed in 7.3 as it applies to the design and development of products. ISO is simply suggesting that there is no reason not to apply the techniques and methodology used with clause 7.3 to the planning required by clause 7.1.

Product Realization, Continued

ISO 9001:2000, Clause 7.2: Customer-Related Processes

The requirements of Clause 7.2 are found in three subclauses.

7.2.1 Determination of Requirements Related to the Product

7.2.2 Review of Requirements Related to the Product

7.2.3 Customer Communication

Customer-Related Processes, Continued

ISO 9001:2000, Clause 7.2.1: Determination of Requirements Related to the Product

The organization shall determine

a) requirements specified by the customer, including the requirements for delivery and post-delivery activities,

b) requirements not stated by the customer but necessary for specified or intended use, where known,

c) statutory and regulatory requirements related to the product, and

d) any additional requirements determined by the organization.

Requirements—Clause 7.2.1

Clause 7.2.1 requires the organization to determine four levels of requirements for the new product:

a) Customer requirements for the product (including delivery and post-delivery activities)
b) Requirements beyond those stated by the customer that are necessary for the specified or intended use of the product
c) Any laws or regulations that apply to the product
d) Any other requirements to be imposed by the organization

Application Information—Clause 7.2.1

This clause requires the organization to determine the complete set of requirements for the new product. All four levels of requirements designated in this clause will not apply to every product. In some cases there will be no customer requirements. In many cases statutory and regulatory requirements will not apply. There remain many categories of products to which all of the levels of requirements apply. Some examples are aerospace products, automobiles, ships, specialized equipment, military equipment, medical equipment, computers, television sets, food products, and so forth. They can even apply to something as generic as an automobile tire or the gasoline we buy at the corner service station.

Requirement 1 The organization must determine what requirements the customer has established for the new product. This could be none, or it could be a set of precise specifications that must be met. ISO also requires the organization to determine any peripheral customer requirements, like shipping, installation, test at the customer's site, or other post-production activities.

Requirement 2 The organization has to look beyond customer requirements (if any) to determine other requirements necessary considering the specified or intended use—when that information is known. For example, a customer has specified an all-terrain vehicle with special capabilities. The organization is aware that the vehicle is primarily to be used in the tropics, but arctic expeditions are also planned. Although the customer did not require it, the organization imposed its own requirement to enable operation across a temperature range compatible with the vehicle's intended use, specifying an operating range of −70 to +140 degrees Fahrenheit rather than the normal −40 to +140. This new internal requirement will be addressed during the design of the vehicle.

Requirement 3 The organization must determine if any laws or regulations apply to the product and, if so, what they require. It is difficult enough to keep track of applicable legal and regulatory requirements in one country, but doing business on an international basis raises this issue to a whole new level. An organization dealing with international customers must determine what relevant statutory and regulatory requirements exist wherever its product are sold.

Requirement 4 The organization must determine what additional requirements are to be placed on the product from within the organization. This sounds strange. The reader may say that if the organization can determine internal requirements, then they must already have been determined. ISO is not attempting to put us into a loop here. The fact is, especially in larger organizations, requirements may come from any of several departments. Manufacturing may, for example, require that the product be designed in such a way that assembly is facilitated. A test department may have a requirement for attaching test adapters. The point is, before the product is committed to design and development, all the requirements must be known. Only by doing so can the organization avoid costly and disrupting redesign at some later stage.

At this point the organization should be aware of all requirements associated with the new product.

Customer-Related Processes, Continued

ISO 9001:2000, Clause 7.2.2: Review of Requirements Related to the Product

The organization shall review the requirements related to the product. This review shall be conducted prior to the organization's commitment to supply a product to the customer (e.g. submission of tenders, acceptance of contracts or orders, acceptance of changes to contracts or orders) and shall ensure that

a) product requirements are defined,
b) contract or order requirements differing from those previously expressed are resolved, and
c) the organization has the ability to meet the defined requirements.

Records of the results of the review and actions arising from the review shall be maintained (see 4.2.4).

Where the customer provides no documented statement of requirement, the customer requirements shall be confirmed by the organization before acceptance.

Where product requirements are changed, the organization shall ensure that relevant documents are amended and that relevant personnel are made aware of the changed requirements.

NOTE In some situations, such as internet sales, a formal review is impractical for each order. Instead the review can cover relevant product information such as catalogues or advertising material.

Requirements—Clause 7.2.2

The previous clause, 7.2.1, required the organization to determine the product requirements. Clause 7.2.2 picks up at that point, and requires the organization to do four things concerning the requirements.

1. Review the requirements prior to any commitment to produce the product. The purpose of the review is to ensure that:
   a) All necessary product requirements are defined
   b) Contract or order requirements that differ from previously expressed requirements are resolved
   c) The organization has the ability to meet the requirements
2. Maintain records of the results of the requirements review, and actions arising from it
3. In the absence of documented customer requirements, confirm requirements before order/contract acceptance
4. If product requirements are changed, ensure that relevant documents are updated, and that relevant employees are made aware of changes

Application Information—Clause 7.2.2

Once the organization has identified all the product requirements, from all sources, ISO requires a formal review of the requirements to make certain that the organization knows exactly what the product is to be, and to avoid any potential contractual miscues. The idea is that before an organization enters into an agreement to produce a product for a customer, it should:

- Know what the product must be in every detail
- Eliminate any misunderstandings or information gaps between the organization and the customer concerning the product and contract or order
- Satisfy itself that the organization is capable of providing a product that meets the requirements

Requirement 1 Before agreeing to any order or contract, the organization must review all the product requirements, from all sources, to:

a) Be certain that all the requirements are in hand and understood.
b) Resolve real or apparent conflicts in the requirements, or differences between the requirements and those earlier expressed. Any uncertainty between the organization and the customer concerning requirements must be eliminated prior to contract/order acceptance.
c) Ensure that the organization has the capability to produce the required product before it agrees to the contract or order. It is not unusual for organizations to default on a contract because it determined too late that a product was beyond its capabilities and resources. ISO 9000 seeks to avoid this type of failing.

Requirement 2 The organization must keep and maintain records of the product requirement reviews and of any actions resulting from the reviews. This requirement is linked to clause 4.2.4. These records will be invaluable in the event of subsequent contractual disputes, and will also be useful to auditors as evidence of conformance with the standard.

Requirement 3 In the event that the customer provided no product requirements, or if the requirements were not documented, the organization must confirm the requirements it believes to be relevant with the customer. This is sometimes called reverse contracting, and involves documenting the organization's requirement set and obtaining formal customer approval. This should always be documented. Failure to do this is an invitation to contractual problems down the road.

Requirement 4 When product requirements are changed, the organization must ensure that all relevant documentation (i.e., specifications, drawings, test instructions, and so on) reflect those changes. In addition, the organization's employees associated with the product must be made aware of the changed requirement(s). This requirement should apply equally before contract/order acceptance and at any other time a requirement is changed.

Clause 7.2.2 has a note attached that concedes that *in some situations, such as internet sales, a formal review is impractical for each order.* We would suggest that in many sectors such situations are the norm. Such reviews are impractical in virtually all retail sales and in most categories of services. The note goes on to say that if the reviews required above are impractical, then *the review can cover relevant product information such as catalogues or advertising material.* The point is that the organization is required to mutually agree with the customer that the product being sold is the product the customer wants to buy. Even that cannot be done in every case, but the effort will pay off in customer satisfaction.

Customer Related Processes, Continued

ISO 9001:2000, Clause 7.2.3: Customer Communication

The organization shall determine and implement effective arrangements for communicating with customers in relation to

a) product information,

b) enquiries, contracts or order handling, including amendments, and

c) customer feedback, including customer complaints.

Requirements—Clause 7.2.3

Clause 7.2.3 requires the organization to determine and implement the means to effectively communicate with customers in relation to product information, inquiries, contracts and order handling, and customer feedback, including complaints.

Application Information—Clause 7.2.3

Requirement Clause 7.2.3 requires the organization to set up an effective communication system with its customers. It would seem that organizations would do this naturally, without the need for another requirement. Many organizations devote a great deal of attention to **customer communications**, but unfortunately many more seem to feel that the less customer communication the better. In our present competitive environment, any organization that depends on its customers for its very existence must be in tune with them. Failure to do so can lead to rapid extinction. Specifically, clause 7.2.3 is concerned with customer communication in three areas:

a) The exchange of product information. This entails
   - Customers telling the organization what they need in terms of concepts, specifications, or requirements
   - The organization telling its customers about its products and capabilities
   - Communicating to ensure that the customer and the organization understand each other in terms of products and contracts or orders

b) Handling customer inquiries, and providing an effective means of two-way information flow concerning contracts, order handling, and contract/order amendments. Until recently these things were done in face-to-face meetings, by telephone, or by mail. Technology makes many more communication media available today, including Web sites, e-mail, facsimile, voice/video conferencing, and others. Organizations should take advantage of those deemed to offer effectiveness and value to the organization and convenience for customers.

c) Channels for customer feedback. Customer input is vital at the beginning of a new product cycle so that the product ultimately produced is what the customer wanted. Customer input is equally important after the product has been delivered. The final product validation and the final judgment on the product's quality and fitness for the purpose intended is something only the customer can provide. Organizations need that information to fuel the continual improvement effort and to learn from experience. Some customer input inevitably comes in the form of complaints. While the organization may prefer positive comments, customer complaints should be seen as gems to be mined. Management should ensure that customer complaints are input to the PDCA Cycle.

Product Realization, Continued

ISO 9001:2000, Clause 7.3: Design and Development

The requirements of Clause 7.3 are found in seven subclauses.

7.3.1 Design and Development Planning

7.3.2 Design and Development Inputs

7.3.3 Design and Development Outputs

7.3.4 Design and Development Review
7.3.5 Design and Development Verification
7.3.6 Design and Development Validation
7.3.7 Control of Design and Development Changes

The thrust of all of clause 7.3 is that the design and development of products is not something that should be done ad hoc. Rather, the **design and development** process should be logically structured and rigorously adhered to. The first step should be planning. The second, determining the design inputs. That is followed by the actual design stage, which has specific outputs. Design and development outputs are then reviewed to ensure that all is understood and in order. Then the design and development steps are verified and validated. That is followed by putting the design under control.

Design and Development, Continued

ISO 9001:2000, Clause 7.3.1: Design and Development Planning

The organization shall plan and control the design and development of product. During the design and development planning, the organization shall determine

a) the design and development stages,
b) the review, verification and validation that are appropriate to each design and development stage, and
c) the responsibilities and authorities for design and development.

The organization shall manage the interfaces between different groups involved in design and development to ensure effective communication and clear assignment of responsibility.

Planning output shall be updated, as appropriate, as the design and development progresses.

Requirements—Clause 7.3.1

The overriding requirement of clause 7.3.1 is that the organization must plan and control the design and development of its products.

1. In the planning stage the organization must determine
   - The stages necessary for the product's design and development
   - The review, verification, and validation appropriate to each of the design and development stages
   - The responsibilities and authorities necessary for the design and development processes

2. The organization must manage the interfaces between different groups involved in the design and development to
   - Ensure effective communication between the groups
   - Establish clear assignment of responsibility
3. The organization must update planning output as indicated as the design and development progresses.

Application Information—Clause 7.3.1

As the initial stage of design and development, the organization is required to plan and control the design and development process necessary for the new product. *Plan* means determine what is needed in terms of the design and development effort, and *control* means manage the effort. This is to be achieved by following a set of requirements.

Requirement 1 The organization must determine several things:

a) The design and development stages that are necessary for this new product
b) The review, verification, and validation that is appropriate at each stage of the design and development process
c) The design and development responsibilities and authority assignments that are necessary

Requirement 2 Design and development of the product will usually involve more than one department or group. Primary responsibility is customarily with an engineering group, but manufacturing, quality assurance, marketing, and accounting should also have input. The organization is required to "manage" the interfaces between the groups. This is to ensure that all relevant groups are heard, understood, and that their inputs are acted upon when appropriate. It is also to maintain the established lines of responsibility. It is not unusual for groups that should be playing a support role to attempt to insert themselves into the line management responsibility. An excellent example is that until a decade or so ago, the accounting departments of some automakers actually controlled the selection of components used, including tires and other items critical to safety and handling, in order to save a few cents on a car that sold for $10,000 or more. (That is analogous in today's environment to an insurance company employee with no medical training telling an HMO doctor that the test he or she wants to perform on a patient is not needed.)

On the other hand, it is not unusual for line departments that should have a voice in design and development to be ignored. The best example of this is the engineering department refusing to accept design and development input, even though it could improve product quality and reduce cost. ISO's intent is for management to manage these interfaces so that effective communication between the groups is accomplished, while maintaining the intended responsibility.

Requirement 3 As the design and development process moves forward, it is likely that changes to the original plan will be desirable or even necessary. Should that be the case, the organization must update the plans. This implies that the planning output is documented, otherwise what is there to update? Although there is no explicit requirement for it, we believe that design and development plans should be documented.

Design and Development, Continued

ISO 9001:2000, Clause 7.3.2: Design and Development Inputs

Inputs relating to product requirements shall be determined and records maintained (see 4.2.4). These inputs shall include

a) functional and performance requirements,
b) applicable statutory and regulatory requirements,
c) where applicable, information derived from previous similar designs, and
d) other requirements essential for design and development.

These inputs shall be reviewed for adequacy. Requirements shall be complete, unambiguous and not in conflict with each other.

Requirements—Clause 7.3.2

Clause 7.3.2 requires the organization to do three things concerning design and development inputs.

1. Determine the design and development inputs relating to product requirements, including
   a) Functional and performance requirements
   b) Applicable statutory and regulatory requirements
   c) Information derived from previous similar designs where applicable
   d) Any other requirements essential to the design and development of the new product
2. Maintain records of inputs relating to product requirements
3. Review these inputs to assure that they are complete, unambiguous, and not conflicting

Application Information—Clause 7.3.2

Requirement 1 The organization must determine all the inputs relating to requirements for the new product. This requirement is closely akin to those of clause 7.2.1. The design and development inputs from all sources are to be determined, and they must include:

- Product functional requirements, that is, what the product is supposed to do.
- Product performance requirements, namely, performance specifications.
- Any laws or regulations placing requirements on the product that must be taken into account during design and development.

- Any applicable information derived from previous similar designs that should be observed as product requirements (lessons learned).
- Any other requirements essential to the new product's design and development. These requirements may be in many categories, including time-to-market, produceability, repairability, cost, materials, reuse of appropriate designs, compatibility with existing manufacturing processes, environmental concerns, and so on.

Requirement 2 Records of the assembled design and development inputs related to product requirements must be maintained in accordance with clause 4.2.4.

Requirement 3 The organization must review the inputs to ensure that they are adequate for the product's design and development, and that they are unambiguous (to the reviewers and employees who will have to use them). The review must also ensure that no requirement is in conflict with any other. Frequently design and development projects find that requirements for one thing or another are missing, and that requirements may be interpreted in different ways (ambiguous). It is also not unusual to have conflicting requirements. The review of design and development inputs is intended to catch such problems before they cause trouble. Of course, that is not enough. The problems must then be eliminated before the design process starts.

Design and Development, Continued

ISO 9001:2000, Clause 7.3.3: Design and Development Outputs

The outputs of design and development shall be provided in a form that enables verification against the design and development input and shall be approved prior to release.

Design and development outputs shall

a) meet the input requirements for design and development,
b) provide appropriate information for purchasing, production and for service provision,
c) contain or reference product acceptance criteria, and
d) specify the characteristics of the product that are essential for its safe and proper use.

Requirements—Clause 7.3.3

1. Design and development outputs must:
   a) Meet the design and development input requirements
   b) Provide information needed for purchasing product materials, manufacturing the product, and/or service provision

   c) Either contain or make reference to product acceptance criteria
   d) Specify product characteristics necessary for safe and proper use
2. Design and development outputs must be in a form that allows verification against the inputs.
3. Design and development outputs must be approved prior to release.

Application Information—Clause 7.3.3

The previous clause, 7.3.2, gathered all the inputs needed for the design and development of the new product. Clause 7.3.3 is focused on the output of the design and development process, and requires the outputs to be crafted and handled in such a way that subsequent product realization steps have the approved design and development output information they need. This information is typically in the form of specifications, drawings, instructions, verification and test methodology, and acceptance criteria.

Requirement 1 The organization must ensure that design and development outputs:

a) Have addressed and satisfied all inputs. All design and development inputs from all sources (customer, government agencies, the organization itself) must be satisfied in the corresponding outputs.
b) Contain all the information needed by the subsequent product realization processes, including purchasing, manufacturing, service provision, and verification. Purchasing has to know what materials to buy. Manufacturing has to know what processes to use, how to set up production lines, special tooling to be used, skills required, and how the product is to be tested and verified. If the product is a service, the service providers need complete instructions and lists of equipment and materials required.
c) Contain product acceptance criteria, or provide reference to it. The people responsible for accepting or rejecting the product must have the information upon which to base that decision. Acceptance criteria may be contained directly in the design and development output, or the output may provide reference to another document containing the acceptance criteria.
d) Address any characteristics of the product that are essential for its safe and proper use. For safety, many products have to be used and maintained appropriately. For example, automobile tires are designed to be operated within specified load, speed, and pressure ranges. If the tire becomes overloaded, its pressure becomes too low, or if it is operated above its specified design speed, the tire may well self-destruct with disastrous consequences. ISO requires that such concerns have to be addressed for every product. Designers will use the design and development output data to make the product as safe as they know how. When complete safety is still not assured, and it often cannot be, the output data guides the development of instructions for proper use and warning labels.

Requirement 2 Although clause 7.3.3 does not address it, the design and development output clearly has to be documented. There is simply no other way to do it. The clause requires that the output data be *in a form that enables verification* against the design and development inputs. In other words, there must be a means of verifying that the outputs satisfy all input requirements. Because of the manner in which some organizations go about design and development, it is sometimes impossible to discern from the outputs whether or not all of the inputs have been addressed. This requirement is intended to cause the organization to specifically relate outputs to corresponding inputs, or in some other manner provide for verification. It is not difficult. It is simply a matter of discipline.

Requirement 3 As outputs flow from the design and development process, they must be reviewed by management (see 7.3.4) and approved before they are released to the subsequent product realization processes. Use of outputs before they are fully approved by management can result in wasted resources, and is to be prevented. Note that the requirement for approval-prior-to-release is another clue that the design and development outputs must be documented.

Design and Development, Continued

ISO 9001:2000, Clause 7.3.4: Design and Development Review

At suitable stages, systematic reviews of design and development shall be performed in accordance with planned arrangements (see 7.3.1)

a) to evaluate the ability of the results of design and development to meet requirements, and
b) to identify any problems and propose necessary actions.

Participants in such reviews shall include representatives of functions concerned with the design and development stage(s) being reviewed. Records of the results of the reviews and any necessary actions shall be maintained (see 4.2.4).

Requirements—Clause 7.3.4

Clause 7.3.4 requires the organization to do three things:

1. Conduct systematic reviews of design and development work. The reviews are to be performed in accordance with planned arrangements (relating to clause 7.3.1), and are to be done at stages deemed suitable to the organization. The purpose of the review is to evaluate the ability of the design and development results to meet requirements, and to identify problems encountered and the actions necessary to overcome them.
2. Include as participants representatives from each of the functions concerned with the stage(s) of design and development being reviewed.

3. Record the results of the design and development reviews and any necessary actions planned or taken. Maintenance of these records must be in accordance with clause 4.2.4.

Application Information—Clause 7.3.4

This clause is intended to prevent design and development being done in an unstructured, uncontrolled fashion. Through the requirements of clause 7.3.4, order and discipline is mandated for the design and development process.

Requirement 1 The organization must plan for and conduct formal, systematic reviews of the design progress at designated stages of the design and development process. Planning for these reviews is done under the requirements of clause 7.3.1. The function of the review is to

a) Evaluate the ability of the design and development results to meet the input requirements. That is, will the product as specified at this design and development stage meet customer, legal, regulatory, and organizational requirements?

b) Identify problems with the design, and propose actions to overcome them. If an input requirement has not been addressed, or if it is not being met, this information will be revealed by the reviews. Management has the responsibility to direct the design team to address an overlooked or misunderstood input requirement, or to establish an action to find a means of meeting the requirement.

Requirement 2 The organization must include members of all the functions concerned with the design and development stage(s) being reviewed as review participants. This means that the review is not just between management and the design staff. If manufacturing is (or if it can be) affected by the design work in this stage, then manufacturing must be represented. Purchasing must be represented when the review concerns materials or components that will have to be purchased. Representatives from the test function must participate in reviews concerning the testing of the product. This is a vital requirement since it keeps all the appropriate functions in the design loop in order to prevent sometimes disastrous surprises later on, when correcting them can become significantly expensive.

Requirement 3 The organization must keep formal records of the design and development reviews, including actions arising from the reviews, and must maintain them as part of the QMS records as required by the link to clause 4.2.4.

Design and Development, Continued

ISO 9001:2000, Clause 7.3.5: Design and Development Verification

Verification shall be performed in accordance with planned arrangements (see 7.3.1) to ensure that the design and development outputs have met the design and development input requirements. Records of the results of the verification and any necessary actions shall be maintained (see 4.2.4).

Requirements—Clause 7.3.5

This clause has two requirements.

1. The organization must, in accordance with planned arrangements (clause 7.3.1), verify that the design and development outputs meet all design and development input requirements.
2. The organization must record the results of the verification and any actions arising from the verification, and must maintain them as QMS records in accordance with clause 4.2.4.

Application Information—Clause 7.3.5

The dictionary definition of *verify* is *to prove to be true by demonstration, evidence, or testimony; to confirm or substantiate*. ISO is looking for confirmation that the design process has properly done its job.

Requirement 1 Clause 7.3.5 requires the organization to *prove to be true* that the design and development outputs have met all design and development input requirements. Verification requires that the input requirements be tested against design and development outputs to confirm (1) that they have been addressed, and (2) that the requirements have been met.

Requirement 2 The organization must record the results of the verification process, including any actions taken or planned. Verification results and actions are to be QMS records and must be maintained in accordance with clause 4.2.4.

Design and Development, Continued

ISO 9001:2000, Clause 7.3.6: Design and Development Validation

Design and development validation shall be performed in accordance with planned arrangements (see 7.3.1) to ensure that the resulting product is capable of meeting the requirements for the specified application or intended use, where known. Wherever practicable, validation shall be completed prior to the delivery or implementation of the product. Records of the results of validation and any necessary actions taken shall be maintained (see 4.2.4).

Requirements—Clause 7.3.6

This clause has two requirements.

1. The organization must perform validation of the design and development to ensure that the product is capable of meeting the input requirements for the product's specified application or intended use. This validation is to be completed prior to delivery or implementation when possible.

2. The organization must record the validation results and any actions arising from them, and must maintain them as QMS records in accordance with clause 4.2.4.

Application Information—Clause 7.3.6

Clause 7.3.5 above required the organization to verify the design and development outputs as having responded to and meeting the input requirements. The requirements of this clause, 7.3.6, are similar. They require the organization to prove the design and development to be valid by ensuring that the product itself meets the requirements for the product's specified application or intended use.

Requirement 1 The organization must perform design and development validation in accordance with the planned arrangements arising from clause 7.3.1. Validation is to be done prior to delivery or implementation of the product wherever practicable. Such validations are usually done with mock-ups, bread-boards, or preproduction units, but can be product from the production stream. Validation testing is to confirm that specifications are met and that the product is capable of meeting the requirements for its specified or intended use. Manufacturers do not always know to what use their products might be subjected.

Requirement 2 The organization must record the results of the validation process. Any action items (taken or planned) must also be recorded. These results and action items are considered QMS records, and must be maintained in accordance with clause 4.2.4. Note: This is the point at which the design is considered ready for full-scale production, and a *freeze* is typically applied on the design documents. After this, design documents may be changed only through a formal "change control" process, which we find in clause 7.3.7.

Design and Development, Continued

ISO 9001:2000, Clause 7.3.7: Control of Design and Development Changes

Design and development changes shall be identified and records maintained. The changes shall be reviewed, verified and validated, as appropriate, and approved before implementation. The review of design and development changes shall include evaluation of the effect of the changes on constituent parts and product already delivered.

Records of the results of the review of changes and any necessary actions shall be maintained (see 4.2.4).

Requirements—Clause 7.3.7

This clause has four requirements.

1. Design and development changes must be identified.
2. Design and development changes must be reviewed, verified, and validated, as appropriate. This is to include evaluation of the change's effect on constituent parts of the product and on product already delivered.

3. Design and development changes must be approved before implementation.
4. The organization must maintain records of change reviews and any resulting action items in accordance with clause 4.2.4.

Application Information—Clause 7.3.7

After verification and validation of the design and development results, ISO restricts the introduction of changes to the design documentation (specifications, drawings, bills of material, and test and acceptance criteria) to those that have been formally reviewed, verified and validated, and approved by appropriate authority.

Requirement 1 Changes to design and development documentation must be identified. This is usually done via an engineering change proposal (ECP) form which carries the affected document name and number, along with a change number or letter. For example, the second change proposal affecting "Flat Panel Matrix," Drawing 18922 might be identified as Flat Panel Matrix, Drawing 18922, Revision 002.

Requirement 2 Design change proposals must be reviewed by the organization to ensure that the change is required, and verified and validated to prove that the change will accomplish the intended result.

Change reviews must evaluate possible effects on the remaining constituent parts of the product. Changing the design of one part of the product may, for example, impose more mechanical or electrical stress on other parts. This will lead to future problems, and must be dealt with before the change is approved. Change reviews must also explore what the change means in terms of product already delivered. A change to improve a product that already meets customer requirements is one thing. However, if the change is to improve safety performance, the organization may decide to issue a recall, or at least a bulletin.

Design and development change reviews ensure that the proposed change is valid, while preventing frivolous or unnecessary changes.

Requirement 3 Design changes must have the approval of the appropriate organizational authority before they are implemented. That means that an authorized manager must sign the change document. It is sometimes necessary to obtain customer approval also.

Requirement 4 The organization must maintain records of the results of design and development change reviews, and any resulting actions. These records are to be maintained in accordance with clause 4.2.4.

Product Realization, Continued

ISO 9001:2000, Clause 7.4: Purchasing

The requirements of Clause 7.4 are found in three subclauses.

7.4.1 Purchasing Process
7.4.2 Purchasing Information
7.4.3 Verification of Purchased Product

The intent of clause 7.4 is to establish controls on the purchasing element of the product realization process, the benefit of which is better control of purchased product quality.

Purchasing, Continued

ISO 9001:2000, Clause 7.4.1: Purchasing Process

The organization shall ensure that purchased product conforms to specified purchase requirements. The type and extent of control applied to the supplier and the purchased product shall be dependent upon the effect of the purchased product on subsequent product realization or the final product.

The organization shall evaluate and select suppliers based on their ability to supply product in accordance with the organization's requirements. Criteria for selection, evaluation and re-evaluation shall be established. Records of the results of evaluations and any necessary actions arising from the evaluation shall be maintained (see 4.2.4).

Requirements—Clause 7.4.1

Clause 7.4.1 requires the organization to:

1. Ensure that purchased product conforms to the specified requirements.
2. Establish supplier controls proportionate to the effect the purchased product can have on product realization processes or the finished product.
3. Establish a supplier-selection criteria.
4. Evaluate and select suppliers based on their ability to meet the organization's requirements.
5. Maintain records of supplier evaluation results and any actions arising from the evaluations. Records must be maintained in accordance with clause 4.2.4.

Application Information—Clause 7.4.1

This clause requires the organization to establish effective and efficient **purchasing processes** for the evaluation and selection of suppliers, and to ensure that purchased products satisfy the organization's requirements. ISO 9004:2000, clause 7.4.1 offers the following to be considered by the purchasing processes:

- Timely, effective and accurate identification of needs and purchased product specifications
- Evaluation of the cost of purchased product, taking account of product performance, price, and delivery (cost-versus-price analysis, in TQM terms)
- The organization's need and criteria for verifying purchased products
- Unique supplier processes
- Consideration of contract administration, for both supplier and partner arrangements
- Warranty replacement for nonconforming purchased products
- Logistic requirements
- Product identification and traceability
- Preservation of product
- Documentation, including records
- Control of purchased product which deviates from requirements
- Access to suppliers' premises
- Product delivery, installation, or application history
- Supplier development
- Identification and mitigation of risks associated with the purchased product

Requirement 1 The organization must ensure that purchased product conforms to the purchase requirements specified. This is usually done by inspection, and may also require testing. Forward-looking organizations form partnerships with key suppliers and certify the supplier to determine and declare conformance. This eliminates, or at least greatly reduces, the need for incoming inspection by the organization.

Requirement 2 The organization must establish controls on its suppliers and purchased product. The degree of control should vary depending upon the significance of potential effect of the purchased product on the product realization processes or the finished product. In other words, the more important the purchased product to the production processes or the finished product, the more stringent the controls should be.

Requirement 3 The organization must establish criteria for evaluating, re-evaluating and selecting suppliers. The primary objective is to select suppliers who have the ability to supply conforming product on time, every time. Most of the bulleted items just listed are appropriate considerations for supplier evaluation and selection criteria. Other considerations include:[23]

- Supplier experience
- Supplier performance relative to competitors

- Product quality
- Delivery performance
- Response to problems
- Assessment of supplier's ability to satisfy requirements
- Supplier references and customer satisfaction record
- Financial viability
- Service, installation, and support capability
- Response to inquiries, requests for quotations
- Supplier awareness of and compliance with applicable legal and regulatory requirements
- Logistics factors, including location
- Supplier's community relations

ISO does not specify that the criteria be documented, but if it is not, it really does not exist in any practical sense. There are so many elements in the effective supplier evaluation and selection criteria, and so many people must employ them, that it must be documented. Registrars expect to see documented procedures for supplier evaluation and selection.

Requirement 4 The organization must evaluate and select suppliers in accordance with its procedures and criteria.

Requirement 5 The organization must maintain records of supplier evaluation/re-evaluation and any actions that result. The records must be maintained in accordance with clause 4.2.4.

Purchasing, Continued

ISO 9001:2000, Clause 7.4.2: Purchasing Information

Purchasing information shall describe the product to be purchased, including where appropriate

a) requirements for approval of product, procedures, processes and equipment,

b) requirements for qualification of personnel, and

c) quality management system requirements.

The organization shall ensure the adequacy of specified purchase requirements prior to their communication to the supplier.

Requirements—Clause 7.4.2

This clause requires the organization to:

1. Develop purchasing information that describes the product to be purchased, including where appropriate
   a) Requirements for approval of product and procedures, processes, and equ used by the supplier
   b) Requirements for qualification of supplier personnel
   c) Quality management system requirements
2. Ensure the adequacy of specified purchase requirements prior to transmitt supplier.

Application Information—Clause 7.4.2

Clause 7.4.2 requires the organization to employ a rather formal purchasing ir on system. Such a system imposes discipline on the organization to be certair pur-chased product specifications and requirements are complete and accurate.

Requirement 1 Purchasing information must adequately describe the prc uct to be purchased and, where required, must also

- Specify any requirements for approval of the product and approval of procedures, processes, and equipment used by the supplier. In some cases the organization may find it necessary to evaluate and approve the supplier's processes, procedures, and equipment. This is particularly important where the organization has had similar requirements imposed by the customer. When product approval requirements are included in the purchasing information, both the supplier and the organization understand and accept the basis for acceptance of the purchased product.
- Specify any requirements for the certification or qualification of supplier personnel in the purchasing information provided to the supplier. This might be the case, for example, if the organization is under a NASA or military contract stipulating that only operators who are certified to the "Hi-Rel" (high reliability) soldering standard may work on the product.
- Specify in the purchasing information any requirement for the supplier to operate and maintain a quality management system. In some instances organizations require suppliers to meet quality management system requirements. In many cases such a requirement would be for ISO 9000 registration, although some industries have developed similar QMS standards, tailored to the industry. For example, suppliers to Ford, General Motors, and DaimlerChrysler in the United States are required to be registered to, and be operating a QMS conforming to, the QS 9000 standard. The aerospace industry uses the AS 9000 QMS. When such requirements apply, they must be included in the purchasing information.

Requirement 2 The organization must ensure the adequacy of purchase requirements before they are communicated to the supplier. It is customary in many industries to

ISO INFO

We have found that rather than allowing the paperwork to flow from one group to the next for approval, it is far better to assemble the concerned parties, and go through the approval and release process in a single, coordinated effort. By using this technique, the time required to approve the purchasing information is dramatically reduced—sometimes from weeks to hours.

Goetsch & Davis

submit purchasing information to quality assurance review prior to transmittal to the customer. However it is managed, inputs from several departments, including engineering, manufacturing, contracts, and purchasing are necessary.

Purchasing, Continued

ISO 9001:2000, Clause 7.4.3: Verification of Purchased Product

The organization shall establish and implement the inspection or other activities necessary for ensuring that purchased product meets specified purchase requirements.

Where the organization or its customer intends to perform verification at the supplier's premises, the organization shall state the intended verification arrangements and method of product release in the purchasing information.

Requirements—Clause 7.4.3

This clause levies two requirements on the organization:

1. The organization must establish and implement the inspection and other activities necessary to ensure that purchased products meet specified requirements.
2. Where it is intended that purchased product verification is to take place at the supplier's premises, the intended verification arrangements and the method of product release must be included in the purchasing information.

Application Information—Clause 7.4.3

This clause contains two requirements for the organization. Their purpose is to ensure that the organization and the supplier both understand how and where the purchased product is to be verified to purchase requirements.

Requirement 1 The organization must determine how the purchased product is to be verified to purchase requirements, and must implement the means to accomplish the verification. This will normally involve inspection, and frequently requires parametric testing. The clause does not specifically state that this information must be communicated to the supplier, but it should be. The supplier needs to know how the organization intends to verify its product. It is not unusual for these details to be worked out jointly between the supplier and the organization.

Requirement 2 When it is intended that verification be performed at the supplier's plant, the organization must state that in the purchasing information, along with the intended method of product release and other arrangements that may be required. Failure to advise the supplier initially may cause delays and contractual conflicts later on because verification at the supplier's premises will require support from the supplier in terms of people, equipment, facilities, and so on.

Product Realization, Continued

ISO 9001:2000, Clause 7.5: Production and Service Provision

The requirements of clause 7.5 are found in five subclauses.

7.5.1 Control of Production and Service Provision
7.5.2 Validation of Processes for Production and Service Provision
7.5.3 Identification and Traceability
7.5.4 Customer Property
7.5.5 Preservation of Product

The previous clauses under clause 7 have taken us through planning for the new product, determination and review of product requirements, product design and development of the product, and purchasing related to the product. We are now at the point in the product realization process where all of these requirements come together on the manufacturing floor or the service provider's workplace, to be assembled as a finished product.

ISO INFO

The best rule for working with suppliers is to provide more *information rather than* less. *More information than the supplier needs may not help, but too little will certainly hurt.*

Goetsch & Davis

Production and Service Provision, Continued

ISO 9001:2000, Clause 7.5.1: Control of Production and Service Provision

The organization shall plan and carry out production and service provision under controlled conditions. Controlled conditions shall include, as applicable

a) the availability of information that describes the characteristics of the product,
b) the availability of work instructions, as necessary,
c) the use of suitable equipment,
d) the availability and use of monitoring and measuring devices,
e) the implementation of monitoring and measurement, and
f) the implementation of release, delivery and post-delivery activities.

Requirements—Clause 7.5.1

This clause requires the organization to *plan and carry out* production of product and provision of service under controlled conditions, and then it provides, in subclauses a through f, several examples of items included in *controlled conditions.*

Application Information—Clause 7.5.1

The manufacturing process, or the service provision process, is a critical step in the provision of product and must not be left to an ad hoc approach to build the product or supply the service.

Requirement 1 The organization must plan the production or service provision. That means that a disciplined effort is to be undertaken to plan each step of the process, defining the processes to be employed, stipulating tools and equipment, determining the work space requirements, defining the skills required and selecting the right people, establishing schedules, providing for purchased materials and other resources, and so on. Nothing is to be left intentionally to chance, although the plans may have to be modified as experience with the new product is gained.

Requirement 2 Production and service provision must be carried out under controlled conditions. As applicable to the product or service, the following are required:

- Availability of all information necessary to adequately describe the product's characteristics (specifications, drawings, functional descriptions).
- Availability of work instructions for the process operators. These are the working level procedures used by operators. Whether or not many of these require documentation under ISO 9000 is left to the organization to determine. The test is whether the organization can ensure the absence of deviations and nonconformities

without them. If the organization cannot ensure that, then they must be documented. Smart organizations always document them.

- Availability and use of suitable equipment. This could be something as simple as having the right size screwdriver at the workstation, or something as complex as a walk-in altitude and temperature chamber required for product testing. Whatever equipment the plan requires must be available.
- Availability and use of monitoring and measuring devices. We have the same comment as in the last bulleted item. These devices may be anything from a twenty-dollar micrometer to equipment worth millions of dollars. If it is in the plan, it must be available. Otherwise it should not be in the plan.
- Implementation of monitoring and measurement. This applies to both the product and the processes used in its production. Critical processes need to be controlled through monitoring and measuring to ensure consistency of product and the minimization of waste. As for monitoring and measuring the product itself, key production stages should be identified for product verification as it passes through the production processes, and as a finished product.
- Implementation of release activities. This normally entails a final inspection and test of the product to ensure that specified characteristics and performance have met the requirements. Upon satisfactory completion of this step, the product can be released for delivery to a customer.
- Implementation of delivery and post-delivery activities. Whatever agreements have been made for product delivery must be executed accordingly. The same is true for any post-delivery requirements, such as installing and checking out the product at the customer's premises.

Production and Service Provision, Continued

ISO 9001:2000, Clause 7.5.2: Validation of Processes for Production and Service Provision

The organization shall validate any processes for production and service provision where the resulting output cannot be verified by subsequent monitoring or measurement. This includes any processes where deficiencies become apparent only after the product is in use or the service has been delivered.

Validation shall demonstrate the ability of these processes to achieve planned results.

The organization shall establish arrangements for these processes including, as applicable

a) defined criteria for review and approval of the processes,

b) approval of equipment and qualification of personnel,

Production and Service Provision, Continued

ISO 9001:2000, Clause 7.5.2: Validation of Processes for Production and Service Provision

c) use of specific methods and procedures,
d) requirements for records (see 4.2.4), and
e) revalidation.

Requirements—Clause 7.5.2

This clause has four requirements.

1. The organization must validate any production and service provision processes where the resulting output cannot be verified by subsequent testing. This includes processes where product or service deficiencies become apparent only after the product is in use or the service has been delivered.
2. Validation of such processes must demonstrate the ability of the process to achieve planned results, that is, to produce product that conforms to the requirements even though not measurable at that point.
3. As applicable, the organization must establish arrangements for these processes, including
   a) Defined process review and approval criteria
   b) Approval of equipment used and qualification of processs operators
   c) Use of specific process validation methods and procedures
   d) Requirements for records of validation
   e) Revalidation as necessary

Application Information—Clause 7.5.2

As the production processes hand off semicomplete product from one process to the next, it is often impossible or impractical to verify the product. In some cases even the finished product cannot be verified short of destructive testing. In both of these instances it is necessary to ensure that the processes are designed and operated in a manner that ensures the product, even though it cannot be verified, does meet requirements. That is the thrust of clause 7.5.2.

Requirement 1 The organization must validate these processes. That will entail validation of process design on paper, testing, and monitoring to ensure that process parameters stay within their normal ranges, and that established procedures are followed. It may also require destructive testing of samples as appropriate.

Requirement 2 Such validation must demonstrate the ability of the process to achieve planned results, that is, do what it is supposed to do.

Requirement 3 For such processes, the organization must establish arrangements to include, as applicable:

- Defined process review and approval criteria. This is done before the process is put into use. Although the clause does not specify it, this has to be documented.
- Approval of process equipment and qualification of operators. This is to assure that the equipment specified is capable, and that operator qualification is known and met before the process is activated.
- Use of specific methods and procedures to operate the process. There can be no doubt that these *specific methods* (work instructions) *and procedures* must be documented.
- Requirements for records of validation to be maintained under clause 4.2.4. Records of process control results, such as statistical process control charts and parametric data, should also be included.
- Revalidation when required, for example, after experiencing nonconforming product, or after process improvements have been implemented.

Production and Service Provision, Continued

ISO 9001:2000, Clause 7.5.3: Identification and Traceability

Where appropriate the organization shall identify the product by suitable means throughout product realization.

The organization shall identify the product status with respect to monitoring and measurement requirements.

Where tracebility is a requirement, the organization shall control and record the unique identification of the product (see 4.2.4).

NOTE In some industry sectors, configuration management is a means by which identification and traceability are maintained.

Requirements—Clause 7.5.3

This clause carries four requirements for the organization when **traceability** is required.

1. The product must be uniquely identified throughout the production process.
2. The product must be identified as subject to applicable monitoring and testing, and must be identified as to completion of such monitoring and testing.
3. The product's unique identification must be controlled.
4. Product identification records must be maintained under clause 4.2.4.

Application Information—Clause 7.5.3

In some industry sectors it is commonly required that products be traceable back to the plant (or plants) where they were produced, and through the processes, including testing and monitoring processes, involved in their production. In fact, the purchased materials and components going into the product may also carry traceability requirements. The Space Shuttle offers an example. Every component of the Space Shuttle is traceable back to the raw materials used. If a Space Shuttle valve fails, the unique identification on the failed valve (a serial number, for example) allows traceability to the plant where it was assembled, the production lot, and relevant dates, processes, and tests. Records of the valve's testing at each level of assembly may be reviewed. The component parts of the valve can be traced to their origin, and more production history may be reviewed. The purpose is that if this one valve fails, NASA wants to know why, and when that is determined, if the same failure mode is likely in other valves of the type. Suppose NASA found that the valve failed because of a faulty o-ring, a sealing device. If analysis of the o-ring found that it had been improperly installed, then NASA would probably require at the least that all valves of this type manufactured in the same lot (or batch) be disassembled and inspected for the same assembly defect. This is possible because of traceability. This is a very exotic example, but traceability is required in many, far less glamorous products. Using the automobile tire example again, all tires carry serial numbers. From these it is possible to trace any tire to the date and location of its manufacture. This information is useful to the manufacturer for warranty issues, and is useful also in the event of safety issues. Should data from the field indicate that a particular production run of tires is experiencing an unacceptable failure rate, it is possible to issue recall notices to customers. Without traceability that would not be possible.

Traceability adds cost to products. Therefore, it is used only for products where there is a clear reason for doing so. This clause is applicable for products that require traceability.

Requirement 1 When traceability is a requirement, the organization must uniquely identify the product at the earliest possible stage of production. **Identification** may be by serial number, barcode, or other suitable means. The product must carry that identification through the production processes, including inspection and test processes.

Requirement 2 The uniquely identified product must also carry its current monitoring and measurement (test) status as it proceeds through the production processes. This may be done by an attached "traveler tag" or by other suitable means, listing the required tests and inspections, along with verification of those accomplished.

Requirement 3 The organization must control product identification. *Control,* in this instance means that the identification must be accomplished through procedures that ensure that each product item is identified, that the identifier carries all the information necessary (such as model, date, location, etc.), and that no two products can be given the same identification. As the note to clause 7.5.3 indicates, this is often achieved by configuration management procedures. Whether a part of a larger configuration management system or not, the procedures used for identifying product should be documented. Product identification is fundamentally a bookkeeping issue, and it cannot be done properly without documented procedures.

Requirement 4 The organization must maintain product identification records under clause 4.2.4.

This clause seems more likely to be excluded than others. If the organization has no customer, legal, or regulatory requirement for product traceability, and if it offers no advantage to the organization or its customers, then a legitimate exclusion to the clause should be available under clause 1.2.

Production and Service Provision, Continued

ISO 9001:2000, Clause 7.5.4, Customer Property

The organization shall exercise care with customer property while it is under the organization's control or being used by the organization. The organization shall identify, verify, protect and safeguard customer property provided for use or incorporation into the product. If any customer property is lost, damaged or otherwise found to be unsuitable for use, this shall be reported to the customer and records maintained (see 4.2.4).

NOTE Customer property can include intellectual property.

Requirements—Clause 7.5.4

This clause has four requirements.

1. The organization must exercise care with customer property with which it is entrusted.
2. The organization must identify, verify, protect and safeguard customer property.
3. The organization must notify the customer of lost, damaged, or otherwise unsuitable customer property.
4. Customer property records must be maintained according to clause 4.2.4.

Application Information—Clause 7.5.4

There are numerous reasons for **customer property** to be placed in the organization's hands. In the service sector an organization may provide a warehousing operation for a customer. Everything in the warehouse belongs to the customer. A customer may contract the operation and maintenance of a plant or facility to an organization. All the property belongs to the customer. In the manufacturing sector it is not unusual for customers to provide certain components to be incorporated in the finished product. Customers often provide specialized equipment needed for the production or test of products being supplied by the organization. This is to say that it is not uncommon for customer-owned property to be left in an organization's care. When that is the case, Clause 7.5.4 requires nothing more than appropriate treatment of customer property.

Requirement 1 The organization must exercise care with customer property. This requirement at first seems almost condescending, or unnecessary. Every organization should understand that. However, having stated it as a requirement, ISO has armed the registrar to withhold or terminate registration should the organization fail to conform.

Requirement 2 The organization must identify, verify, protect, and safeguard customer property. This is really a requirement to manage all customer property in the organization's possession. Customer property should always be identified by name, type, serial number, and be listed in a current inventory of customer property. The organization should always verify that the property it received from the customer is the property the customer believes it provided. Verification may also entail inspection and testing to verify that it was received undamaged and is in good working order. Customer property must be protected and safeguarded from all hazards. For the most part this is obvious, but it may require segregating customer property from like organization property to prevent its unauthorized use.

Requirement 3 The organization must notify the customer of any lost, damaged, or otherwise unsuitable customer property. It is important that the customer be notified of loss, damage, or unsuitability as soon as the organization makes that determination.

Requirement 4 The organization must maintain records of customer property under clause 4.2.4. As the clause is written, it seems that ISO only wants records of loss, damage, and unsuitability reports. We believe the real requirement also includes records of all customer property transactions.

ISO has appended a note to this clause stating that customer property can include intellectual property. ISO 9004:2000, clause 7.5.3 defines intellectual property as including specifications, drawings, and proprietary information. Computer software programs are often considered to be intellectual property also.

Production and Service Provision, Continued

ISO 9001:2000, Clause 7.5.5: Preservation of Product

The organization shall preserve the conformity of product during internal processing and delivery to the intended destination. This preservation shall include identification, handling, packaging, storage and protection. Preservation shall also apply to the constituent parts of a product.

Requirements—Clause 7.5.5

This clause requires that the organization *preserve the conformity* of the product as it is being processed for delivery and being delivered to the customer.

Application Information—Clause 7.5.5

The organization is responsible for the preservation of its products until delivered to the customer. The organization should have defined and implemented processes for handling, packaging, storage, preservation, and delivery of product.[24]

Requirement The organization must handle, package, store, and deliver product in such a manner that the conformity of the product and its constituent parts will be preserved. The processes used for protecting product from damage must be appropriate for the type of product. The product may have special handling and protection requirements. Electronic media is one example. Products containing hazardous materials are another.

Product Realization, Continued

ISO 9001:2000, Clause 7.6 Control of Monitoring and Measuring Devices

The organization shall determine the monitoring and measurement to be undertaken and the monitoring and measuring devices needed to provide evidence of conformity of product to determined requirements (see 7.2.1).

The organization shall establish processes to ensure that monitoring and measurement can be carried out and are carried out in a manner that is consistent with the monitoring and measurement requirements.

Where necessary to ensure valid results, measuring equipment shall

a) be calibrated or verified at specified intervals, or prior to use, against measurement standards traceable to International or national measurement standards; where no such standards exist, the basis used for calibration or verification shall be recorded;
b) be adjusted or re-adjusted as necessary;
c) be identified to enable the calibration status to be determined;
d) be safeguarded from adjustments that would invalidate the measurement result;
e) be protected from damage and deterioration during handling, maintenance and storage.

In addition, the organization shall assess and record the validity of the previous measuring results when the equipment is found not to conform to requirements. The organization shall take appropriate action on the equipment and any product affected. Records of the results of calibration and verification shall be maintained (see 4.2.4).

When used in the monitoring and measurement of specified requirements, the ability of computer software to satisfy the intended application shall be confirmed. This shall be undertaken prior to initial use and reconfirmed as necessary.

NOTE See ISO 10012-1 and ISO 10012-2 for guidance.

Requirements—Clause 7.6

All the requirements for control of **monitoring and measuring** devices used by the organization are found in this clause.

1. The organization must determine the monitoring and measurement required for the product.
2. The organization must determine the monitoring and measuring devices needed.
3. The organization must establish processes to ensure that monitoring and measurement can be, and are, carried out in a manner consistent with the requirements.
4. Where necessary to ensure valid results, the organization must
   a) Calibrate or verify measurement devices at specified intervals, or prior to use. This calibration or verification must be traceable to international or national measurement standards. If such standards do not exist, then the basis for calibration or verification must be stated and recorded.
   b) Adjust or readjust measurement equipment as necessary.
   c) Provide suitable identification of measurement equipment to enable calibration status to be determined.
   d) Safeguard against adjustments that would invalidate the calibration and measurement results.
   e) Protect measurement equipment from damage and deterioration.
5. When measuring equipment is found to be out of calibration or otherwise not in conformance with requirements, the organization must assess the validity of previous measurements taken with this equipment, and record the results of the assessment.
6. When measuring equipment is found to be out of calibration or otherwise not conforming to requirements, the organization must take appropriate action on both the equipment and any affected product.
7. The organization must maintain records of the results of calibration and verification under clause 4.2.4.
8. When computer software is used in connection with monitoring and measurement, the organization must confirm its ability to satisfy the intended application prior to initial use, and must reconfirm as necessary.

Application Information—Clause 7.6

The control of monitoring and measuring resources involves selecting the right devices for the task, maintaining calibration of the devices, and keeping them in good working order.

Requirement 1 The organization must determine what kind of monitoring and measurement is required for the product. What tests must be conducted to verify that the product meets its requirements? That is the first step, because without that knowledge one can go no farther.

Requirement 2 The organization must specify (and secure, if not already on hand) the monitoring and measurement equipment needed to satisfy Requirement 1.

Requirement 3 The organization must establish processes to ensure that monitoring and measurement (1) can be carried out as planned, and (2) are carried out in a manner that is consistent with requirements.

Requirement 4 The organization must keep measuring equipment calibrated, properly adjusted, and maintained in order to ensure that measurement results are valid. To accomplish this it is necessary for the organization to:

a) Calibrate equipment at specified intervals, or prior to use. The proper intervals are usually specified by the device manufacturer, and are typically six months or one year. Calibration must be traceable to international or national standards where such exist. In the United States it is customary to calibrate through calibration laboratories with traceability to the National Institute for Standards and Technology (NIST). Other nations have similar institutions. In those rare instances where no standard exists for the measurement, the organization must state the basis for calibration or verification, and record that information.
b) Keep measurement equipment properly adjusted, while at the same time avoiding adjustments that would invalidate calibration.
c) Uniquely identify measurement equipment so that calibration data may be associated with the particular instrument.
d) Safeguard against adjustments that would invalidate calibration. This is often accomplished through the use of tell-tale seals applied to relevant adjustment points. It is impossible to make an adjustment without breaking such seals, thus providing an alert that calibration has been invalidated.
e) Protect monitoring and measurement equipment from damage and deterioration. This kind of equipment is usually delicate, and does not fare well with rough handling or improper storage. Care must be taken to ensure that it is properly handled, stored, and maintained.

Requirement 5 Should the equipment be found to be out of calibration, or otherwise not in conformance with requirements, the organization must

1. Assess and record the validity of previous measuring results by the device in question. The object is to determine if earlier measurements were valid, or if they must be repeated.
2. Maintain records of calibration and verification results in accordance with clause 4.2.4.

Requirement 6 When measurement equipment is found to be out of calibration, the organization must take appropriate action concerning both the measurement equipment and the affected product. That means that the equipment must be repaired and/or calibrated before it may be used again, and, usually, the affected product parameter(s) must be remeasured.

Requirement 7 One could interpret clause 7.6 as requiring calibration and verification records only following out-of-calibration incidents. Such is not the case. Complete, routine calibration records of all relevant equipment must be maintained. This is one of the processes the organization must establish under the second paragraph of the clause. It will be difficult to keep equipment calibrated without such records, and even more difficult to convince registrars. Clause 4.2.4 applies.

Requirement 8 The organization must confirm computer software's ability to satisfy the intended application when used in the monitoring and measuring processes. This confirmation must be accomplished prior to initial use, and reconfirmed as necessary.

Note: ISO advises that the organization see ISO 10012-1 and ISO 10012-2. These are guidance documents applicable to measurement equipment. ISO 10012-1 is entitled *Quality Assurance Requirements for Measuring Equipment—Part 1: Metrological Confirmation System for Measuring Equipment.* ISO 10012-2 is entitled *Quality Assurance for Measuring Equipment—Part 2: Guidelines for Control of Measurement Processes.*

Measurement, Analysis, and Improvement—Requirements of Clause 8

Clause 8 is concerned with demonstrating conformity of the product and the QMS, and continual improvement through processes to

- Monitor and measure both the production processes and the product
- Prevent delivery of nonconforming product
- Ensure the continued suitability and effectiveness of the QMS
- Continually improve the QMS and the product

The requirements of Clause 8 are found in its five two-digit clauses:

8.1 General
8.2 Monitoring and Measurement
8.3 Control of Nonconforming Product
8.4 Analysis of Data
8.5 Improvement

With clause 8 we have arrived at the fourth and final block of the process-based quality management system. This section of ISO 9001 is intended to demonstrate that the QMS and delivered products conform to requirements, and that both are subjected to continual improvement. External input from customers is coupled with internal measurement and analysis to develop data that is fed back to the process-based quality management system's clause 5 block, Management Responsibility. This feedback is used by management in their Plan-Do-Check-Adjust (PDCA) Cycle to determine what QMS adjustments may be required. That signals the start of a new cycle through the clause 5 through clause 8 blocks. Refer to Figure 4-1 and Figure 4-10.

Measurement, Analysis and Improvement

ISO 9001:2000, Clause 8.1: General

The organization shall plan and implement the monitoring, measurement, analysis and improvement processes needed

a) to demonstrate conformity of product,
b) to ensure conformity of the quality management system, and
c) to continually improve the effectiveness of the quality management system.

This shall include determination of applicable methods, including statistical techniques, and the extent of their use.

Requirements—Clause 8.1

This clause requires the organization to

1. Determine the appropriate methods for monitoring, measurement, analysis and improvement related to the product and to the QMS in order to ensure
   a) Conformity of product
   b) Conformity of the QMS
   c) Continual improvement of the QMS
2. Plan and implement these monitoring, measurement, analysis, and improvement processes

Application Information—Clause 8.1

Requirement This clause represents an overview which will become focused in the succeeding section 8 clauses. It lays out the general requirement that the organization must determine methods best suited to its business and the interests of its customers to ensure that when product is delivered to customers it meets requirements, and that the quality management system conforms to ISO 9001 and is continually being improved. Once these methods are determined, the organization must plan for and implement the processes required to fulfill the methods.

ISO has made the valuable point in this clause that it may be appropriate for an organization to employ statistical techniques as a part of its monitoring, measurement, and analysis methods. We agree, and commend TC 176 for presenting the issue. Statistical process control (SPC) has been demonstrated as the best way to ensure that processes continue to operate within the limits of their normal variation envelopes. SPC can instantly flag nonnormal variations within the processes, preventing the production of nonconforming product and alerting operators to the fact that something has gone wrong. There are many other statistics-based TQM tools that organizations can employ, including:

- Pareto charts to separate the important few factors from the trivial many
- Check sheets to display collected data in a manner that is easily interpreted
- Histograms to chart the frequency of occurrence (when dealing with process variability)
- Scatter diagrams to determine correlation between two variables
- Run charts to display process results over time
- Control charts (SPC) to determine whether process variation is the result of the process' inherent capability, or caused by some controllable external factor

For a comprehensive, easily understandable treatise on all of the total quality tools and SPC, the reader is directed to Chapters 15 and 18 of our book, *Quality Management: Introduction to Total Quality Management for Production, Processing, and Services,* published by Prentice-Hall, ISBN: 0-13-011638-6.

Measurement, Analysis and Improvement, Continued

ISO 9001:2000, Clause 8.2: Monitoring and Measurement

The Requirements of Clause 8.2 are found in four subclauses:

8.2.1 Customer Satisfaction
8.2.2 Internal Audit
8.2.3 Monitoring and Measurement of Processes
8.2.4 Monitoring and Measurement of Product

ISO INFO

The origin of what is called statistical process to control (SPC) goes back to 1931 and Dr. Walter Shewhart's book The Economic Control of Quality of Manufactured Product. *Shewhart, a Bell Laboratories statistician, was the first to recognize that industrial processes themselves could yield data, which, through the use of statistical methods, could signal that the process was in control or was being affected by special causes (causes beyond the natural, predictable variation).*[25]

Monitoring and Measurement, Continued

ISO 9001:2000, Clause 8.2.1: Customer Satisfaction

As one of the measurements of the performance of the quality management system, the organization shall monitor information relating to customer perception as to whether the organization has met customer requirements. The methods for obtaining and using this information shall be determined.

Requirements—Clause 8.2.1

This clause contains two requirements for the organization:

1. The organization must monitor customer feedback related to customer requirements being met.
2. The organization must determine how the customer feedback is to be obtained and how it is to be used.

Application Information—Clause 8.2.1

ISO requires the organization to collect and utilize customer satisfaction information as a component of its monitoring and measuring activity, a purpose of which is, as we saw in clause 8.1, demonstration of conformity of the product and the processes associated with manufacture.

Requirement 1 The organization must monitor customer satisfaction information concerning whether the product satisfies customer requirements. Customers may find that product has failed to fulfill their stated requirements. They may also find that when a new product is put to use, even though their stated requirements have been fully met, the product cannot do what they intended because of an unanticipated requirement. A customer who was initially satisfied with the product may later determine if one or more requirements were changed, the product could be more useful. As competitors introduce new competing products, previously satisfied customers may find them more attractive than the organization's product. All of these represent customer satisfaction information related to product requirements. The first, whether or not the product met the stated requirements, is the most critical under the ISO clause. However, the other three, while not suggesting nonconformance with requirements, are equally important to the organization if it wants to retain its customers. Securing customer feedback on the first is vital to continued ISO 9000 registration. The other three are vital for the organization's survival.

Requirement 2 The organization must determine the methods to be employed in obtaining customer satisfaction feedback, and determine how the customer feedback is to be used by the organization. Although this leaves a lot to the organization's discretion, and is open to interpretation by the registrars, there is really no other practical approach. There is clearly no one-size-fits-all method for obtaining and using customer feedback.

ISO intends, as will registrars, for the organization to treat the use of customer feedback seriously, and to take maximum advantage of it in terms of determining whether customer requirements have been met.

ISO has directed this clause at feedback from external customers, the ones who buy and use the product. The organization would be well advised also to pay critical attention to *internal customer satisfaction.* Internal customers are the ones who accept the output product of a preceding process. They are often the first to determine that the semifinished product has a problem, or if it could be processed differently, then succeeding processes would be easier. This is valuable information and, compared with external customer feedback, is much easier to obtain.

There are many ways to obtain customer satisfaction feedback. They include:

- Direct telephone surveys
- Direct mail-out surveys
- Use of firms specializing in customer satisfaction surveys
- Management and employee contact with customers
- Customer visits
- Warranty data
- Customer service contacts
- Organization-sponsored user groups

All organizations will find several of these methods useful in collecting customer feedback data.

Monitoring and Measurement, Continued

ISO 9001:2000, Clause 8.2.2: Internal Audit

The organization shall conduct internal audits at planned intervals to determine whether the quality management system

a) conforms to the planned arrangements (see 7.1), to the requirements of this International Standard and to the quality management system requirements established by the organization, and
b) is effectively implemented and maintained.

An audit program shall be planned, taking into consideration the status and importance of the processes and areas to be audited, as well as the results of previous audits. The audit criteria, scope, frequency and methods shall be defined. Selection of auditors and conduct of audits shall ensure objectivity and impartiality of the audit process. Auditors shall not audit their own work.

Monitoring and Measurement, Continued

ISO 9001:2000, Clause 8.2.2: Internal Audit

The responsibilities and requirements for planning and conducting audits, and for reporting results and maintaining records (see 4.2.4) shall be defined in a documented procedure.

The management responsible for the area being audited shall ensure that actions are taken without undue delay to eliminate detected nonconformities and their causes. Follow-up activities shall include the verification of the actions taken and the reporting of verification results (see 8.5.2).

NOTE See ISO 10011-1, ISO 10011-2 and ISO 10011-3 for guidance.

Requirements—Clause 8.2.2

Eight requirements are found in this clause:

1. The organization must conduct internal QMS audits at planned intervals.
2. The audits must be planned, and must take into account status, importance, and previous audit results of the processes and areas to be audited.
3. Audit criteria, scope, frequency, and methods must be defined.
4. Selection of auditors and conduct of audits must ensure objectivity and impartiality.
5. The organization must ensure that auditors do not audit their own work.
6. Responsibilities and requirements for planning and conducting audits and reporting audit results must be defined in a documented procedure.
7. Management of the audited process or area must take follow-up actions without undue delay.
8. Follow-up activities must include verification of the actions taken and reporting on verification results.

Application Information—Clause 8.2.2

Audits are covered in Chapter 7, Registration and the Audit Process. The reader is invited to turn to Chapter 7 for a more thorough description of the **internal audit**.

Requirement 1 ISO requires that the organization conduct internal QMS audits to determine that the QMS:

- Conforms to planned arrangements, that is those planned under clause 7.1
- Conforms to the requirements of ISO 9001
- Conforms to the QMS requirements established by the organization, namely the quality policy, objectives, plans, and procedures
- Is effectively implemented and maintained.

These internal audits must be conducted according to a schedule planned in advance by the organization.

Requirement 2 Internal audits must be planned in advance, and must take into account the status, importance, and previous audit results of the processes and areas to be audited. The intent is that the most critical processes and functional areas get the most audit attention in terms of frequency and intensity, especially if previous audits have shown them to be problematic. *Previous audits* include both internal audits and those conducted by the registrar. No QMS audit is to be undertaken without thorough planning.

Requirement 3 Audit planning must include definition of the audit criteria, scope, frequency, and methods. Audit *criteria* are the set of policies, procedures, and/or requirements against which audit findings are compared. Audit *scope* defines the boundaries of the audit. Audit *frequency* defines how often the process or area is to be audited. Audit *methods* refer to how the audit is to be conducted. (See Chapter 7.)

Requirement 4 The organization must select auditors and conduct internal audits in ways that ensure objectivity and impartiality. Ideally, auditors should be ambivalent regarding the possible outcomes of the audit. We do not expect the auditor to approach an audit with the aim of flunking the auditee. Neither should we expect the auditor to approach the audit with the aim of passing the auditee. No feelings one way or the other is the stance desired for the auditor. Similarly, the audit plan should be developed with no inherent, preconceived point of view. Both the plan and the auditor should be neutral as regards the outcome. That assures objectivity and impartiality. This is admittedly easier to achieve in the registrar's audits, but must be a goal of internal audits as well.

Requirement 5 The organization must ensure that no internal auditor audits his or her own work. Earlier versions of the standard required that internal audits be *carried out by personnel independent of those having direct responsibility for the activity being audited.*[26] That means that the auditor of the activity being audited could not be affiliated with the activity. It goes beyond the prohibition of auditing one's own work, and prevents an employee of an activity from auditing anything connected with the activity. We consider the 1994 version to be better suited to assuring objectivity and impartiality.

Requirement 6 The organization must develop and implement a documented procedure for planning and conducting audits and for reporting audit results. This is one of only six QMS procedures that are explicitly required to be documented. This is pointed out to highlight the importance ISO attaches to the conduct of internal audits.

It is important to understand that this procedure must be more than just a document to assign responsibilities and requirements for planning and conducting audits. The procedure should cover all aspects of planning and conducting audits, reporting on them, and maintaining audit records in accordance with clause 4.2.4.

Requirement 7 Management of the activity audited must take timely action to eliminate any reported discrepancies or nonconformities. *Timely,* in this case, means as soon as possible. Under no circumstances should a reported nonconformity stay on the books until the next audit, which may be conducted by the registrar. (The typical pattern is for internal audits to occur between registrar audits.) It normally falls to the management representative to keep track of audit discrepancies, and to ensure that the management of the activity involved expedites the necessary follow-up action.

Requirement 8 The organization must verify that the follow-up actions taken effectively eliminate the nonconformities detected by the audit, and must report the verification results to interested parties, including top management and usually the registrar. Records of actions taken and verification results must be maintained in accordance with clause 4.2.4. Note: Clause 8.2.2 refers the organization to three ISO documents for guidance related to internal audits but does not identify them except by document number. They are:

> ISO 10011-1, *Guidelines for Auditing Quality Systems—Part 1: Auditing.*
>
> ISO 10011-2, *Guidelines for Auditing Quality Systems—Part 2: Qualification Criteria for Quality Systems Auditors.*
>
> ISO 10011-3, *Guidelines for Auditing Quality Systems—Part 3: Management of Audit Programs.*

ISO plans to revise these three documents in a single document, ISO 19011, *Guidelines On Quality and/or Environmental Management Auditing.*[27]

Monitoring and Measurement, Continued

ISO 9001:2000, Clause 8.2.3: Monitoring and Measurement of Processes

The organization shall apply suitable methods for monitoring and, where applicable, measurement of the quality management system processes. These methods shall demonstrate the ability of the processes to achieve planned results. When planned results are not achieved, correction and corrective action shall be taken, as appropriate, to ensure conformity of the product.

Requirements—Clause 8.2.3

This clause requires the organization to do three things with regard to monitoring and measurement of the processes used in the production of the product.

1. The organization must apply suitable methods for monitoring and, when applicable, measurement of the QMS processes.
2. The organization must choose methods that demonstrate the ability of the processes to achieve the intended (planned) results.
3. When the intended results are not achieved, the organization must take appropriate corrective action to ensure conformity of the product.

Application Information—Clause 8.2.3

The objective of clause 8.2.3 is to establish process monitoring capability, and, where it is possible, process measurement systems—to ensure both that the processes meet production requirements and ultimately that the products realized through these processes conform to requirements.

Requirement 1 The organization must apply suitable methods for monitoring QMS processes. Many methods are available, including statistical process control. The organization has to determine the best methods for its unique processes, and implement them. The object of process monitoring is to prevent the production of nonconforming product or constituent parts of the product.

It is sometimes possible to apply measurement to processes for the same purpose. When that is the case, the organization must also apply suitable measurements. Measurements can be in the form of direct measurement of process parameters, or measurements related to process output.

Requirement 2 The organization must carefully select monitoring and measurement methods that are capable of demonstrating the ability of processes to produce conforming product. Monitoring and measurement that do not provide information verifying process capability are of little use, and may be a waste of time and effort.

Requirement 3 The organization must take appropriate corrective actions when the processes fail to produce the planned results. These actions must ensure conformity of any delivered product. Such corrective actions may include changes to the process to prevent further nonconformances, or repair or rework to the product to achieve conformance. Corrective action may also include scrapping the nonconforming product.

Monitoring and Measurement, Continued

ISO 9001:2000, Clause 8.2.4: Monitoring and Measurement of Product

The organization shall monitor and measure the characteristics of the product to verify that product requirements have been met. This shall be carried out at appropriate stages of the product realization process in accordance with the planned arrangements (see 7.1).

Evidence of conformity with the acceptance criteria shall be maintained. Records shall indicate the person(s) authorizing release of product (see 4.2.4).

Product release and service delivery shall not proceed until the planned arrangements (see 7.1) have been satisfactorily completed, unless otherwise approved by a relevant authority and, where applicable, by the customer.

Requirements—Clause 8.2.4

This clause establishes three requirements for the organization.

1. At appropriate stages of the production process, the organization must monitor and measure characteristics of the product to verify that requirements are met. This is to be in accordance with the planning for product realization done under clause 7.1.
2. The organization must maintain evidence of product conformity with acceptance criteria, and these records must indicate the person(s) authorizing release of the product. Records are to be maintained in accordance with clause 4.2.4.
3. The organization must not release product or deliver service until the planned arrangements (clause 7.1) have been satisfactorily completed. The only exception to this requirement is when release is otherwise approved by a relevant authority in the organization and, where applicable, by the customer.

Application Information—Clause 8.2.4

The intent of this clause is that product be monitored and measured at appropriate stages of the production process as planned under clause 7.1, and that except for extraordinary circumstances, product not be delivered unless and until it has been shown to meet product requirements.

Requirement 1 The organization planned for the production and verification of the product under clause 7.1. Under clause 8.2.4, the organization must monitor and measure the characteristics of the product as it planned to do. That means that monitoring and measuring must take place at the production stages earlier defined, and that it must be demonstrated that product requirements have been met.

Requirement 2 As product undergoes monitoring and measuring, including verification testing for conformity with acceptance criteria, records must be maintained. These records must identify the person(s) authorizing release of product.

Monitoring data and test measurements recorded at each applicable production stage, along with signatures authorizing product release, are considered to be evidence of product conformity, and are to be maintained in accordance with clause 4.2.4.

Requirement 3 Provision is made in clause 8.2.4. for the extraordinary situation in which, even though it has not satisfactorily completed all of the planned monitoring and measurement, there is an overriding circumstance requiring the product to be released. This happens occasionally, for various reasons. There is usually an urgent need to put the product into operation, even though it is not fully meeting requirements. When that happens, the product may be released by a relevant authority in the organization. The organization should also seek authorizing signature(s) from the customer. Under no circumstances should such product be released without the customer's full knowledge and agreement.

Monitoring and Measurement, Continued

ISO 9001:2000, Clause 8.3: Control of Nonconforming Product

The organization shall ensure that product which does not conform to product requirements is identified and controlled to prevent its unintended use or delivery. The controls and related responsibilities and authorities for dealing with nonconforming product shall be defined in a documented procedure.

The organization shall deal with nonconforming product by one or more of the following ways:

a) by taking action to eliminate the detected nonconformity;

b) by authorizing its use, release or acceptance under concession by a relevant authority and, where applicable, by the customer;

c) by taking action to preclude its original intended use or application.

Records of the nature of nonconformities and any subsequent actions taken, including concessions obtained, shall be maintained (see 4.2.4).

When nonconforming product is corrected it shall be subject to re-verification to demonstrate conformity to the requirements.

When nonconforming product is detected after delivery or use has started, the organization shall take action appropriate to the effects, or potential effects, of the nonconformity.

Requirements—Clause 8.3

This clause contains six requirements related to the handling of product that has failed to conform to the acceptance criteria.

1. The organization must identify and **control nonconforming product** to prevent unintended delivery or use.
2. The organization must define the controls for nonconforming product, and related responsibilities and authorities in a documented procedure. The organization must use one or more of the following three ways of dealing with nonconforming product:
   a) Correcting the nonconformity
   b) Authorizing its use, release, or acceptance if concession is obtained from a relevant organizational authority and, where applicable, by the customer
   c) Taking action to prevent its intended use or application
3. The organization must maintain records of the nature of nonconformities, and of any actions taken, including concessions obtained. These records are to be maintained in accordance with clause 4.2.4.

4. The organization must demonstrate conformity of previously nonconforming product that has been corrected. This must be done by re-verification to the same acceptance criteria.
5. If it is discovered that product is nonconforming subsequent to delivery or start of use, the organization must take action that is appropriate to the effects, or potential effects, of the nonconformity.

Application Information—Clause 8.3

It is inevitable that some manufactured product will not satisfy the acceptance criteria, while other units going through the same processes meet the requirements. Significant problems can result if the nonconforming product is not promptly identified and put under controls to prevent its being delivered to a customer. Sometimes nonconforming product can be reworked to correct the deficiency, and returned to conformance. Sometimes it is not practical to do that. Occasionally the customer's need for the product is so urgent that a concession may be granted to permit delivery of product which does not fully meet requirements. This is often done with the understanding that the product will be subsequently repaired or replaced. Clause 8.3 is concerned with these issues.

Requirement 1 The organization must identify and control any nonconforming product. Identification must be visible to anyone, and should be in a media that is not easily removed or separated from the product. Quality assurance rejection stickers are frequently used. Any associated traveler tags should also reflect the rejected status. Control of nonconforming product usually means that it is placed in a secure area designated for the purpose of holding nonconforming product. The purpose of identifying it and placing it under control is to prevent the product from being inadvertently delivered or put to use.

Requirement 2 In order to accomplish a consistent and effective means of dealing with nonconforming product, ISO requires the organization to develop and implement a documented procedure. The nonconforming product procedure must define the controls and the related responsibilities and authorities, including the designation of authority for release of nonconforming product. (This is the fourth of six documented procedures explicitly required by ISO 9001.)

Requirement 3 The organization must deal with nonconforming product in one or more of the following ways:

a) The defect causing the nonconformity may be reworked or repaired. Once done, the formerly nonconforming product is ready for re-verification.
b) A relevant authority in the organization (with customer approval if applicable) may authorize its use, acceptance, or release under concession. This should only be done in exceptional circumstances.
c) Action may be taken to preclude its original intended use or application. This ordinarily means scrapping or destroying the nonconforming product.

Requirement 4 The organization must keep records of nonconforming product, including the types and nature of nonconformities, actions taken to repair, deliver under concession, or to preclude its delivery or use. These records are to be maintained in accordance with clause 4.2.4.

Requirement 5 If nonconforming product is reworked or repaired, before it can be delivered or put to use, the organization must re-verify it. Only after demonstrating conformance to the acceptance criteria is the product to be released.

Requirement 6 In some instances it may be discovered that product already delivered and in use is nonconforming. This can happen if the acceptance criteria are defective, and the organization failed to detect nonconformance. Should this happen, for any reason, the organization must take action appropriate to the effects, or potential effects, of the nonconformance. If there is a safety issue, then the organization may have to issue a warning or an immediate recall. If the effects are benign, there is less pressure on the organization to take action. Obviously, the effects can cover a wide spectrum, and the organization should keep the customer's interest in the forefront of any relevant decision making.

Measurement, Analysis and Improvement, Continued

ISO 9001:2000, Clause 8.4: Analysis of Data

The organization shall determine, collect and analyze appropriate data to demonstrate the suitability and effectiveness of the quality management system and to evaluate where continual improvement of the effectiveness of the quality management system can be made. This shall include data generated as a result of monitoring and measurement and from other relevant sources.

The analysis of data shall provide information relating to

a) customer satisfaction (see 8.2.1),

b) conformity to product requirements (see 7.2.1),

c) characteristics and trends of processes and products including opportunities for preventive action, and

d) suppliers.

Requirements—Clause 8.4

This clause requires the organization to do two things with respect to **analysis of data:**

1. The organization must determine, collect, and analyze appropriate data to demonstrate that its quality management system is (and continues to be) suitable and operates effectively.
2. The organization must determine, collect, and analyze appropriate data in order to evaluate how its QMS effectiveness might be improved.

The data collected and analyzed is to come from monitoring and measuring done by the organization, and any other relevant sources. Data analysis must provide information relating to:

a) Customer satisfaction (see 8.2.1)
b) Product conformity (see 7.2.1)
c) Characteristics and trends of processes and products including opportunities for preventive action
d) Suppliers

Application Information—Clause 8.4

Through monitoring and measuring of processes, product, and customer satisfaction, and determination of product requirements, the organization has the data ingredients for producing a wealth of valuable information. By analyzing that data, it can be determined whether or not the QMS, as it is structured and implemented, is performing effectively and efficiently in terms of its plans, objectives, and other defined goals. Another expected result of data analysis is that opportunities for continual improvement of the QMS and the product will be identified. This clause requires the organization to accomplish this data analysis.

Requirement 1 The organization must determine what data to collect, collect the data, and analyze it to validate the continued suitability and effectiveness of the QMS. The question is, is the QMS operating according to its plans, achieving its objectives and goals, and producing conforming product? Analysis of the data will reveal the answers.

Requirement 2 The organization must determine what to collect, collect the data, and analyze it to evaluate opportunities for continual improvement of the QMS. Two points follow: (1) For the most part this will be the same data as used for Requirement 1, although the analysis may be different. (2) Clause 8.4's first paragraph seems to be directed to the QMS only for evaluation of improvement opportunities. We believe that ISO intends for the analysis to yield product improvement opportunities also. (Note that there is no other clause specifically relating to analysis of data for product improvement.) This contention is supported by subclauses b and c which require the data analysis to provide information relating to *conformity to product requirements,* and *characteristics and trends of processes and products including opportunities for preventive action* . . . ISO clearly expects the product to benefit from the required data analysis. We would have preferred to see the words *continual improvement* instead of (or in addition to) *preventive actions,* but it seems that ISO has still not quite come to the realization that, by definition, if a process or a product is prevented from taking on an unwanted characteristic or doing something undesirable, the process or product has been improved.

Requirement 3 The organization must ensure that its data analysis provides information relating to:

a) Customer satisfaction. Clause 8.2.1 requires the organization to monitor customer satisfaction information. This clause requires that it be analyzed.

b) Conformity to product requirements. Clause 7.2.1 requires the organization to determine product related requirements. This clause requires that actual process and product data be compared against the requirements to determine QMS performance.

c) Characteristics and trends of processes and products, to include opportunities for preventive action. As we have said, this should also be targeted at continual improvement for both processes and products.

d) Suppliers. This means supplier performance. ISO expects organizations to collect and analyze data, allowing selection of the suppliers who best support the organization and the QMS.

Measurement, Analysis and Improvement

ISO 9001:2000, Clause 8.5: Improvement

The requirements of clause 8.5 are found in three three-digit clauses.

8.5.1 Continual Improvement
8.5.2 Corrective Action
8.5.3 Preventive Action

Improvement, Continued

ISO 9001:2000, Clause 8.5.1: Continual Improvement

The organization shall continually improve the effectiveness of the quality management system through the use of the quality policy, quality objectives, audit results, analysis of data, corrective and preventive actions and management review.

Requirements—Clause 8.5.1

The organization must continually improve the effectiveness of its quality management system through the use of all the QMS elements and activities.

Application Activities—Clause 8.5.1

One of the most significant changes between the 1994 and 2000 versions of ISO 9000 is the addition of the requirement of 8.5.1 for continual improvement of the QMS. The concept of continual improvement is new, and perhaps radical, to organizations not already operating under TQM. It is difficult for them to accept the notion that no process or product is ever perfect, and therefore, should be improved. However, the concept is so powerful that during the 1960s, 1970s, and early 1980s, entire industries have shifted from the United States, where continual improvement was not practiced, to Japan where continual improvement was a well integrated way of industrial life. Today the emphasis is on the competitiveness and hence the survival of competing organizations. Given two firms starting equally in all respects, the one that relentlessly pursues continual improvement of its processes and products will quickly capture market share from the organization that is content with the status quo. Organizational survival should be an incentive for continual improvement. That is part of the motivation, but benefit to the customer is at least as important, as it should be. Customers benefit from better, higher quality, longer lasting, more capable products at lower costs. The organization benefits from enhanced competitiveness and higher profits. If there has ever been a win-win situation, it is continual improvement.

Requirement The organization must apply the concept of continual improvement to improve the effectiveness of its quality management system. It is to do this through the use of:

- The quality policy (clauses 5.1 and 5.3). Top management sets the stage for everything done under the QMS through the quality policy.
- Quality objectives (clause 5.4.1). Quality objectives are in harmony with the quality policy and the corporate strategy. Quality objectives can be used to establish improvement goals and to reveal weaknesses.
- Audit results, internal and external. Audit results may point directly to QMS elements that need improvement.
- Analysis of data (clause 8.4). Data analysis reveals areas of needed improvement, and verifies improvements that are made.
- Corrective and preventive actions (clauses 8.5.2 and 8.5.3). As ISO has defined corrective and preventive actions, both can be improvements.
- Management review (clause 5.6). Management review meetings are the vehicle through which many continual improvement opportunities are first presented. Management's responsibility is to sincerely seek these opportunities and to support those which offer authentic continual improvement.

Improvement, Continued

ISO 9001:2000, Clause 8.5.2: Corrective Action

The organization shall take action to eliminate the cause of nonconformities in order to prevent recurrence. Corrective actions shall be appropriate to the effects of the nonconformities encountered.

Improvement, Continued

ISO 9001:2000, Clause 8.5.2: Corrective Action

A documented procedure shall be established to define requirements for

a) reviewing nonconformities (including customer complaints),
b) determining the cause of nonconformities,
c) evaluating the need for action to ensure that nonconformities do not recur,
d) determining and implementing action needed,
e) records of the results of action taken (see 4.2.4), and
f) reviewing corrective action taken.

Requirements—Clause 8.5.2

This clause contains three requirements for the organization.

1. The organization must take action to eliminate the cause of nonconformities in order to prevent their recurrence.
2. Corrective actions taken must be appropriate to the seriousness of the effects of the nonconformities encountered.
3. The organization must develop and implement a documented corrective action procedure that defines the requirements for
   a) Reviewing nonconformities
   b) Determining the cause of nonconformities
   c) Evaluating the need for action to ensure that nonconformities do not recur
   d) Determining and implementing action needed
   e) Recording results of actions taken
   f) Reviewing corrective action taken

Application Information—Clause 8.5.2

Corrective action should be used as a component part of the continual improvement program. ISO defines *corrective action* as *action taken to eliminate the cause of a detected nonconformity*.[28] Eliminating the *cause* of a nonconformity represents improvement.

Requirement 1 When a nonconformity is encountered, the organization must take action to eliminate the cause of the problem. To do that, the root cause of the problem must first be determined. Determining root causes requires a different approach from the typical reaction to production problems, namely, working on the symptoms. Use of the Total Quality tools, root cause analysis, and corrective action project teams is recommended. Training in the tools and techniques may be required.

Requirement 2 Corrective actions taken by the organization must be appropriate to the significance of the impact, or potential impact, of the nonconformance. That does not mean that some nonconformities are not to be corrected, but rather that the urgency of action and the extent of resources committed is to be based on the nonconformity's significance. ISO is looking at priorities. Some of the aspects of the nonconformity's impact to be evaluated include product performance, safety, dependability, customer satisfaction, and cost.

Requirement 3 The organization must develop and implement a documented corrective action procedure defining the requirements associated with items a through f of this clause. (See above in Requirements—Clause 8.5.2.) This is the fifth of six documented procedures explicitly required by ISO 9001:2000.

Improvement, Continued

ISO 9001:2000, Clause 8.5.3: Preventive Action

The organization shall determine action to eliminate the causes of potential nonconformities in order to prevent their occurrence. Preventive actions shall be appropriate to the effects of the potential problems.

A documented procedure shall be established to define requirements for

a) determining potential nonconformities and their cause,
b) evaluating the need for action to prevent occurrence of nonconformities,
c) determining and implementing action needed,
d) records of results of action taken (see 4.2.4), and
e) reviewing preventive action taken.

Requirements—Clause 8.5.3

This clause contains three requirements for the organization.

1. The organization must determine action needed to eliminate causes of potential nonconformities in order to prevent their occurrence.
2. Preventive actions taken by the organization must be appropriate to the effects of potential problems.
3. The organization must develop and implement a documented preventive action procedure to define the requirements for
   a) Determining potential nonconformities and their causes
   b) Evaluating the need for action to prevent occurrence of nonconformities
   c) Determining and implementing the needed preventive action

d) Records of actions taken and their results

e) Reviewing preventive actions taken

Application Information—Clause 8.5.3

The organization should make **preventive action** a component part of the continual improvement program. ISO defines *preventive action* as *action to eliminate the cause of a potential nonconformity or other undesirable potential situation.*[29] While the requirements of clause 8.5.3 sound similar to those of clause 8.5.2, ISO is dealing here with an entirely different approach. The corrective actions taken under 8.5.2 are in response to the detection of an actual nonconformity. Something went wrong, and the organization reacted to it. Clause 8.5.3 deals with actions taken to preclude nonconformities where they have never existed. Preventive action is proactive, before the fact. The trick with preventive action is determining what *may* go wrong. The following can point toward potential nonconformance:[30]

- Risk analysis
- Review of customer needs and expectations
- Market analysis
- Management review output
- Outputs from data analysis
- Relevant QMS records
- Lessons learned from past experience
- Processes that provide early warning of impending out-of-control operating conditions

Requirement 1 The organization must determine actions necessary to prevent potential nonconformities by eliminating their causes. To do this is to prevent the problem before it happens, and that is the objective of this clause.

Requirement 2 In similar fashion to *corrective actions,* ISO wants preventive actions to be "appropriate to the effects of the potential problems." Suppose, for example, that the organization discovered that its product had the potential for a problem that could endanger the user of the product. The organization should at once apply maximum attention and resources to preventing the realization of the problem. It is a matter of the logical use of priorities. The potential problems having the most significant consequences should get the most urgent attention, while other potential problems of less significance can be eliminated as time and resources permit.

Requirement 3 This is the last of six documented procedures explicitly required by ISO 9001:2000. The requirement is for a procedure addressing the following:

a) Determining nonconformities that *may* occur, and their causes

b) Evaluating the consequences of potential nonconformities, and establishing the priority for preventive action

c) Determining appropriate action to prevent the potential nonconformity, and taking that action
d) Maintaining records of the results of preventive actions taken
e) Reviewing preventive actions taken to determine success and to ensure closure on the potential nonconformities

CASE STUDY

ISO 9000 in Action

The executive management team approved Jake Butler's recommendation for establishing a temporary office to get all of the company's documentation up to date and placed in electronic files. Now he faced the next and even bigger challenge. Butler needed to convince higher management that in order to win ISO 9001 certification according to the 2000 standard, the company would need to implement the Total Quality Management or TQM philosophy or, at least, eight of its main principles.

In reviewing the ISO 9001:2000 standard, Butler could not find the term TQM anywhere. But the concept, without the actual name, practically leaped off the pages at him. Throughout the document, Butler was confronted over and over again with such concepts as customer service, leadership, employee involvement, process approach, system approach to management, continual improvement, factual decision making, and mutually beneficial supplier relationships. Selling all of these concepts to higher management as a package was not going to be easy. Butler had already let the CEO know that there were major differences between the previous ISO 9000 standards and the 2000 versions. He had explained that continual improvement would be a necessity. But even Butler had been surprised to learn in detail just how much of the Total Quality approach to doing business had been built into the new standard.

His plan was to convince higher management that now would be the right time to fully implement the Total Quality approach and to make ISO 9001 certification just a part of the overall process, albeit a large part. To do this, he would need to get the executive management team more knowledgeable of Total Quality and more committed to its basic principles. Butler was sure the convincing needed to be done by a third party. There had already been enough convincing done on his part, and there would be more before the process was over. Consequently, he decided to visit a nearby university to arrange an on-site "executive training" session. Butler's idea was, "If the company's top managers can be convinced to go along with the implementation of Total Quality, everything else will fall in line. If not, we are going to be hard pressed to succeed in our ISO 9000 certification effort."

Working with the heads of the industrial engineering and business departments, Butler was able to sketch out a plan for a one-day seminar covering everything an executive level manager needed to know conceptually about Total Quality and its various elements. Now all he had to do was convince the executive management team to participate in the seminar and hope the university's personnel were good salespeople.

SUMMARY

1. The philosophical approach of ISO 9000:2000 is much different than that of earlier versions. ISO 9000 is now closely aligned with the Total Quality approach to doing business as indicated by the adoption of eight of the main principles of Total Quality: customer focus, leadership, employee involvement, process approach, systems approach to management, continual improvement, factual decision making, and mutually beneficial supplier relationships.
2. The new requirements of ISO 9000:2000 fall into the following nine categories: continual improvement, increased emphasis on the role of top management, consideration of legal and regulatory requirements, establishment of measurable objectives for all functional units, monitoring of customer satisfaction as a measure of system performance, increased attention to resource availability, determination of training effectiveness, measurements extended to system, processes, and products, and analysis of collected data on the performance of the QMS.
3. The new ISO 9001 structure is as follows: Clause 1—Scope, Clause 2—Normative Reference, Clause 3—Terms and Definitions, Clause 4—Quality Management System (QMS), Clause 5—Management Responsibility, Clause 6—Resource Management, Clause 7—Product Realization, and Clause 8—Measurement, Analysis, and Improvement. The actual requirements are set forth in clauses 4 through 8.
4. The instructional scheme used to explain standards herein is as follows: (1) a verbatim statement of the actual clause from ISO 9001, (2) an explanation of what the clause requires, and (3) application information relating to the clause in question. When reading the actual ISO 9001 standard, it is a good idea to apply the following approach. Read the actual clause verbatim. Formulate an explanation of the requirements and write them down. Then ask yourself how this requirement actually applies to your organization.

KEY TERMS AND CONCEPTS

Analysis of data
Continual improvement
Control of nonconforming product
Corrective action
Customer communication
Customer focus
Customer property
Deming Cycle
Design and development
Disposition of records
Documented procedures
Factual approach to decision making
Identification
Identification and traceability
Internal audit
Involvement of people
Leadership
Management responsibility
Management review
Measurement, analysis, and improvement
Monitoring and measurement
Mutually beneficial supplier relationships
PDCA Cycle
Planned intervals

Preventive action
Process approach
Product realization
Protection
Purchasing process
Quality management system (QMS)
Quality plan
Quality planning
Quality objectives
Quality policy
Resource management
Retention time
Retrieval
Shewhart Cycle
Storage
System approach to management
Total Productive Maintenance

REVIEW QUESTIONS

1. Explain the philosophical changes that were introduced into ISO 9000 with the 2000 release.
2. Summarize the new requirements in ISO 9000.
3. List the various clauses that make up the structure of the ISO 9001:2000 standard and what each clause covers.
4. Explain as briefly as possible the requirements of Clause 4.1: General Requirements.
5. What are the requirements relating to the quality manual?
6. What does ISO 9001 mean by the term *customer focus*?
7. Describe the various requirements of the quality policy.
8. What does ISO 9001 mean by the term *product realization*?
9. What are the requirements relating to customer communication?
10. During design and development planning, the organization is required to make certain determinations. What are they?
11. What are the requirements relating to design and development outputs?
12. What are the requirements relating to design and development review?
13. Clause 7.5.1 lists several controlled conditions. What are they?
14. If your company receives customer property that is damaged, what should it do?
15. Clause 8.5.1. requires continual improvement. How are organizations supposed to ensure continual improvement?

APPLICATION ACTIVITIES

1. Create a fictitious company and develop a quality policy for the company.
2. Develop a set of realistic quality objectives for the fictitious company in Activity 1. Develop a comprehensive annotated outline for a quality manual for the fictitious company in Activity 1. You do not have to develop the actual manual, just the outline showing everything that would be in it (like the table of contents for a book).

ENDNOTES

1. ISO, Frequently Asked Questions, FAQ 020, 11/1999.
2. Ibid, FAQ 017, 11/1999.
3. ISO 9004:2000, clause 4.1.
4. ISO 9001:2000, clause 4.2.1d.
5. ISO 9001:2000, clause 4.2.1d.
6. ISO 9004:2000, clause 4.2.
7. ISO 9000:2000, clause 3.2.9.
8. Ibid, clause 3.7.5.
9. ISO 9004:2000, clause 5.1.1.
10. ISO 9000:2000, clause 2.6.
11. ISO 9001:2000, Introduction, clause 0.2.
12. ISO 9004:2000, clause 5.3.
13. ISO 9000:2000, clause 3.2.8
14. ISO 9000:2000, clause 3.2.9.
15. Brian L. Joiner, *Fourth Generation Management: The New Business Consciousness* (New York: McGraw-Hill, Inc., 1994), p. 243.
16. ISO 9004:2000, clause 5.5.2.
17. Goetsch and Davis, *Understanding and Implementing ISO 9000 and ISO Standards* (Upper Saddle River, N.J.: Prentice-Hall, 1998), p. 62.
18. ISO 9001:2000, clause 0.2.
19. Mary Walton, *The Deming Management Method* (New York: The Putnam Publishing Group, 1986), pp. 86–88.
20. ISO 9001:2000, clause 0.2, Figure 1.
21. Seiichi Nakajima, *Introduction to TPM: Total Productive Maintenance* (Cambridge, Massachusetts: Productivity Press, 1988), p. xviii.
22. ISO 9004:2000, clause 6.4.
23. Adapted from ISO 9004:2000, clause 7.4.2.
24. ISO 9004:2000, Clause 7.5.4.
25. Goetsch and Davis, *Quality Management: Introduction to Total Quality Management for Production, Processing, and Services,* 3rd Edition (Upper Saddle River, NJ: Prentice-Hall, 2000), p. 556.
26. ISO 9001:1994, clause 4.17.
27. ISO 9004:2000, p. 57.
28. ISO 9000:2000, clause 3.6.5.
29. ISO 9000:2000, clause 3.6.4.
30. Adapted from ISO 9004:2000, clause 8.5.3.

CHAPTER FIVE

The Quality Management System (QMS)

MAJOR TOPICS

- Definition: What Is a Quality Management System?
- Quality Management System References in ISO 9000
- Management Responsibility
- Elements of a Quality Management System: Developing the QMS
- Quality Management System Structure
- Continual Improvement As a Part of the Quality Management System
- Documenting the Quality Management System
- Verifying the Integrity of the Quality Management System
- Review and Evaluation of the Quality Management System

DEFINITION: WHAT IS A QUALITY MANAGEMENT SYSTEM?

ISO arrives at a definition of **quality management system** by first breaking the term into its constituent parts: *quality, system,* and *management system.* ISO 9000, clause 3.1.1 defines *quality* as the "degree to which a set of inherent characteristics fulfills requirements." Clause 3.2.1 defines *system* as a "set of interrelated or interacting elements." Clause 3.2.2 defines *management system* as a "system to establish policy and objectives and to achieve those objectives." Clause 3.2.3 puts these together to define *quality management system* as a "management system to direct and control an organization with regard to quality."

Although they may not be formally designated as such, every organization will make use of several management systems. For example, a financial management system keeps track of the organization's financial status and controls spending. A business management system may control the kinds of business opportunities the organization pursues. An ISO 14000 organization will utilize an environmental management system to manage its environmental aspects, impacts, and objectives. Most organizations use some sort of management system, whether it is considered to be a quality management system or not, to assure the quality of products and services. ISO 9000 firms deliberately and carefully develop and utilize a suitable management system to control all elements and activities of the organization that can have an impact on the quality of product or service. The result is called a quality management system.

The authors have developed the following definition, which, we believe, conforms to the requirements and the intent of ISO 9000:

> *The quality management system is comprised of all the organization's policies, procedures, plans, resources, processes, and delineation of responsibility and authority, all deliberately aimed at achieving product or service quality levels consistent with customer satisfaction and the organization's objectives. When these policies, procedures, plans, etc. are taken together, they define how the organization works, and how quality is managed.*

QUALITY MANAGEMENT SYSTEM REFERENCES IN ISO 9000

References to the *quality management system* relative to developing, implementing, and operating and improving a QMS may be found in:

ISO 9000:2000: Providing Fundamentals and Vocabulary

Introduction	Background information
Clause 0.1	Background
Clause 0.2	Quality management principles as the basis for the QMS
Clause 1	Scope
Clause 2	Fundamentals of quality management systems
Clause 2.1	Rationale for quality management systems
Clause 2.2	Distinguishing between requirements for *QMS* and requirements for *products*
Clause 2.3	Approach to developing and implementing
Clause 2.4	Process-based QMS
Clause 2.6	Role of top management within the QMS
Clause 2.7.1	Contribution of documentation to QMS evaluation
Clause 2.7.2	Types of QMS documentation
Clause 2.8	Evaluating the QMS
Clause 2.8.1	Evaluating processes within the QMS
Clause 2.8.2	Auditing the QMS
Clause 2.8.3	Management review of the QMS
Clause 2.8.4	The organization's self-assessment of the QMS

Clause 2.9	Continual improvement of the QMS
Clause 2.11	Focus of the QMS
Clause 2.12	Relationship between the QMS and excellence models
Clause 3.2.3	Definition of

ISO 9001:2000: Requirements of the Standard

Introduction	Background information
Clause 0.1	Adaptability to the organization's unique needs
Clause 0.2	Promoting the process approach to QMS
Clause 1	Establishing scope and applicability
Clause 4	(All of section 4) Establishing QMS requirements
Clause 5	(All of section 5) Top management's responsibility to the QMS
Clause 6	(All of section 6) Providing resources needed to implement, maintain, and continually improve QMS effectiveness
Clause 7.1	Product realization planning consistent with QMS requirements
Clause 8.2.1	Customer satisfaction as a measurement of QMS performance
Clause 8.2.2	Internal audits to determine continued suitability and effectiveness
Clause 8.2.3	Monitoring and measuring QMS processes
Clause 8.4	Data analysis to demonstrate QMS suitability, effectiveness, and continual improvement
Clause 8.5.1	Continual improvement of QMS effectiveness

ISO 9004:2000: Guidance and Advice for Performance Improvement

Clause 0.1	Expansion on ISO 9001, clause 0.1
Clause 0.2	Essentially identical to ISO 9001, clause 0.2
Clause 0.3	ISO 9004's wider range of QMS objectives relative to ISO 9001
Clause 1	Scope of QMS related guidelines beyond ISO 9001 requirements
Clause 4	(All of section 4) Advice and guidance for a more effective QMS
Clause 5	(All of section 5) Advice and guidance related to ISO 9001, clause 5
Clause 6	(All of section 6) Advice and guidance related to ISO 9001, clause 6
Clause 8.2.1.1	Advice and guidance related to ISO 9001, clause 8.2.1
Clause 8.2.1.3	Advice and guidance related to ISO 9001, clause 8.2.2
Clause 8.2.1.5	Advice and guidance relative to self-assessment
Clause 8.2.2	Advice and guidance relative to ISO 9001, clause 8.2.3
Clause 8.4	Advice and guidance relative to ISO 9001, clause 8.4
Clause 8.5.1	Advice and guidance relative to ISO 9001, clause 8.5.1
Annex A	Guidance for self-assessment

MANAGEMENT RESPONSIBILITY

There are two levels of **management responsibilities** in connection with ISO 9000. The first level relates to the responsibilities specified in the section 5 clauses of ISO 9001. The second level relates to the responsibilities that are not specified but are necessary to assure that the QMS functions effectively. Regarding the unspecified responsibilities, any ISO 9001 clause that states something like "the organization shall establish . . ." can be interpreted as an implicit management responsibility. Although the statement does not mention management, only management has the authority and power to establish. ISO 9001:2000 makes it clear that top management, meaning the person or group of people who direct and control an organization at the highest level[1], is responsible for the development, implementation, operation, and continual improvement of the quality management system. This position corresponds to the Total Quality philosophy; namely, that without leadership and commitment from the top, success is impossible. Everyone in the organization must understand that, with both Total Quality Management and ISO 9000 quality management systems, ultimate responsibility rests with the chief executive. This responsibility cannot be delegated. If the quality management system involved only one department within the organization, it might be possible for the managers of just that one department to develop a QMS and maintain a commitment to it. But the fact is, the QMS will apply to *all activities throughout the organization.*

This means every department (or every activity) and every person in the organization. The only person who has the authority to commit every department in the organization is the chief executive. In addition to commitment, there is a symbolic role that the senior manager must accept if all employees, at all levels, are to share his or her commitment, not only to the quality management system, but also to the customer orientation the QMS supports. By being visibly involved in the activities of the QMS, the senior manager leads by example. Employees are inclined to follow the lead of a chief executive who sets a good example.

ISO 9001 explicitly makes the following requirements the responsibility of top management:

- Communicating to employees the importance of meeting customer requirements (clauses 5.1 and 5.2)
- Communicating the importance of meeting statutory and regulatory requirements (clause 5.1)
- Establishing [and adhering to] the quality policy (clauses 5.1 and 5.3)
- Ensuring that quality objectives are established [and achieved] (clauses 5.1, 5.3, and 5.4.1)
- Conducting management reviews of the QMS (clauses 5.1, 5.3, and 5.6)
- Ensuring the availability of required resources (clause 5.1e)
- Ensuring that planning of the QMS is appropriately carried out (clause 5.4.2)
- Ensuring that responsibilities and authorities are defined and communicated throughout the organization (clause 5.5.1)

- Appointing the management representative (clause 5.5.2)
- Ensuring establishment of appropriate internal communication system (clause 5.5.3)

Implicit top management requirements of ISO 9001 include:

Clause	Management's Implicit Responsibilities
4.1	To establish and maintain the QMS and continually improve its performance
4.2.2	To establish and maintain a quality manual
4.2.4	To establish and maintain quality records
6.2.2 b	To provide competence-and-awareness training
6.3	To provide and maintain the needed infrastructure
6.4	To determine and manage the appropriate work environment
7.6	To establish and use appropriate monitoring and measuring processes to provide evidence of conformity

Many other ISO 9001 requirements contain "the organization shall . . ." statements, and it is safe to assume that all of these place an implicit responsibility on the various levels of management even though the actual work may be done by others. Ultimately, top management is responsible for everything that is done, and for ISO 9000 to produce the desired results, top management, as well as all other levels, must be active participants.

From the explicit and implicit management responsibilities listed above, one can conclude that ISO 9000 expects management, including top management, to be involved in virtually everything connected with the quality management system. This *is* the ISO position. Management is responsible for the success of the QMS; compliance with legal, regulatory, and customer requirements; and conformance to the QMS—regardless of whether the responsibility is explicitly or implicitly stated. Success cannot be achieved without employees at all levels willingly supporting the QMS and its objectives. Likewise, just as is the case of Total Quality Management, an ISO 9000 quality management system simply cannot be successful without complete commitment and participation of top management. Senior managers must state their commitment and demonstrate it by actively participating in the ISO 9000 effort. People follow by observing; so when employees see their leaders involved in QMS activities, they will accept that a commitment does, in fact, exist. This point, though lost on many managers, is critical for the successful implementation and operation of a quality management system. ISO states the following:

> Leadership, commitment and the active involvement of the top management are essential for developing and maintaining an effective and efficient quality management system to achieve benefits for interested parties.[2]

ELEMENTS OF A QUALITY MANAGEMENT SYSTEM: DEVELOPING THE QMS

ISO 9000:2000, clause 3.2.1 defines a system as a "set of interrelated or interacting elements." The quality management system is made up of several components, or elements. These elements are necessary to support the structure of a QMS that conforms to the requirements of ISO 9000. See Figure 5-1. QMS elements include those explained in the following sections.

Management Commitment Management commitment must be present at the start and sustained over time. If management is not committed to the objectives of ISO 9000 and actively involved in related QMS activity, no chance exists for successful integration and operation of a quality management system. Top management's commitment flows into the quality policy, and must be supported throughout the organization (clauses 5.1, 5.3).

The importance of demonstrating a continuous commitment to quality by management cannot be overstated. This is a proven concept from Total Quality Management. Employees tend to focus their energy on the things that are important to management, and they understand that the things that are important to management do not always correspond with what managers say. Employees almost universally understand that if managers spend a lot of time or money on something, that is what is important. Employees take cues from observing what management does more than from what it says. If you want your employees to pay attention to something, let them see you involved in it.

Customer Focus The needs and expectations of current and future customers must be determined and fulfilled if the organization is to be successful. This requires a full time customer focus by the organization. Quality and teamwork expert Peter R. Scholtes explains the concept of *customer focus* as follows:

> Whereas Management by Results begins with profit and loss and return on investment, Quality Leadership starts with the customer. Under Quality Leadership, an organization's goal is to meet and exceed customer needs, to give lasting value to the customer. The return will follow as customers boast of the company's quality and service. Members of a quality organization recognize both external customers, those who purchase or use the products or services, and internal customers, fellow employees whose work depends on the work that precedes them.[3]

Customer focus should be a key part of the quality policy (clauses 5.2, 5.1, 5.5.2c, 5.6.2b, 7.2, 8.2.1, 8.4).

Conforming Quality Policy The quality policy crafted by management, or under management's direction, must be in harmony with the organization's vision and strategy for the future, and is the guiding document that establishes the overall principles and sense of direction for the quality management system. It should reflect top management's commitment to ISO 9000 and management's quality philosophy. Organizational customer focus should be supported by the policy (clauses 4.2.1, 5.1, 5.3).

Articulating the organization's **quality policy** is the first step in developing the quality system. It is important to have the quality policy defined and documented in order to

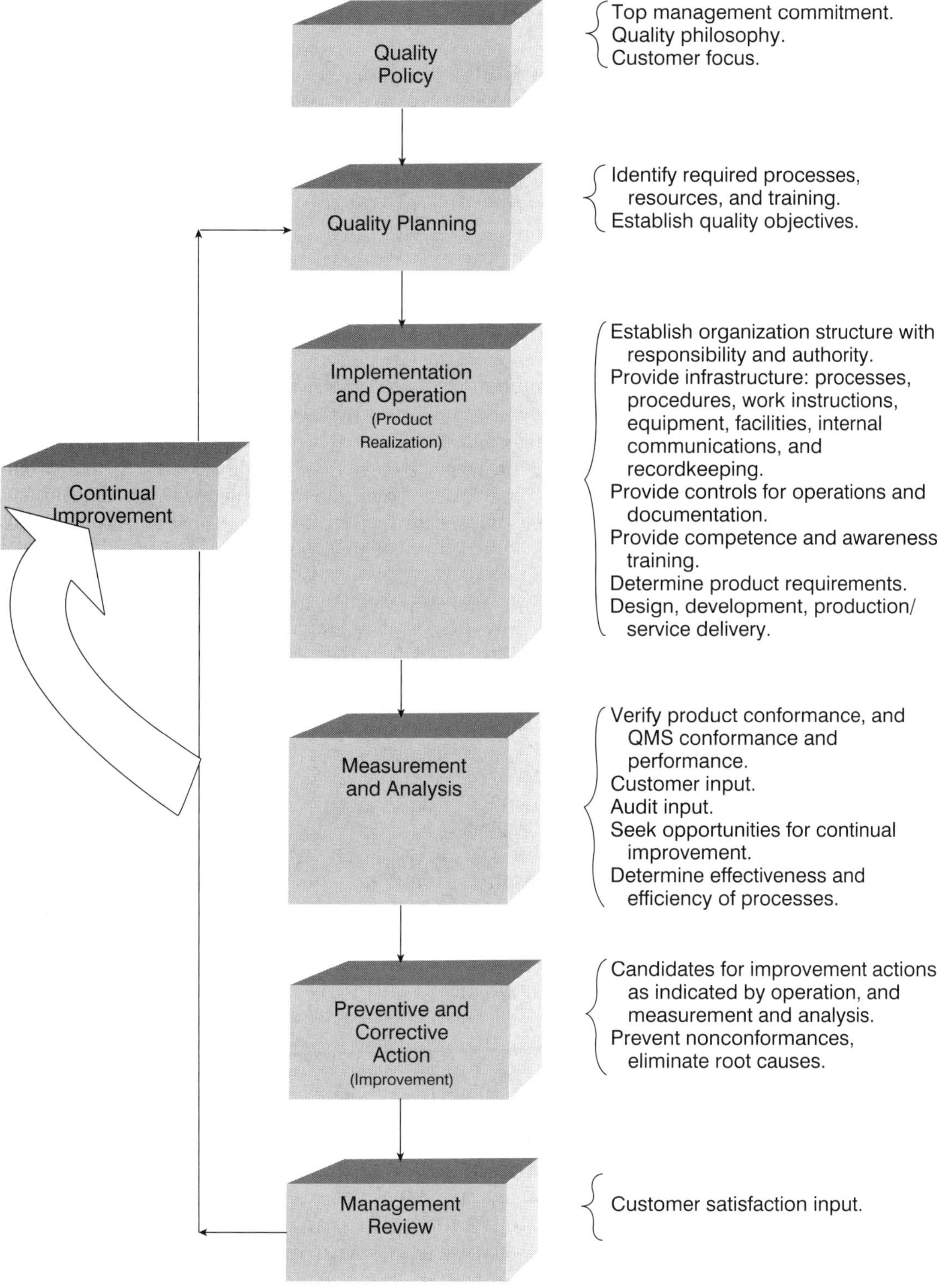

Figure 5–1
ISO 9000 QMS Model

ISO INFO

The organization's senior manager should never attempt to delegate responsibility for developing or deploying the quality policy, nor expect anyone else to be committed to the quality policy if he or she is not.

ensure that the quality system is harmonized with the aims and policies of the organization. It somewhat analogous to the corporate vision which acts as a beacon for guiding all activities within the organization.

The sample quality policy of Figure 5–2 was written to respond to the guidance provided in ISO 9004:2000, clause 5.3. It is stated in the context of a manufacturing organization. A quality policy should be short, certainly no more than one page, and should portray a lofty ideal. It should contain no specifications or other details. The quality policy should be a general statement of how the organization's top management sees its quality reputation, a broad vision of the level of quality to be achieved, the quality objective at the highest level, the approach to be used to achieve quality objectives, and the role of the organization's employees in that pursuit. It must state the organization's commitment to comply with requirements of, and to continual improvement of the QMS.

ISO 9001:2000, clause 5.1, explains that the responsibility for establishing a quality policy rests with the "top management," but this does not mean that the CEO should draft the policy in isolation. It is better for the senior manager and those directly reporting to that manager to develop the quality policy statement as a team. This will result not only in a better document, having input and discussion from all the senior managers, but will promote buy-in by all departments.

There is no advantage to be gained in trying to cover everything in the quality policy. If the policy statement is too specific, it will have to be rewritten every time there is an organizational change such as introducing a new product, adding a new customer, complying with a new specification, or devising a new process. A critical rule of thumb is *never* say anything in the quality *policy* with which you do not intend to comply. For example, if LCT had no plans to go to a team format, the tenth bulleted item in Figure 5.2 should not have been included. Keep it short, broad in its application, and somewhat lofty. Specifics are dealt with elsewhere in the quality management system.

Quality Planning For an effective QMS, the organization must identify the processes, responsibilities, resources, and training necessary to meet product requirements and strategic goals. Planning also includes the establishment of quality objectives that are consistent with the corporate vision (assuming such exists) and the quality policy and are supportive of continual improvement (clauses 5.4, 4.1).

Developing and documenting quality objectives is a significant element of quality planning. It is management's responsibility to define and document the organization's quality objectives. ISO 9001, clauses 0.2 (PDCA), 4.2.1, 5.1c, 5.3c, and 5.4.1 speak to this subject.

Quality Policy

QP - 02
Rev C

Being recognized for unfailing customer satisfaction is our foremost objective.

Meeting the expectations of our external and internal customers is the primary task of every ***LCT*** employee. We will always strive to deliver products that meet or exceed the requirements and expectations of our customers by:

- Listening to and acting on input from our customers, external and internal.
- Maintaining harmony between our corporate vision, principles and strategies, and our quality policy.
- Translating our quality philosophy into tangible actions to improve our products and the way we do business.
- Living up to and surpassing our commitment to conform to the requirements of ISO 9000 and our Quality Management System.
- Committing our company to the relentless pursuit of perfection through continual improvement of our products and processes.
- Rigorously controlling all our processes.
- Never releasing product that does not conform to customer requirements and expectations.
- Enlisting the involvement of experts—all our employees, partners, and suppliers—in our quest for continual improvement.
- Providing training for employee growth and skill enhancement.
- Bringing teamwork to all our tasks.
- Full, complete, and timely communication throughout the company.

J.M. Martinez, CEO

April 20, 2001
Date

Oct. 19, 2001
Review-By Date

Figure 5–2
Sample Quality Policy Statement

In a Total Quality organization, quality objectives are written into the strategic and tactical goals, and they need only be referenced. If your organization does not already have such objectives, they need to be developed (see Figure 5–3) in order to:

- Provide direction
- Establish commitment
- Measure progress
- Provide continual improvement

In a Total Quality organization all objectives flow from the corporate *vision statement,* which describes what the organization hopes to become in the long term. Accompanying the vision should be two other documents: the first is the *mission statement,* which describes what the organization does, its purpose for being. The second is what we call the *guiding principles,* which establish the rules of conduct the organization will observe while carrying out its mission. Directly from the vision is developed the mid-to-long-term strategy for achieving it. The strategy is articulated in terms of broadly stated objectives (we call them *broad objectives*) that relate to the entire organization. These

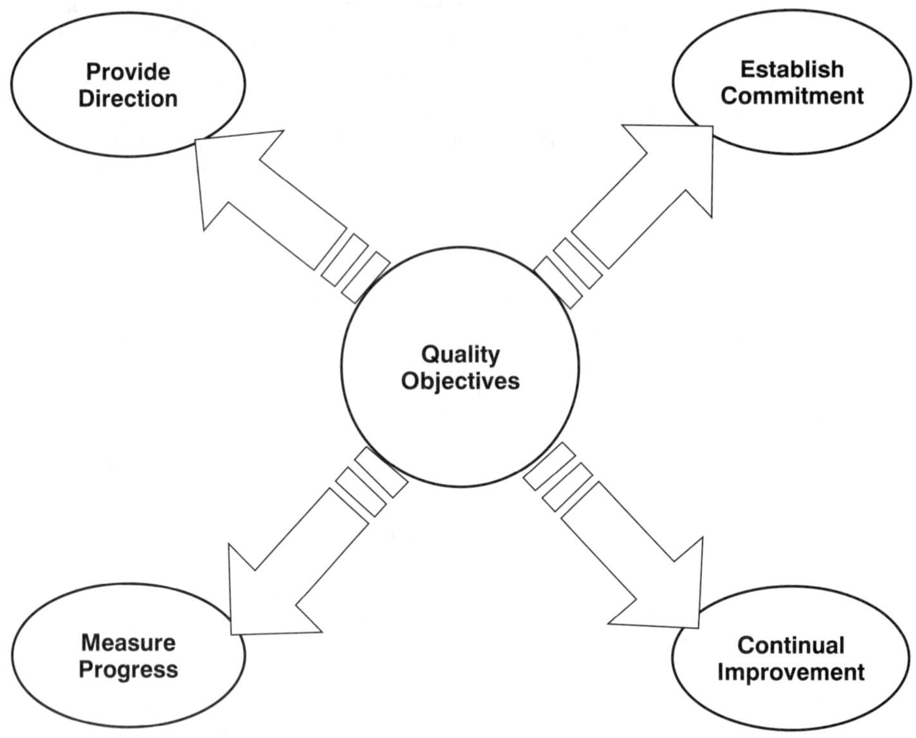

Figure 5–3
Purposes of Quality Objectives

broad objectives tell how the vision is to be achieved. From the strategic-level objectives flow the specific *tactical objectives* that may relate to departments, activities, or even individuals. These specific objectives tell the organization exactly *what* must be accomplished to satisfy the strategy that leads to the vision. Refer to Figure 5–4. Quality objectives will be of both these broad and specific types.

An organization that already has a set of objectives fashioned in accordance with the above guidelines will certainly be in a good position with regard to ISO's requirement for objectives. They need only be referenced in the new quality QMS documentation. An organization which does not have a vision statement and strategic and tactical objectives clearly needs them whether ISO requires them or not. No organization should be left to drift without a clear understanding of its course, and without instructions on how to get to some destination. Moreover, every person in the organization must understand the course, strategy, and tactics in order to align his or her activities to support achieving the vision.

Any organization interested in quality will have quality-related objectives. Whatever the organization's other objectives, some must speak to *customer satisfaction,* and must include dependability, performance, and fitness-for-use of products; *continual improvement* of processes, products, or services and people; *societal and environmental* considerations; and *efficiency* that results in lower costs. These are the objectives of interest to ISO 9000. Refer to ISO 9001, clause 1.1, and ISO 9004, clause 0.1. If your organization already has such objectives, you may be able to format them for ISO, or just simply reference the document in which they appear. If your organization does not have a set of quality objectives, then it is management's responsibility to develop, document, and deploy them.

For ISO 9000 purposes, your objectives should be at the strategic level (i.e., showing *how* you intend to achieve your vision). If you use the lower level tactical objectives, you will find yourself updating your ISO 9000 documentation constantly as these specific objectives are achieved and new ones are added. Keep in mind though, it is still necessary to develop the lower level objectives; just do not include them in the ISO documentation. Tactical objectives will be necessary for two different reasons:

1. They will give direction and purpose to your quality system, providing guidance to your employees and making the measurement of progress possible.
2. Even though not in the primary ISO documentation, these objectives will be referenced, and auditors (both internal and external) will use them and documented progress relating to them as evidence that your organization's quality system is or is not functioning properly.

Having established a commitment, and having developed a quality policy and quality objectives, it remains for senior management to complete the elements of the quality management system by doing the following:

- Providing a supporting organizational structure (ISO 9001, clause 5.5.1; ISO 9004, clause 5.5.1)
- Providing the necessary processes (ISO 9001, clauses 4.1, 5.1, 6.3, 7.1; ISO 9004, clauses 4.1, 5.1.1, 5.4.2, 6.1.1, 7.1.1)

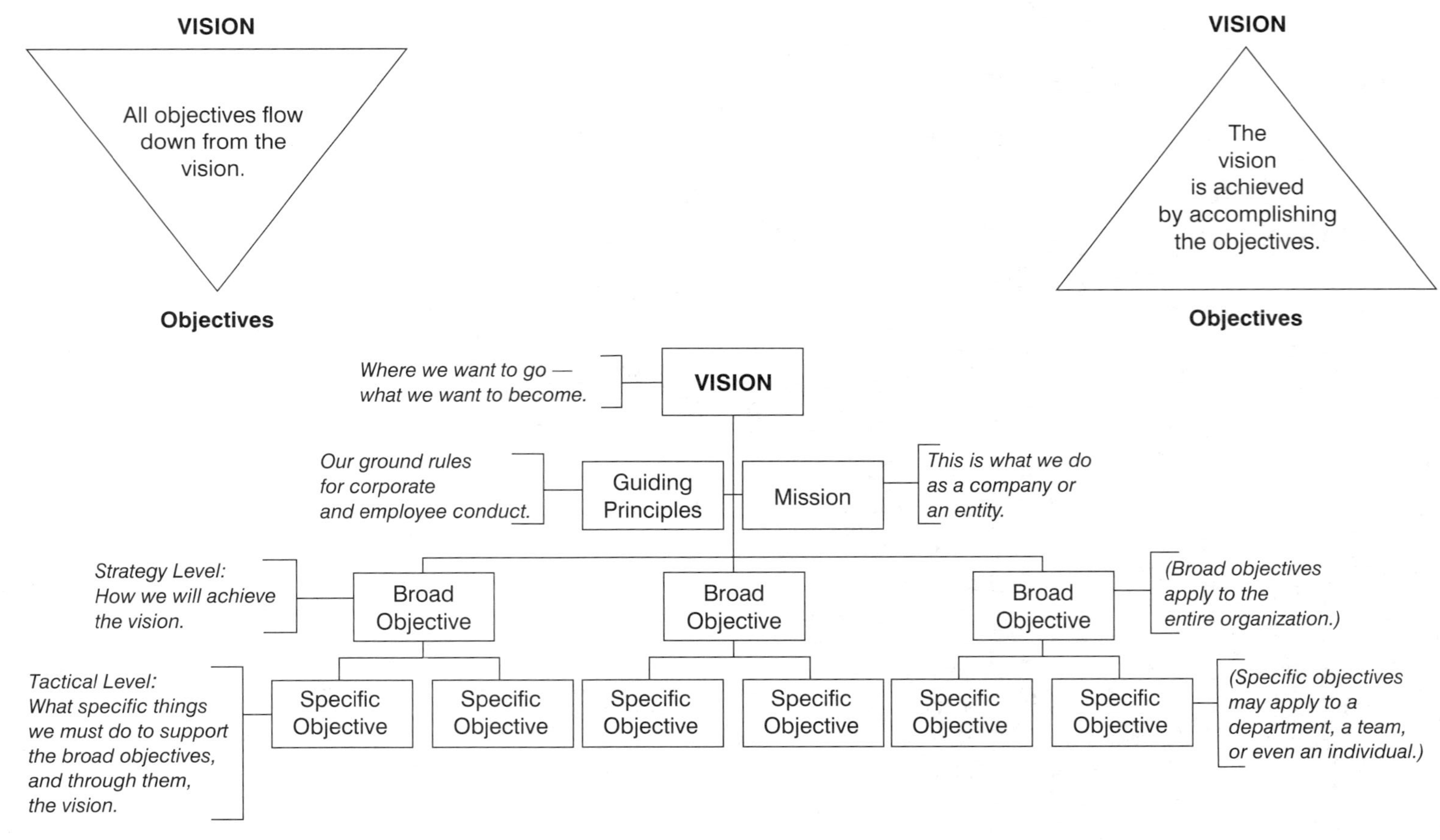

Figure 5–4
Hierarchy of Vision, Guiding Principles, Mission, Strategic Objectives, and Tactical Objectives

ISO ISSUE

What Would You Do?

John Starr is facing a dilemma. His company is moving toward ISO 9000 registration, and doing so in an attempt to eventually adopt TQM. But the company's CEO, Branch Markum, does not fully understand what will be required. For example, Markum wants to delegate management's responsibility for developing the company's quality policy to Starr. Although Starr is the company's quality manager, he does not think he should develop the quality policy. If you were John Starr, how would you explain management responsibility to Branch Markum?

- Providing the necessary procedures (ISO 9001, clauses 4.2.1, 4.2.2; ISO 9004, clauses 5.1.2, 4.2)
- Providing documentation for all of these elements as may be required (ISO 9001, clause 4.2.1; ISO 9004, clause 4.2)
- Providing the means for effective internal communications
- Providing controls for operations and documentation
- Providing the required resources (people, machines, facilities, money) (ISO 9001, clause 4.1, 6; ISO 9004, clause 6.1)
- Appointing the ISO 9000 management representative (ISO 9001, clause 5.5.2; ISO 9004, clause 5.5.2)
- Providing complete information to all employees to assure understanding (ISO 9001, clause 5.1, 5.5.1, 5.5.3; ISO 9004, clause 5.1.1, 5.1.2, 5.5.1, 5.5.3)
- Periodically reviewing QMS performance and effectiveness (and opportunity for improvement) (ISO 9001, clause 5.6; ISO 9004 clause 5.6)
- Demonstrating an unwavering commitment to quality, the QMS, and ISO 9000
- Creating an environment in which continual improvement thrives

Through the totality of these, quality management is implemented.

Organizational Structure The organization's hierarchy and the responsibility and authority assignments for each relevant level and their interrelation must be clearly defined and understood by employees. This is part of the infrastructure supporting implementation and operation of the QMS (clause 5.5.1).

The *supporting organizational structure* required by ISO 9000 is whatever you choose it to be, provided that the structure appears reasonable in terms of its potential for implementing quality management in your organization's individual situation. You must tell your registrar how your organization is structured. Be prepared to be challenged by auditors if the design of the structure does not convey obvious visible support for a quality management system. The absence of clarity on this issue suggests that the structure probably does not support a QMS, and should be redesigned. The structure must define *responsibility,* lines of *authority,* and the *interrelationships* of the people who manage, perform, and verify work affecting quality. In particular, this must be the case for people *who need the organizational freedom and authority to:*

- Prevent the occurrence of nonconformities
- Identify and record product, process, or QMS problems
- Recommend, initiate, or provide solutions for quality problems or improvements
- Verify the implementation of solutions
- Cause the cessation of further production, delivery, or installation of product until any deficiency in product or process is corrected

Distilling this requirement to its essence, the organization must have an organization chart with supporting text that clearly supports quality, and with responsibilities and lines of authority that empower all employees to do what they must do to ensure that product/service quality measures up to the quality policy. No employee should ever have to wonder if he or she has the authority to stop a process that has gone awry. It must also be clear to employees that they have the freedom to provide solutions or recommendations for solutions to problems, and to contribute to the accomplishment of quality objectives. The organizational structure has to be compatible with what is set forth in the quality policy.

The organizational chart must match the actual organization, in that there are people in the designated slots, and in that day-to-day operations match the chart. You cannot have the structure and the quality policy saying one thing, while in actual practice the organization does something else. With ISO 9000 say what you do, and do what you say.

Processes In every organization it is management's responsibility to provide the resources for doing the organization's work. The processes through which your organization's work is accomplished are provided, whether good or bad, by management. Whatever they are at this point in time, management has to accept responsibility for them. If they are not capable of producing the level of quality your customers demand, management is at fault, not employees who operate the processes. ISO 9000 correctly considers processes to be the heart of the quality system (see Process Approach, clause 0.2 of ISO 9001 and ISO 9004.) Since the overall quality system is management's responsibility, it behooves management to become familiar with the organization's processes, determine their capability, and if they are not capable, improve them—or replace them with capable processes. Process characterization should be accomplished for all key processes, certainly for all of those that can affect the quality of the product. The next step requires the development of procedures for all processes. This cannot be done without first gaining a thorough knowledge of the processes in question (clauses 4.1, 5.1e, 6.1, 6.3, 7.1).

Procedures To satisfy ISO 9001, clause 4.2.1, two kinds of procedures are required. First are procedures that directly respond to the 51 major clauses of ISO 9001. These are typically part of the organization's quality manual. As we indicated earlier, it is possible to create these procedures by parroting ISO 9001's words, changing such phrases as "the organization shall" to "we will." Of course it is permitted to go beyond this approach to assure a complete understanding and proper coverage of your unique situation. However, organizations are advised to take the simplest approach available.

The second set of procedures relates directly to the processes used for accomplishing the organization's work. These should be thought of as work instructions or work methods. ISO 9000:2000 has taken an abrupt departure from the standard's historic course of explicitly requiring many of these procedures to be documented, or requiring documentation for any procedure which could impact quality. Under the 2000 release, ISO 9000 explicitly requires only six procedures to be documented:

Clause 4.2.3	Procedure for Documentation Control
Clause 4.2.4	Procedure for Control of Records
Clause 8.2.2	Procedure for Internal Audits
Clause 8.3	Procedure for Control of Nonconforming Product
Clause 8.5.2	Procedure for Corrective Action
Clause 8.5.3	Procedure for Preventive Action

Implicit requirements for documented procedures are based on phrases such as these:

> [Clause 4.2.1] The quality management system documentation shall include . . . (d) documents needed by the organization to ensure the *effective planning, operation and control of its processes.* [Emphasis added.]

The implication is that if the effective *planning, operation,* or *control* of its processes cannot be ensured in the absence of documented procedures, then the organization *shall* provide the relevant documented procedures.

> [Clause 4.2.1, NOTE 2] The extent of quality management system documentation can differ from one organization to another due to
>
> a) the size of the organization and type of activities,
> b) the complexity of processes and their interactions, and
> c) the competence of personnel.

The implication is a) the larger the organization, b) the more complex the processes and their interactions, or c) the less competent the employees, the greater the need to document procedures and work instructions.

> [Clause 7.1] In planning product realization, the organization shall determine the following, as appropriate: . . . b) the need to establish processes, *documents,* and provide resources specific to the product.

The implication is that rather than ISO requiring the documentation, the organization must determine the need for documentation. In actual practice, the organization must not only make the determination, but also convince the registrar that the determination is valid. Subsequently, the first time there is a nonconformance which might have been avoided if a documented procedure had been used, documenting becomes a requirement per clause 4.2.1, above.

> [Clause 7.5.1] The organization shall plan and carry out production and service provision under controlled conditions. Controlled conditions shall include, as applicable . . . b) the *availability of work instructions as necessary.*

The implication is that if work instructions (procedures) are required for carrying out a task under controlled conditions (meaning doing it the same every time), then they must be available. To be available, procedures have to be documented.

> [Clause 7.5.2—on the validation of production processes] The organization shall establish arrangements for these processes including, as applicable . . . c) *use of specific methods and procedures.*

The implication is that for these processes, specific methods and procedures may be required. That most likely means that they will have to be documented.

> [Clause 7.6—on the control of monitoring and measuring devices] The organization shall *establish processes* to ensure . . .

The implication is that documented processes will be required since it is unlikely that processes related to monitoring and measuring can be consistently applied unless they are documented.

Although at first blush ISO 9000:2000 seems to *require* fewer documented procedures than the 1994 version, there is not likely to be a significant reduction except in very small organizations employing the least complex processes. The real difference is that instead of ISO deciding which procedures should be documented, those decisions are left to the organization and the registrar.

ISO 9001, clause 4.2.2 lists these procedures as a part of the quality management system. As such they are to be provided by management. This does not mean that management must develop the procedures. Rather, management must ensure that the procedures are developed. As any practitioner of Total Quality will tell you, the best people to develop procedures are the employees who actually use them. Management involvement consists of directing that the procedures be developed, and by facilitating their development through the provision of resources.

For many organizations procedure development is one of the major efforts in preparing for ISO 9000 registration. This is because many organizations don't fully understand their processes, and so have to begin with process characterization. Then the related procedures have to be developed and published. These are not jobs that can be handed off to consultants. An organization must do the job itself, perhaps with some assistance from an outside consultant.

Once the processes are really understood, and there are process flow diagrams that accurately depict the actual processes (not the way they should be, but the way they are), then procedures can be developed. An important consideration for the procedure writers is that procedures should be as simple and clear as possible, consistent with the process in question. Procedures should be written from the viewpoint of someone who has never operated the process before, but has to do so with the help of the procedure being written. The object of the procedure is consistency. That is, if the procedure is valid and is easily comprehended, then regardless of who operates the process, the output should always be the same.

The registrar will spend a lot of time going through process flow diagrams and procedures. After registration, the registrar will conduct periodic audits to make certain that your organization is meticulous in following the published procedures. This is another good reason to keep them as simple as possible while still ensuring consistency. If procedures are complicated or unwieldy every audit will become a nightmare.

Control of Operations and Documentation Control systems are required for organizations to perform operations consistently. QMS control of production and service operations is accomplished primarily through the use of proven documented procedures for the processes that can have an impact on product or service quality, and by ensuring that procedures are rigorously followed. To support this the organization must have a documentation control system that assures (1) proper procedures are issued for use and (2) changes follow the established approval process (clauses 7.5.1, 4.2.1, 4.2.3).

As any organization preparing for registration can attest, documentation is a primary consideration of ISO 9000. Certainly, the documenting of organizational structure, work processes, and procedures are not exceptions. This emphasis on documentation is in complete accord with the principles of Total Quality Management. Organizational structures must be fully understood so that all know how much latitude they have in dealing with quality issues. It is necessary to know how work processes actually function in order to improve them, and to notice when something has changed. Consistency of performance is a requirement if the process is to produce goods or services that meet requirements every time. ISO 9004, clause 4.1 says that the quality management system should function in such a manner as to ensure that:

- The system is understood, managed, and improved in effectiveness and efficiency
- Processes are effectively and efficiently controlled

None of this can happen without good documentation. The QMS cannot be understood nor can customer needs and expectations be consistently met without well-crafted documentation (see Figure 5–5) of the:

- Quality policy
- Organization's structure (delineating responsibility, authority, and interrelation of work functions)
- Process inputs, flows, and outputs
- Procedures by which the processes are operated
- Expectations in terms of quality objectives

Figure 5–5
Five Components of Good Documentation

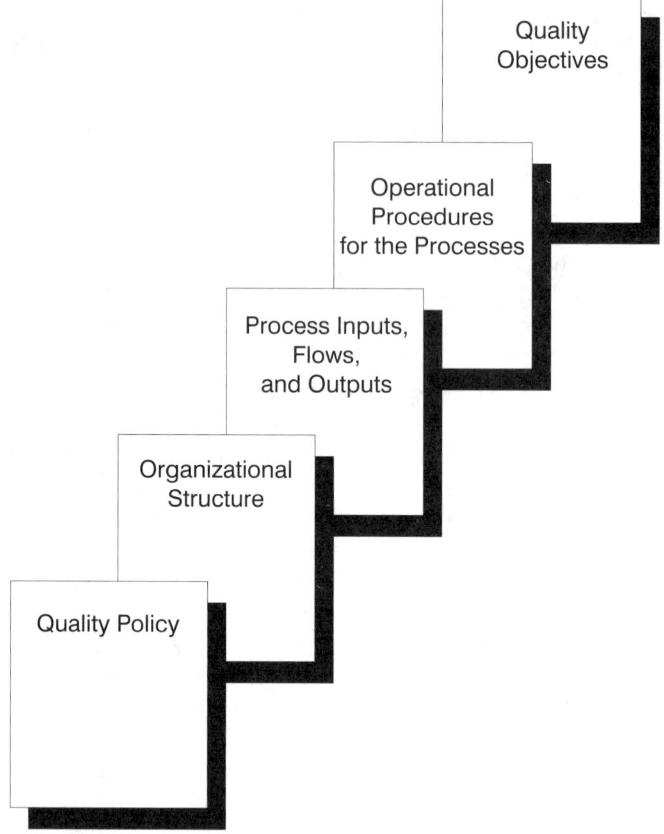

Beyond just the logic of documentation, there is the additional rationale that you cannot achieve ISO 9000 registration without having the appropriate documentation—and proving that it is routinely and consistently followed. ISO 9001, clause 4.2.2 requires the development of a quality manual for the organization. The quality manual must cover the applicable requirements of ISO 9001. As explained earlier, this requirement is most easily satisfied by changing the ISO 9001 expression *The organization shall* . . . to *We will* . . ., and republishing it as the basic quality manual. Beyond this, the quality manual must include the organization's quality policy and documented quality procedures (or make reference to them), description of interaction between QMS procedures, and details and justification for exclusion of any clause 7 requirements. The quality manual should also define the structure of the documentation used in the quality system. The quality manual, then, becomes either the home for documentation required for the QMS, or a directory from which employees (and auditors) may easily locate the documentation they need. It is also a good idea to let the quality manual be the format guide for all QMS documentation by making the approved formats a part of the manual.

It is interesting that so many business organizations try to operate without the benefit of documented processes, procedures, and structures. There is simply no way that such organizations can stay competitive in the global marketplace. The authors are told over and over by organizations that have secured registration that the most valuable aspect of ISO 9000 was that it forced the issue of documentation. With ISO 9000, an organization must either spend the time, money, and effort to develop documentation and use it, or forget applying for registration. This effort remains a management responsibility, but should involve employees not in the management ranks, especially those who operate the processes in question.

Internal Communication For the QMS to operate efficiently, the organization must have an effective internal communication system that keeps all employees and management informed of the needs and performance of the QMS. This is part of the QMS infrastructure (clause 5.5.3).

Total Quality Management has taught us that a primary function of management is communicating the aspirations, goals, code of conduct, plans, and performance results of the organization so that all stakeholders can align their efforts with those of the organization. This is not a trivial task, requiring as it does continual reinforcement and review through all the channels that a clever management team might devise. It is of critical importance to the Total Quality organization, and it is equally important for the ISO 9000 organization, whether it intends to go on to the Total Quality level or not. Just a few years ago some senior managers would have been appalled at the thought of sharing information on goals, performance, and plans below the level of middle managers. Employees, in the old days, were to do what they were told, like so many soft-skinned robots. These were the days when "managers thought and employees worked." Those days have passed. We have learned that employees need all relevant information if they are to perform to the best of their ability and for the maximum benefit to the organization. As for the old concern about employees sharing information with the competition, competitors have a good idea of what you are up to anyway, and besides, they have their hands full with their own efforts. As a sports analogy, consider the offensive squad of a football team whose game plan is known only to the manager, coach, and quarterback. The other players can react only to their immediate situations on the field. Each will play his own mini-game instead of working as a part of a coordinated effort to advance the ball to the opponent's end zone. Consider the potential difference when the blockers know their assignments for every play, and receivers have their pass routes down cold. The analogy holds true in the world of business. If you want to win in today's competitive marketplace, you must share the whole game plan with employees at all levels. The requirement to communicate with all employees is found in ISO 9001, clauses 5.1a, 5.3d, 5.5.1, and 5.5.3.

When ISO says something must be communicated, that carries with it the need to be understood. In addition to documenting the system, the implication is that it must be explained to all employees, and that management must check to make sure that the understanding of employees is complete and accurate by reviewing results (ISO 9001, clause 5.6).

This is familiar territory for Total Quality organizations. On the other hand, for traditional organizations moving toward ISO 9000 registration, this level of communication probably represents a culture change. In such cases, ISO is serving the purpose of moving the organization in the right direction.

Management of Resources ISO 9001, clause 6.1 requires the organization to determine what resources are necessary to implement, maintain, and continually improve the QMS, and to meet customer requirements. It then requires the organization to provide those needed resources. Resources, of course, come in many flavors. Employees are resources. So are buildings, machines, tools, materials, and money. All of these are tangible; they can be seen and touched. Resources also include some intangibles. Clause 6.2.1 requires that employees whose work can affect quality must be competent on the basis of education, training, skills, and experience. For example, a process might require an operator who has the specialized knowledge, skills, and experience to competently manage the process. Thus, human resources include knowledge, skills, and experience. Organization-provided training is often required to bring employees up to the desired level of competence. The training itself can be considered a resource, or at the very least the result of the combination of several resources, including instructors, classrooms, training programs, and the time of employees. Finally, resources include the total organizational infrastructure. Clause 6.3 requires the organization to determine its appropriate infrastructure, and then to provide and maintain it. An organizational infrastructure is made up of the resources we have already mentioned, together with the processes, procedures, work instructions, policies, organization structure, and even the underlying organizational culture. Clause 5.1 states that providing all of these necessary resources is a management responsibility.

Providing resources is clearly management's responsibility, regardless of ISO 9000's position. The fact is, only management can commit the resources—human and capital—needed to pursue the organization's mission.

It should be noted that under ISO 9000, "competent personnel" in an organization must include internal auditors who ensure that the documented procedures are adequate and rigorously followed and that required records are being maintained. If these internal auditors perform their duties effectively, the third party (registrar) audits should pose no problem for the organization.

Competence and Awareness Training Management is responsible for ensuring that all employees are knowledgeable about the organization's commitment to ISO 9000 and the QMS, the quality policy, and the quality objectives. It is also responsible for ensuring that relevant operating employees are competent to execute their functions. This is accomplished through training and evaluation as a part of the infrastructure established under the QMS (clauses 6.2.2, 5.1, 5.3).

Product Realization In more familiar language, this is developing, implementing, and operating the processes necessary to:

1. Determine customer requirements for a product (or service)—(clauses 7.1, 7.2)
2. Design and develop the product (clause 7.3)
3. Verify and validate the design (clauses 7.3.4, 7.3.5, 7.3.6)
4. Control design and development changes (usually referred to as configuration management) (clause 7.3.7)
5. Determine and implement the processes necessary to build the product (clauses 7.1, 7.3.3, 7.5.1)

6. Validate production processes (clause 7.5.2)
7. Purchase required materials, components, and so forth (clause 7.4)
8. Manufacture the product (clause 7.5)

Product realization processes are those that conceive of the product in response to customer (or otherwise determined) requirements; define, implement and operate the means of production; and, at the same time, verify/validate the design and the production processes.

Measurement, Analysis, and Improvement Through measurement and monitoring and the analysis of data, both process and product conformity is determined on a continuing basis, and continual improvement opportunities are revealed (clauses 8.2.3, 8.2.4, 8.4, 8.5). The organization needs to

1. Validate product through monitoring and measurement (clauses 8.1, 8.2)
2. Continually improve product and the means of production, including the QMS (clause 8.5)

The ISO 9000 Management Representative ISO 9001, clause 5.5.2 requires that top management appoint a manager to be the ISO 9000 management representative. In addition to whatever normal responsibilities this individual has, as the management representative he or she is to (1) ensure that processes needed by the QMS are established, implemented, and maintained; (2) report to top management on QMS performance/effectiveness and any need for improvement; and (3) ensure the promotion of awareness of customer requirements throughout the organization. A note appended to the clause suggests that the management representative may also have responsibility for liaison with external parties on matters relating to the quality management system.

In the earlier version, ISO 9000-2:1993, clause 4.1.2.3 raised the issue of a potential conflict of interest on the part of the management representative. We would say that the potential is certainly real, and it must be guarded against or the integrity of the QMS may be compromised. For example, consider a situation in which Art Swick, the Vice President of Manufacturing, is assigned the duties of Management Representative for ISO 9000. On the one hand his performance is being measured by on-time delivery, and on the other by assuring that no nonconforming product is shipped. Clearly ISO 9000 would require that he adhere strictly to the latter, but there will be pressure from the on-time delivery standpoint to ship regardless. Even if he plays it straight by the ISO 9000 rules, he will be subjected to pressure to do otherwise, and it will be visible to knowledgeable customers, who may well assume that the organization's quality system is suspect. In selecting the management rep, top management must be careful to avoid the appearance of a conflict of interest.

In addition, there are problems with the management representative premise itself, and we had hoped the 2000 version would eliminate it altogether. One of the first lessons learned in Total Quality Management is that the senior executive cannot delegate responsibility for the TQM implementation or for being its champion. To do so virtually assures failure because only the senior executive is in a position to lead the effort. As can be seen in the previous paragraphs of this section, ISO 9000 supports this view. Why, then, does ISO 9000 require that a management representative be appointed? We do not have an answer to this question, but apparently it stems from the belief that the senior manager is too busy to undertake these duties; duties that, after all, require time "on the floor" to view at first hand how the quality system is working, and to look for opportunities to improve the system. ISO apparently takes the view that this is not the senior manager's domain, and so creates a kind of staff function. We recommend that senior managers appoint themselves; there is nothing in ISO 9000 that prevents them from doing so.

However, in most cases the appointee will probably be at a lower level than the CEO, such as the head of the Quality Assurance department. Such an appointment certainly satisfies the requirement of being able to assess the quality system and to be a point of contact for outside parties who may be interested in learning of the organization's status relative to ISO 9000. However, care must be taken to ensure that there is never a suggestion that the responsibility for ISO 9000 is being delegated by the senior manager. The management representative should serve only as a facilitator who assists the CEO in carrying out *his* or *her* responsibility.

Record Keeping The QMS must support the maintenance of important records to provide evidence of conformity to ISO 9000 requirements and of the effective operation of the QMS. Records will be many and varied. They will be useful to the organization and to the auditors and other interested parties. To accomplish this, the records must be stored in a manner that provides protection and at the same time affords easy retrieval (clauses 4.2.3, 4.2.4).

Management Review Management must periodically review the QMS for continued suitability, adequacy and effectiveness, the need for changes to the QMS, and for continual improvement opportunities and results. Review inputs must utilize data from audits, customers, processes, and product performance. Status for current improvement activity, planned changes, and recommendations for improvement are also to be considered (clauses 5.6.1, 5.6.2).

Management review output must include decisions and actions related to (1) improvement of QMS and processes, (2) improvement of product relative to customer requirements, and (3) any resource needs. Formal records of management reviews must be maintained (clauses 5.6.1, 5.6.3).

ISO 9001, clause 5.6.1 uses the term *at planned intervals*. That means that ISO requires the organization to schedule periodic reviews, but the timing is left to the organization, with approval rights by the registrar. The authors suggest that these reviews be scheduled monthly, perhaps in conjunction with other reviews the organization already holds. Quarterly reviews are too infrequent for continuity. In fact many of the items to be reviewed could well start and end within a quarterly window, and fail to be picked up in the review. Bimonthly may be acceptable in some slow-moving organizations, but we think the monthly schedule is best. We also suggest, although it is not required by ISO 9000, that the format and agenda of the reviews be published and distributed appropriately a few days before each review. This will enable participants to better prepare.

Continual Improvement To conform to ISO 9000:2000, systems must be built into the QMS for identifying potential improvements and implementing them. This is a welcome and much needed change from the earlier version of ISO 9000, which had no requirement for continual improvement. Continual improvement most often occurs through the elimination of root causes of actual or potential nonconformance, but may also be the result of incorporating entirely new processes in place of old ones, adopting new technology, or other strategies. Each stage of the ISO 9000 operating model must be a source of continual improvement ideas (clause 8.5).

Borrowing from Total Quality Management, ISO 9004:2000, clause 8.5.1 states:

> Management should continually seek to improve the effectiveness and efficiency of the processes of the organization, rather than wait for a problem to reveal opportunities for improvement. Improvements can range from small-step ongoing continual improvement to strategic breakthrough improvement projects. The organization should have a process in place to identify and manage improvement activities. These improvements may result in change to the product or processes and even to the quality management system or to the organization.

We have described a number of the elements which make up the quality management system, including the quality policy, the processes, procedures, objectives, and organization structure. Clearly there are many processes and many procedures. These processes come into play at different times. For instance, in a manufacturing example, one set of processes determine the requirements for a product (clause 7.2). Once requirements are understood, another set of processes are used to design and develop the product (clause 7.3). Following product design, the procurement-of-materials processes are activated (clause 7.4), and then the production processes manufacture the product (clause 7.5). As product is realized the processes concerned with measuring and monitoring come into play (clause 7.6, 8.1, 8.2). Data arising from measuring and monitoring are subjected to analysis processes (clause 8.4), and this leads to activation of improvement processes (clause 8.5).

The process categories related to the typical phases of a generic product's development and manufacture are:

1. Determination of requirements
2. Product design and development
3. Process planning and development
4. Purchasing
5. Production, or provision of services
6. Verification
7. Improvement

Note that all of these are process categories that relate to each other in time sequence.

An earlier version of ISO 9000 made these points which are still valid uncler ISO 9000:2000:

- A quality management system consists of a number of elements.
- The quality management system is carried out by means of processes, existing within and across functions.
- For the quality management system to be effective, these processes and their supporting responsibilities, authorities, procedures, and resources have to be defined and deployed in a consistent manner.
- The quality management system needs coordination and compatibility of its processes.
- The quality management system needs definition of the process interfaces.

These statements may be a little fuzzy, but notice that the third bulleted item talks about responsibilities and authorities. This is a reference to the organization chart and its supporting text or job descriptions (if any). The same item also says procedures (for the processes) have to be defined, and resources deployed. The QMS should be structured to cover all the processes and their cross-functional interfaces, interrelationships, lines of responsibility and authority, and required resources as elements in a production model (as in the seven-element model above).

Remember, each of these elements, as well as all the QMS elements discussed in this section, is comprised of several processes.

ISO recommends, and the authors concur, that the best approach for setting up the quality management system is one based on processes. ISO provides a model for a process-based quality management system, which we have simplified in Figure 5–1. This figure illustrates the linkages between the processes represented by ISO 9001 clauses 4 through 8. It also illustrates the significant role played by customers and other interested parties. Of the process approach to the quality management system, ISO states:

> An advantage of the process approach is the ongoing control that it provides over the linkage between the individual processes within the system of processes, as well as their combination and interaction.
>
> When used within a quality management system, such an approach emphasizes the importance of
>
> a) understanding and fulfilling the requirements,
>
> b) the need to consider processes in terms of added value,
>
> c) obtaining results of process performance and effectiveness, and,
>
> d) continual improvement of processes based on objective measurement.[4]

The functional and process interfaces, departmental interactions, responsibilities, authorities, and resources should be clearly documented. All documentation must be in harmony with the quality policy. A word of caution is appropriate at this point. Do not succumb to the trap of trying to document everything to the last detail of specificity. Keep documentation as simple as possible.

Where should this information be placed? It needs to be delineated in your organization charts and their accompanying text, in the process procedures, and in any supporting documentation you may use (such as traveler tags on work in process).

QUALITY MANAGEMENT SYSTEM STRUCTURE

ISO 9000 does not attempt to suggest a structure for the quality management system, since it would be impossible to design a structure to fit all organizations. However, ISO 9001:2000 specifies the fundamental requirements and intents for a QMS, which must be tailored and adapted to the particular organization's operations, culture, and resources. From this the organization is expected to translate the standard's requirements and intents into a **QMS structure** that will support them while meeting the organization's needs. The best starting point for developing a QMS structure is a review of ISO 9000's process-based QMS model. Figure 5–6 has been adapted from ISO's Figure 1, which appears in ISO 9000, ISO 9001, and ISO 9004. In Figure 5–6 the four blocks connected with the large arrows represent the QMS in operation, in a continuous loop, not unlike the Shewhart/Deming Plan-Do-Check-Adjust (PDCA) Cycle. In fact, ISO 9001 connects the model with the PDCA Cycle in clause 0.2.

Each of the four blocks in the operating loop contain specific tasks and processes that must be accomplished in order for the next block in sequence to operate its processes. For example, all of the requirements in ISO 9001's clause 5 fall under the Management Responsibility block at the top of the loop. These include:

1. Arrive at a top management commitment to develop and implement the QMS, and to continually improve it (clause 5.1).
2. Establish a customer focus (clause 5.2).
3. Develop the quality policy (clause 5.3).
4. Plan the QMS and quality objectives (clause 5.4).
5. Delineate and communicate responsibilities, authorities, and their interrelationships, and designating the management respresentative (clause 5.5).
6. Conduct scheduled management reviews of the QMS—the process representing the end of a loop (clause 5.6).

The second block in the process-based QMS model, Resource Management, builds on the top of the Management Responsibility block, and includes all of the tasks and processes in clause 6.

1. Provide the resources needed for the QMS (clause 6.1).
2. Ensure that processes are operated by competent employees (clause 6.2).
3. Provide and maintain the required infrastructure (clause 6.3).
4. Provide an appropriate work environment (clause 6.4).

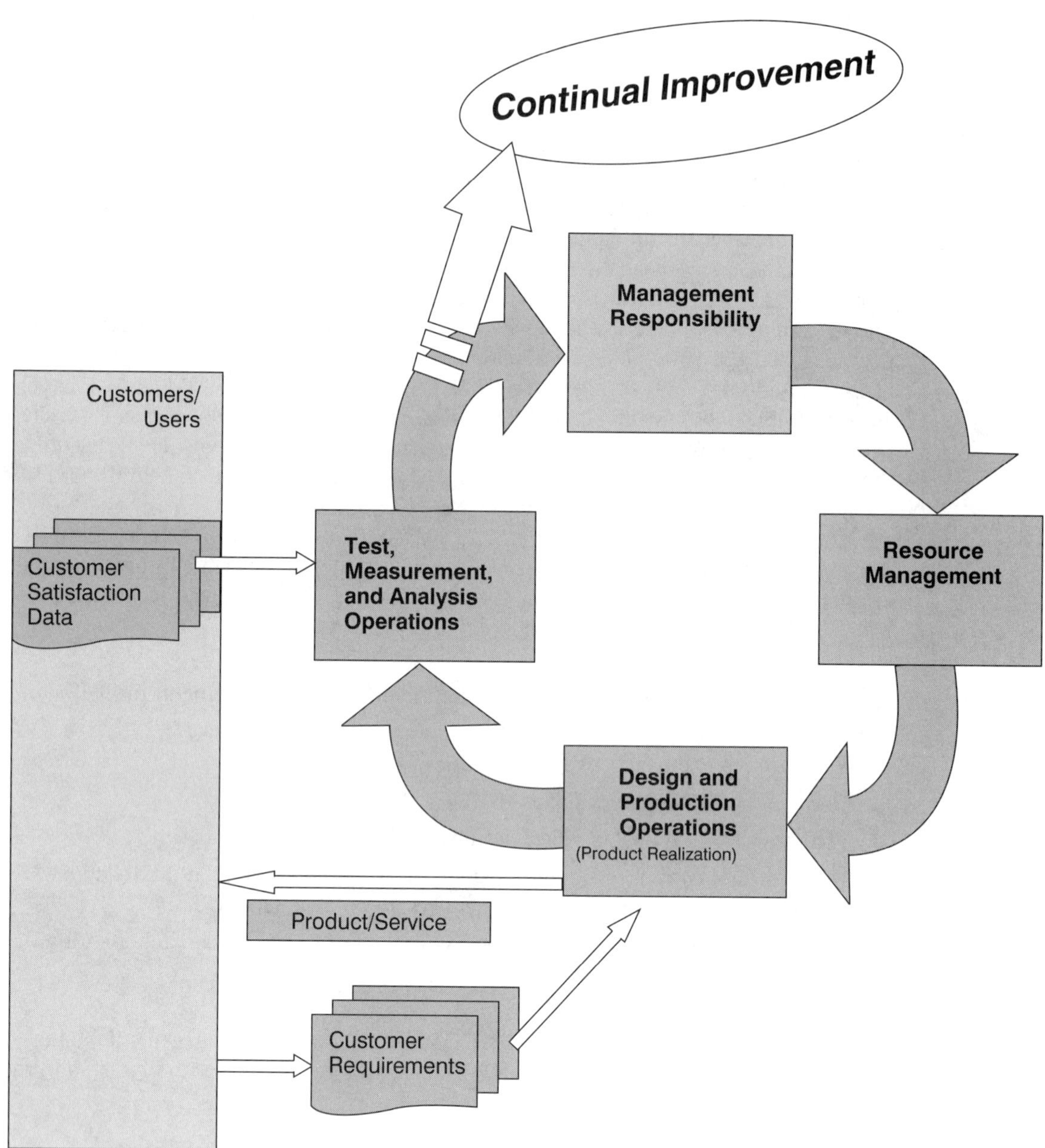

Figure 5–6
Model of a Process-Based QMS

The third block, Design & Production Operations (or Product Realization in ISO-speak) puts to work the combined output of the previous two blocks, with additional input from customers in the form of product requirements, and includes all of the tasks and processes of clause 7.

1. Plan and develop the production processes and required documentation (clause 7.1).
2. Establish product specifications based on customer input (clause 7.2).
3. Design and develop the product (clause 7.3).
4. Purchase and verify required materials for product (clause 7.4).
5. Produce and ship the product (clause 7.5).
6. Provide for the control of monitoring and measuring devices (clause 7.6).

The fourth block, Test, Measurement, and Analysis, operates all the processes of clause 8 to verify that the QMS control loop is operating efficiently, that the products resulting from the loop's processes conform to requirements, and that opportunities for improvement are identified. Measurement and Analysis includes processes to:

1. Demonstrate product conformity, ensure QMS conformity, and continually improve it (clause 8.1).
2. Monitor and measure processes and product (clause 8.2).
3. Control nonconforming product (clause 8.3).
4. Analyze data for QMS effectiveness and improvement opportunities. (clause 8.4).
5. Improve the QMS and product through continual improvement, corrective action, and preventive action (clause 8.5).

The loop next returns to the Management Responsibility box, providing input for the management reviews, which in turn lead to another cycle in a loop that never ends. Viewed in terms of the PDCA Cycle, *Plan* is represented by the Management Responsibility and Resource Management blocks. *Do* is the Design and Production Operations. *Check* is the Measurement and Analysis block, and *Adjust* takes place in the Management Responsibility block as a result of the *Check* data and customer input. Then the loop repeats again, and again, and again, continually improving the QMS and products.

A Typical QMS Structure

As enlightening as Figure 5–6 may be, it is still not a structure, at least from the physical perspective. To build the structure of the QMS in a physical sense, consider Figure 5–7.

The components depicted in that scheme are described in the next sections.

QMS Foundation

Like any physical structure, it starts with a foundation at the bottom, with the upper stories depending on the lower structure to support them. In the case of the QMS structure

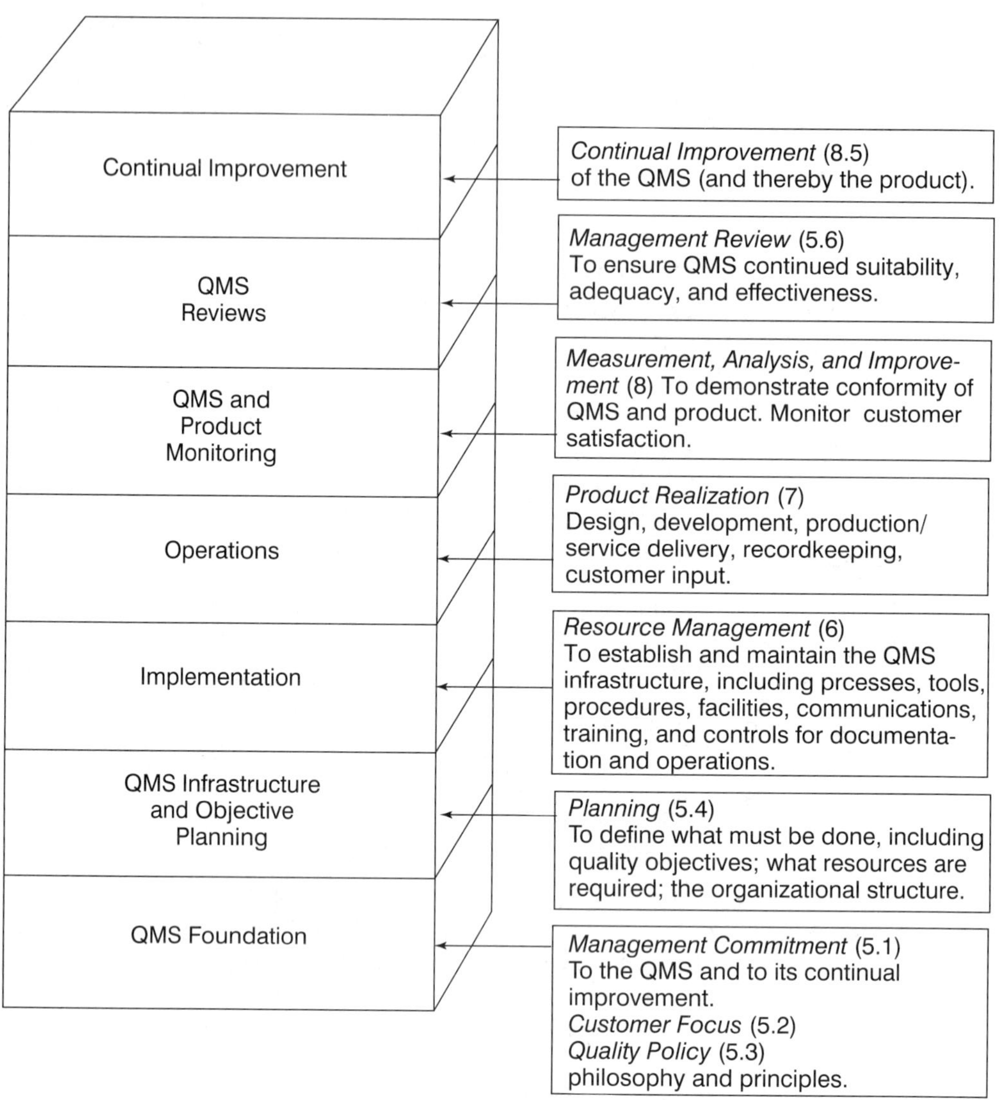

Figure 5–7
Quality Management System Structure

illustrated in Figure 5–7, the foundation consists of the following:

- Management commitment (clause 5.1)
- Customer focus philosophy (clause 5.2)
- Quality philosophy and principles of the quality policy (clause 5.3)

The QMS is envisioned to start with top management's commitment to the quality management system and to continual improvement. This commitment is accompanied by a policy statement that outlines the philosophy (including customer focus) and guiding principles under which the organization intends to operate the QMS. This is the first stage of the QMS structure and provides the foundation upon which the QMS is constructed and operated. The quality policy must be revisited from time to time to assure its continued relevance and adequacy.

QMS Infrastructure and Objective Planning

Built on the foundation is the top management Planning level (clause 5.4), which:

- Establishes quality objectives
- Defines QMS tasks, processes, resources, organization structure

The second stage of the QMS structure deals with identifying and planning for the organization's tasks, processes, resources, organization structure, and quality objectives. After the initial implementation, this stage remains active to deal with changes—in the organization's activities, processes resulting from continual improvement, and the business situation. As such it becomes the Plan stage of the closed-loop Plan-Do-Check-Adjust Cycle. The planning stage effectively establishes what the organization must do to achieve and retain conformance with ISO 9000 and its own QMS.

Implementation

The third story houses ISO's Resource Management activities (clause 6). It converts the plans to reality by establishing and maintaining the QMS infrastructure. The model's third stage provides the tools, procedures, and resources necessary to put the QMS into sustained operation. Implementation is achieved through the following steps:

1. Designation of responsibility and authority relative to the QMS. This may be in the form of an organization chart, job descriptions, or other suitable means of communicating to the employees.
2. Designation of the management representative, who will provide the day-to-day management of the implementation and ongoing functions of the QMS.
3. Providing the necessary human, technological, and financial resources.
4. Establishing and implementing reporting policies and procedures necessary to ensure that top management is aware of QMS performance.
5. Determining training needs and providing required training for relevant personnel.
6. Establishing and implementing policies and procedures for effective QMS-related internal communication.
7. Establishing and implementing policies and procedures relative to documenting the QMS.
8. Establishing and implementing policies and procedures for controlling QMS documentation.

9. Establishing and implementing policies and procedures for controlling the organization's relevant operations and processes.

The Implementation phase puts the QMS framework in place. It will require continual updating when changes, such as employee job reassignments, occur; as the organization's activities change; as training needs change over time; and as policy and procedures evolve through continual improvement.

Operations

The fourth story is the Operations level, which ISO calls Product Realization (clause 7). Operations puts the infrastructure into action designing, developing, and producing the product. The Operations stage, in combination with Implementation, represents the Do stage of the Plan-Do-Check-Adjust Cycle.

QMS and Product Monitoring

At the fifth story the QMS and the products are monitored through ISO's Measurement, Analysis, and Improvement activity (clause 8). Monitoring is intended to:

- Ensure conformity of the QMS to ISO 9001
- Demonstrate that products conform to requirements
- Discover opportunities for improvement
- Capture customer satisfaction input

The QMS structural model's QMS and Product Monitoring stage deals with monitoring the performance of the QMS elements, the system's conformance with ISO 9001, ensuring that products conform to requirements, and monitoring customer satisfaction. This is accomplished through the following steps:

1. Regularly monitoring the key characteristics of the organization's relevant activities and processes.
2. Calibrating and maintaining monitoring equipment according to the organization's procedures.
3. Investigating all nonconformances.
4. Taking corrective and preventive action to prevent recurrence of the same nonconformance.
5. Periodically auditing the QMS to assure to continued conformance to ISO 9001 and the organization's plans, policies, and procedures, as well as to determine if the QMS has been properly implemented. (Audits may be internal or external).

This stage will also be under constant review for updating to improve its own processes or to accommodate changes in the other stages. This stage represents the Check in the Plan-Do-Check-Adjust Cycle.

QMS Management Review

The sixth story of the structure is where the QMS is subjected to management review (clause 5.6). Data generated at the monitoring level, plus any other inputs available, is used to determine whether the QMS, as currently implemented, remains suitable, adequate, and effective. The review process requires the collection of information relevant to the QMS and the presentation of this information to senior management on a scheduled basis, with other inputs as required by the nature of the information. The purpose of the management reviews is to:

1. Ensure the QMS's continued suitability
2. Confirm its adequacy
3. Verify its effectiveness
4. Facilitate continual improvement of the QMS, its processes, and products

From management's review of the QMS, including the processes and equipment employed, it will be determined what is acceptable in its present condition and what needs to be changed. This is the Adjust phase of the Plan-Do-Check-Adjust Cycle.

Continual Improvement

The topmost level of the structure is occupied by Continual Improvement (clause 8.5). At this level the opportunities for improvement of the QMS and the product are realized.

Responsibility and Authority—Within and Between Activities of the Organization

ISO 9004, clause 5.5.1 suggests that "top management should define and then communicate the responsibility and authority in order to implement and maintain an effective and efficient quality management system." All relevant activities, or departments, in the organization should be *defined and documented.* That means the organization's processes should be analyzed and flow-charted, and functional responsibilities and interrelationships documented in a formal manner that is uniform throughout the organization. In particular,

1. Any quality-related responsibilities should be explicitly defined.
2. Any specific responsibility, organizational freedom, or authority held by the activity should be clearly established.
3. Control of interfaces and coordination requirements between activities should be spelled out clearly.
4. Responsibility for quality should be written into the functions of all activities, emphasizing prevention, but accommodating corrective action when it is required.

When these four tasks have been accomplished, the organization will have a *set of documents* that explain the departmental structure, which of these activities is responsible for what, which (or who within) has authority for what, who has the freedom to act when circumstances dictate, who has control over the functional interfaces, what kind of coordination is expected between activities, and everyone's role in quality.

The Organizational Structure As It Relates to Quality

The requirements set forth in the last section cannot be completed without having an organizational structure at least in the formulation stage. ISO 9004, clause 5.5.1 suggests "People throughout the organization should be given responsibilities and authority to enable them to contribute to the achievement of the quality objectives and to establish their involvement, motivation, and commitment." This should be interpreted to mean people in all functions. Lines of authority and communication should be clearly defined. What is asked for here is an organization chart (perhaps with clarifying text) that pays particular attention to functional interfaces with the quality system.

Resources and Personnel

In order to implement and maintain a quality management system compatible with the organization's quality policy and capable of achieving the quality objectives, resources—including people, equipment, and facilities—will be required. This may or may not require increased resources, but whether it does or not, ISO 9001, clause 6.1 specifies that the required resources be identified (meaning in writing) and that the necessary resources be provided by management. In terms of human resources, management must also specify the required level of competence, experience and/or training (clause 6.2.1) for each functional position. The equipment used in these processes, including quality assurance processes, is to be listed by process (clause 6.3). The same applies for computer hardware and software. Management should plan for production demand and schedule resources accordingly. The result of this planning could be a *resource requirements document,* keyed to production demands, that is both quantitative (how many?) and qualitative (what competence level?) for all of the organization's key processes.

Procedures

A major thrust of ISO 9001 is control of processes (see clauses 4.1, 4.2, 5.4.2, 5.5.2, 5.6, 6, 7, 8). It should be noted from the outset that ISO 9000 does not require the use of Statistical Process Control (SPC). In the language of SPC, processes are either *in control* or *out of control.* Although SPC might be appropriate and beneficial in a given situation, when ISO 9000 talks about control of processes, SPC is not what is intended. Rather, ISO 9000 intends that processes be controlled for consistent results through the proper application of documented process procedures. Controlling for consistency is the right thing to do, and ISO makes sure that an organization does it. Beyond this, if the organization *chooses* to apply SPC, it, of course, may do so. Clause 8.1 includes "statistical techniques" as something that may be needed. In addition to those procedures concerned with functional processes, there must be procedures related to the QMS which prevent problems from occurring, and which deal with problems that do occur. There must also be procedures for coordinating the quality-related activities of the various functions. The quality objectives of these procedures must be clearly spelled out. For example, the Engineering Department will have its procedures for the design of new products. The Manufacturing Department will have its procedures for the production of products. In addition, there should be a procedure that requires manufacturing to work with engineering during the development of the new product, the quality objective of which is to ensure the producibility of the design, which, in turn, will ensure higher quality and lower cost through fewer errors and reduced effort. When operational procedures have been documented, they must be strictly adhered to in order to ensure consistency in product output and an emphasis on prevention, both of which will lead to higher quality.

Configuration Management/Control of Documents

The quality management system structure is completed with the application of a document control system, commonly referred to as **configuration management.** The purpose of configuration management as a part of the quality system structure, as stated in ISO 9001, clause 4.2.3, is to:

a) To approve documents for adequacy prior to use
b) To review and update as necessary and reapprove documents
c) To ensure that changes and the current revision status of documents are identified
d) To ensure that relevant versions of applicable documents are available at the point use
e) To ensure that documents remain legible and readily identifiable
f) To ensure that documents of external origin are identified and their distribution controlled
g) To prevent the unintended use of obsolete documents, and to apply suitable identification to them if they are retained for any purpose.

ISO 9001, clause 4.2.3 explicitly requires a documented procedure for this documentation control. This is the first of only six such explicit documented procedure requirements in ISO 9000:2000.

Configuration management can be applied not only to product documentation such as drawings, parts lists, and specifications, but also to the procedures used for every process of the organization. We recommend its application to every policy and procedure document, because it puts discipline into the formatting and release of documents and changes to the documents, and it assures that the proper release version is in use. There are few things worse than building a product to the wrong revision level, or using a procedure that is out of date and no longer valid.

> **ISO INFO**
>
> *"Continuous improvement is fundamental to success in the global marketplace. Companies that are just maintaining the status quo in such key areas as quality, new product development, the adoption of new technologies, and process performance are like a runner who is standing still in a race. Competing in the global marketplace is like competing in the Olympics. Last year's records are sure to be broken this year. Athletes who don't improve continually are not likely to remain long in the winner's circle. The same is true of companies that must compete globally."*[5]

CONTINUAL IMPROVEMENT AS A PART OF THE QUALITY MANAGEMENT SYSTEM

One of the cornerstones of Total Quality Management is the concept of **continual improvement**—not a new concept. Its beginning can be traced to 1950 and the lectures of Dr. W. Edwards Deming to Japanese industrialists. These Japanese leaders based their push into world markets on Deming's ideas of continually improving designs, products, and processes through the use of statistical techniques.

For thirty-five years Japanese businesses forged ahead, virtually unnoticed by western countries, improving the quality of their products, and refining and expanding the teachings of Deming and Joseph Juran—developing along the way the larger concept we now call Total Quality Management. It wasn't until the decade of the eighties that the West began to understand the changes that had occurred in world markets—namely that Japan had used Deming's principles to emerge as a global leader in producing quality products. The time had come for the West to learn from Japan.

Today fewer and fewer companies are exempt from pressures of world markets, what with international trade agreements having removed many of the artificial barriers to competition. Even organizations which do not compete directly with foreign companies find that they have to do a better job with less. This more-with-less trend is likely to be permanent, so all organizations have to find ways to become or remain competitive. Continual improvement should be considered the normal approach to doing business in every organization.

It is interesting to note that in the 1987 version, ISO 9000 did not address continual improvement at all. In the 1994 version, continual improvement was not a major consideration, although one of the guideline documents, ISO 9004-4-1993, was dedicated to quality improvement. Allan J. Sayle suggested that ISO was trying to distance itself from TQM.[6] Our view is that the ISO 9000 committees had not become TQM champions, and the "not invented here" syndrome seemed to be having an effect. In the first edition of this book we said we hoped that future releases of ISO 9000 would clearly endorse continual improvement, and embrace it straight on because it is essential to the objectives of ISO 9000.

Never in our most fervent optimism could we imagine the extent to which ISO 9000:2000 has embraced continual improvement. In the quality policy, the heart, soul and conscience of the quality management system, ISO requires the organization to commit itself to just two things: one is a commitment to comply with requirements, and the other is a commitment to continual improvement. ISO 9000:2000 lists, in clause 0.2, the eight quality management principles. One of them is *Continual Improvement* (clause 5.3). The very first requirement clause of ISO 9001 (4.1) requires the organization to *continually improve* QMS effectiveness. Clause 5.1, the first management responsibility clause, requires top management to provide evidence of its commitment to continual improvement of the QMS. The last requirements section of the standard, clause 8.5, requires the organization to continually improve the QMS (clause 8.5.1). Other clauses requiring the incorporation of continual improvement into the fabric of the QMS include:

5.6 Management Review
6.1 Provision of Resources
8.1 Measurement, Analysis and Improvement, General
8.4 Analysis of Data

In addition, the Introduction clause 0.2 of all three documents, ISO 9000, 9001, and 9004, point out the importance of continual improvement:

> [ISO 9000:2000] Continual improvement of the organization's overall performance should be a permanent objective of the organization.
>
> [ISO 9001:2000 and ISO 9004:2000] When used within a quality management system, [the process] approach emphasizes the importance of continual improvement of processes based on objective measurement.

Moreover, ISO 9000, 9001, and 9004 all use a Figure 1 to illustrate that a major outcome of the process-based QMS is continual improvement. Finally, Annex B to ISO 9004 is a brief tutorial on continual improvement.

The incorporation of continual improvement into ISO 9000 as a firm requirement is welcomed by Total Quality Management proponents, including the authors, but may be less than enthusiastically greeted by many firms currently registered under ISO 9000:1994. Continual improvement and the other TQM elements may be too difficult for their cultures to embrace. Should that be true, their registrations must be invalidated no later than the end of 2003. Worse than that, their competitiveness will erode as their competitors reap the benefits of continual improvement. Their future prospects will be bleak indeed.

DOCUMENTING THE QUALITY MANAGEMENT SYSTEM

The documentation system required by ISO 9000 can be seen as having four distinct levels. Refer to Figures 5–8 and 5–9.

1. Policy
2. Procedure
3. Practice
4. Proof

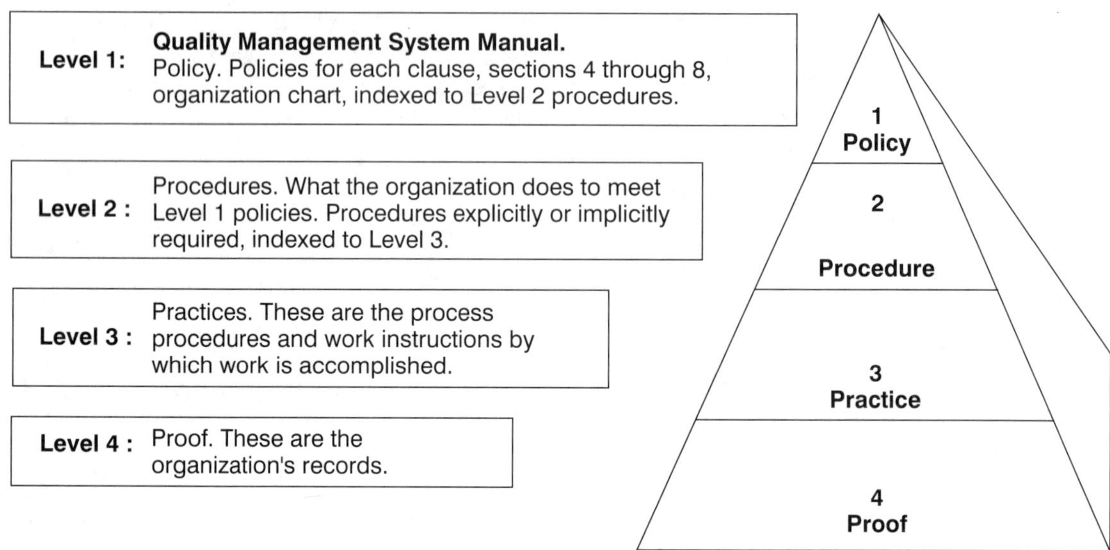

Figure 5–8
QMS Documentation Hierarchy

These four levels of documentation are discussed in the next sections. (Also refer to Figures 4–8 and 4–9.)

Level 1—Policy

Level 1 is the **policy level** that, in essence, declares the organization's quality philosophy and commitment and how it intends to conduct itself. It details the quality issues facing the organization and plans in a general way for improving quality performance. All of this will be in the form of a quality manual.

The quality manual will include the organization's quality policy (clause 5.3), and specific policies corresponding to each clause of ISO 9001, sections 4 through 8. A good starting point for these specific policies is a clause-by-clause duplication of ISO 9001, sections 4 through 8, substituting *will* for *shall, we* or *our* for *organization,* and other words as appropriate. For example, the language "The organization shall establish and maintain . . ." in one of the clauses can be changed to "We will establish and maintain . . ." in the corresponding policy statement.

Organization charts and other forms of documents that define the core elements of the QMS are also included in the QMS Level 1 documentation. They include information on how the documents and activities relate and interact with each other and the organization as well as defining management responsibility and authority for operating the QMS and each of its elements (clauses 4.2.2, 5.5.1).

The quality manual must include details of, and justification for, exclusion of any clause 7 requirement (clause 4.2.2). ISO 9001:2000, clause 1.2 permits requirement

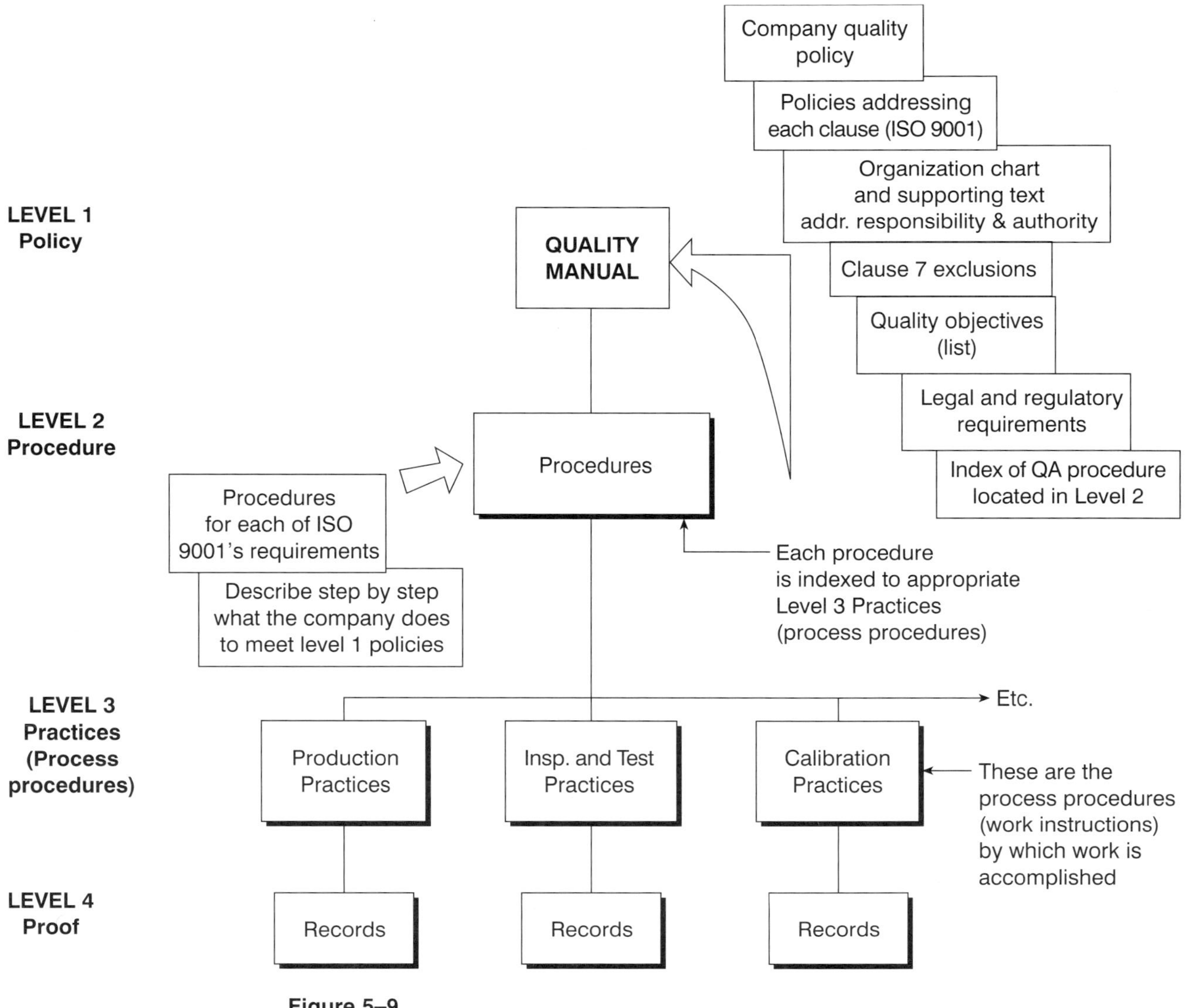

Figure 5–9
ISO 9000 Quality Management System Documentation Model

exclusions within clause 7 only, and only when the organization's nature and product make the requirement irrelevant. An exclusion must not affect the organization's ability to provide product that meets customer and regulatory requirements.

A section in the QMS manual should be reserved to list the current quality objectives. This can be in list format, referring to the actual documentation of the objectives, or it can be the active set of objective documentation. In either case regular updating will be required to account for objectives being achieved and removed from the active list (clause 5.4.1).

It is also a good idea to include the applicable legal and regulatory requirements in the QMS manual, at least in list form with references to actual locations (clause 1.1).

The QMS manual should include a list of procedures that correspond to explicit and implicit ISO 9001 requirements. The procedures themselves are not located in the manual because they would make the manual unwieldy. The list should provide references to the actual location of the procedures so that anyone needing access can find them easily.

A QMS manual constructed according to the foregoing will have these elements:

- Quality policy
- Policies reflecting each of the ISO 9001, section 4 through 8 clauses
- Definition of core elements of the QMS and their interrelationships
- Clear definition of management responsibility and authority for operating the QMS and its elements
- Current list of quality objectives (or the actual objective documentation)
- Copies of legal and regulatory requirements or a list of them with references to actual location
- A list of the organization's procedures related to ISO 9001 clauses, indexed to their Level 2 location

Level 2—Procedures

Level 2, the **procedures level,** describes how the organization operates the QMS. Minimally, it should include procedures to address each procedure requirement of ISO 9001. Clearly, only documented procedures can be included in Level 2 of the QMS documentation. One must be careful here, because while some procedures are explicitly required by the clauses to be documented, others carry only an implicit requirement that can easily be missed until the registrar indicates the error. Beyond the explicitly required procedures, ISO 9001 requires documented procedures where the organization cannot ensure effective operation and control of processes without them (clause 4.2.1). The intent is that the organization must use documented procedures to cover situations where their absence could lead to deviations from QMS policies or the standard. In most organizations, undocumented procedures will lead to all kinds of deviations, but documented procedures will lead to consistency. Properly documented procedures can be easily followed by someone new to the job, and documented procedures are also the only ones that can be controlled. Refer to the discussion of explicit and implicit procedure requirements earlier in this chapter.

Any other quality-related procedures used by the organization should also be included in the Level 2 documentation.

The procedures together describe in a step-by-step fashion what the organization does to meet the Level 1 policies. Ideally there should be procedures related to each of the ISO 9001, section 4 through 8 clauses. The Level 1 list of procedures becomes the table of contents for Level 2. In turn, Level 2 should contain a list of the process procedures (practices) contained in Level 3.

Level 3—Practices

In Level 3, the **practices level,** the process procedures (i.e., the actual work instructions for activities relevant to the QMS) reside. They represent what the employees do in their operational activities. These process procedures, or work instructions, will provide detailed step-by-step instructions dealing with the organization's relevant activities. Organizations involved in Total Quality Management will probably already have these process procedures documented; others may not. Organizations that have not documented their process procedures should do so for consistency of operations. The Level 2 list of process procedures (practices) becomes the table of contents for Level 3. In turn, Level 3 should list the Level 4 forms and records associated with the practices and their location.

Level 4—Proof

Level 4, the **proof level,** is the repository for all forms, records, and the like which represent the objective evidence (proof) that the QMS is, or is not, functioning as it should. In accordance with ISO 9001, these would include as a minimum QMS records such as the following:

- Records from management reviews (clause 5.6.1)
- Records of employee education, training, skills, and experience (clause 6.2.2e)
- Records providing evidence that the processes and product fulfill requirements (clause 7.1d)
- Records of review of customer requirements and related actions (clause 7.2.2)
- Records of design and development reviews and related actions (clause 7.3.4)
- Records of the results of design verification and related actions (clause 7.3.5)
- Records of the results of design validation and related actions (clause 7.3.6)
- Records of results of the review of design and development changes and related actions (clause 7.3.7)
- Records of the results of supplier evaluations and related actions (clause 7.4.1)
- Where applicable, records of unique product identification for traceability (clause 7.5.3)
- Records of measuring results validity (clause 7.6)
- Records of monitoring and measuring device calibration and verification (clause 7.6)

- Records of internal QMS audits (clause 8.2.2)
- Records providing evidence of product conformity (clause 8.2.4)
- Records of the nature of product nonconformities and related actions (clause 8.3)
- Records of the results of corrective actions taken (clause 8.5.2)
- Records of the results of preventive actions taken (clause 8.5.3)

These records will be useful to the organization, and they will be the focus for registrar audits, serving as objective evidence that the organization is or is not in conformance with the ISO 9000 Standard.

Both the registrar and the organization itself have vested interests in compliance with any applicable legal and regulatory requirements. Therefore, it is obvious that in addition to the conformance records listed above, compliance records associated with legal and regulatory requirements should also be maintained as Level 4 documents.

Documentation of the QMS for ISO 9000 will require a significant effort for organizations that are not already operating under the principles of Total Quality Management or government requirements that demand documented processes, procedures, and other similar criteria. ISO 9000 documentation is intended to assist the organization, not to burden it. Procedures should be kept as simple as possible and be consistent with requirements and good judgment. The auditors will require that the organization does exactly as it says it will do in its procedures. Consequently, the best course is to begin simply and add detail later if necessary. Do not make documentation unnecessarily complex.

Clause 4.2.4 establishes the requirement for the control of quality records, and is referenced by each of the clauses indicated in the above bulleted items. Figure 5–9 illustrates this model in graphical form.

VERIFYING THE INTEGRITY OF THE QUALITY MANAGEMENT SYSTEM

For registered organizations two levels of audits are used to verify the integrity of the QMS: **external audits** performed by the registrar and **internal audits** performed by the organization's employees. Registrars will normally perform periodic surveillance audits at least annually. The registration audit, or "conformity assessment," and surveillance audits are explained in Chapter 7.

Clause 8.2.2 establishes the requirement for periodic internal audits. These audits are used to assess the strengths and weaknesses of the QMS[7] and determine (1) if the QMS conforms to ISO 9000 and the requirements of the quality policy and plans, and (2) if the QMS has been properly implemented and utilized. In reference to the second item, the object is to determine whether the QMS been deployed throughout the organization and put into practice and whether the policies, practices, and procedures which comprise the QMS are being rigorously followed.

Internal QMS audits are usually performed by employees of the organization, although the organization can employ outside auditors to perform its internal audits. The two areas of concern when using employees for internal audits are as follows.

- First, the employees must be able to conduct the audit objectively and impartially. This means that internal auditors must be independent of the function being audited. For example, an employee of the finance department could audit the human resources department, but not the finance department.
- Second, the employees selected as internal auditors usually will not be experienced auditors. To achieve maximum benefit from internal audits, the auditors must be trained in auditing techniques, practices, and psychology.

As explained previously, the organization must have a QMS audit program that is supported by procedures. The program and procedures should nominally cover the following:

1. *The organization's activities (processes, functions, departments) to be considered in audits.* All processes and activities which fall within the scope of the QMS (i.e., those which can impact or influence quality) should be included.
2. *The frequency of internal audits.* The frequency with which an activity or process is audited is to be a function of the status and importance of the activity or process, and past audit results and performance.
3. *The responsibilities and requirements for planning and conducting audits and reporting results.* Roles, responsibilities, and requirements of auditors and lead auditors must be clearly established in the internal audit procedure (clause 8.2.2).
4. *The communication of audit results.* Audit results must be communicated to management in accordance with the procedure and the audit plan. ISO 9000 does not specify the method of communicating, but does mention the audit report. Clearly, a written report is required. In addition, we suggest that the report be accompanied by an oral presentation, during which management can obtain any needed clarification from the audit team.
5. *Auditor competence requirements.* Employees carrying out the internal audits should be properly trained. The guidance for auditor qualification criteria is found in ISO 19011, and will be considered daunting by most organizations. At first glance one would assume that the qualification criteria was aimed at external auditors employed by registrar firms, but it is applicable to both internal and external auditors. Fortunately this is only guidance, because in our experience few organizations could meet the education, experience, and training qualifications specified.
6. *Audit methods.* Internal auditors will collect the information they require from the quality records maintained by the organization, through interviews with relevant personnel, and through observation of operations. The information gathered will be audit evidence which will be compared with the audit criteria to develop conclusions. The conclusions will be conveyed to management via a written audit report, perhaps accompanied by an oral presentation to promote understanding.

Considerations for internal QMS audits include the following:

- *Requirements for the audit.* First the audit should focus on clearly defined and documented subject matter. Second, the parties responsible for that subject matter must be identified, with their names documented, and they must be available for the auditors.

- *Objectives and scope of the audit.* The organization's management defines the objectives of the audit. The scope of the audit (boundaries and extent) is determined by the lead auditor, consulting with management, to meet the audit objectives. The objectives and scope should be communicated to personnel involved in the activities to be audited.
- *Objectivity, independence, and competence.* Members of the audit team should be independent of the activities they audit, and they should be objective and without bias or conflict of interest. Audit team members should have the knowledge, skills, and experience to perform audits of assigned activities.
- *Due professional care.* Auditors must exercise the care, diligence, skill, and judgment expected of any auditor. Of utmost importance is confidentiality and discretion between auditor and persons interviewed and management. There should be no attribution (e.g., "John said . . ."). No information obtained during the audit should be disclosed to any third party or anyone outside the organization, unless required by law.
- *Systematic procedures.* To ensure reliability and consistency, quality audits should be conducted in accordance with these principles and the guidelines given in ISO 19011.
- *Audit criteria, evidence, and findings.* Audit criteria normally are the policies, practices, procedures, and requirements against which auditors compare the evidence collected during the audit. The audit criteria for a particular audit should be selected by management and the lead auditor, and communicated to the auditees prior to the audit. Relevant information should be collected during the audit, and it should be analyzed, interpreted, and recorded as audit evidence in order to determine whether the audit criteria are met. Audit evidence should be of such quality and quantity (reliable, and, ideally, from multiple sources) that multiple auditors working independently would reach the same conclusions.
- *Reliability of audit findings and conclusions.* Since QMS audits take place in a discreet, short period of time, any evidence gathered can be no more than a sample. The audit process should be designed to give the auditors and management confidence in the reliability of the audit findings and conclusions. It is necessary that auditors recognize the uncertainty of audit findings and conclusions, and take this into account in planning and conducting the audit.
- *Audit report.* A written **audit report** should be submitted to management. The audit report may, at the discretion of the lead auditor and management, include the following:
  1. Identification of the processes or activities audited
  2. Audit objectives and scope
  3. Audit criteria
  4. Period covered by the audit, and dates of the audit

5. Identification of audit team members
6. Identification of auditee participants
7. Statement of confidentiality
8. Distribution list for the audit report
9. Summary of the audit process and any obstacles encountered
10. Audit conclusions

Note that the audit report does not normally include solutions for any corrective actions required. It is usually up to the organization or relevant activity to determine what corrective action should be taken.

It is interesting to note that ISO 9000 is silent on the subject of subcontractor audits. In the world of TQM, a company's subcontractors (let's call them suppliers) are considered to be a part of the overall organization. Depending upon the extent of the partnership reached between the company and its suppliers, auditing the suppliers is considered the norm. Supplier audits are treated like internal audits that go beyond the company's own four walls. Since the audits customarily relieve the intensity of inspections historically demanded of suppliers, many prefer the audit system. However, in Western countries it is a rare supplier that serves only one company. If all of a supplier's customers demanded the right to conduct audits, the supplier could spend all its time being audited to different requirements and specifications.

An obvious solution to the multiple-audits dilemma is to have all of your suppliers certified to ISO 9000. This is happening at an increasing pace. The question is, will you be absolved of responsibility for your suppliers' product or service quality? Certainly not by the certifying agent. They certify only that the supplier does what it said it would do. From this one can see that there is *consistency* in the supplier's product or service, but consistency at what level of quality? Nor can you expect absolution from the judges who really count—your customers. Customers do not care what lengths you have to go to in order to assure them of high quality, or suitability for use. If you have to flood your supplier's place of business with auditors to do that, so be it. In the final analysis, you will always be responsible for the quality of your products and services. If that quality can be adversely affected by your supplier, you must have a control mechanism. Covering the supplier with your internal audits may continue to be the appropriate preventive measure, even though it is not required by ISO 9000.

ISO INFO

The primary technique in dealing with audit uncertainty is finding corroborating evidence. A single piece of evidence carries a low reliability factor, but when supporting evidence turns up once, twice, or more times the reliability increases correspondingly.

Goetsch & Davis

ISO INFO

In order to relieve suppliers of the requirement to adhere to multiple standards and too many audits, the U.S. Auto Industry adopted to the single QS 9000 Standard. Ford, Chrysler and General Motors together developed a set of add-on requirements to ISO 9000 and now demand that their suppliers conform to the broader standard. This eliminates the multiple audit/multiple standard situation. Consequently, suppliers have generally welcomed QS 9000. At the same time, the Big Three have seen their supplier audit/inspection work load reduced as third party auditors have come on-line.

To summarize this section on internal auditing, remember the following:

What You Have To Do	ISO 9001 Clause
Conduct periodic internal audits.	8.2.2
Have documented internal audit procedures.	8.2.2
Schedule your internal audits.	8.2.2
Ensure that internal auditors are independent of the area audited.	8.2.2
Record internal audit findings.	8.2.2/4.2.4
Put these records under formal control.	4.2.4
Present audit findings to area personnel and management.	8.2.2
Insist area management take timely corrective action.	8.2.2
Have documented procedures for corrective actions.	8.5.2
Conduct follow-up verification to confirm corrective action.	8.2.2
Record follow-up verification results.	8.2.2
Put follow-up audit records under formal control.	4.2.4

For more information on the audit process, see Chapter 7.

REVIEW AND EVALUATION OF THE QUALITY MANAGEMENT SYSTEM

ISO 9001, clause 5.6 requires that management periodically review the QMS for continued suitability, adequacy, and effectiveness, and assess opportunities for continual improvement and the need for changes to the QMS, including the quality policy and quality objectives. One might question the necessity for these reviews once the QMS is in place and accomplishing its objectives. However, nothing remains static for long. The organization's products or services will change over time to meet the changing demands of customers. Those changes will result in changes to processes and procedures, and may eliminate some problems while creating new ones. The organization will find ways to improve its products, processes, and procedures, resulting in the need for new QMS documentation. Technology will also change, perhaps making it possible to improve performance by adopting the new technology. The legal and regulatory agencies may add new product requirements or modify existing ones. With these and many other changes taking place, the need for periodic management reviews of the QMS is critical.

ISO 9000 does not presume to establish a review schedule, or even suggest one. ISO 9001, clause 5.6.1, uses the phrase "at planned intervals." ISO 9004, clause 5.6.1, states, "The frequency of review should be determined by the needs of the organization." Management has to determine an appropriate interval and define it in a schedule that is understood by the relevant employees. Although the reviews should be comprehensive, ISO has held the view that not all elements of the QMS need to be reviewed at once. The 2000 release is quiet on this issue. Therefore, it may be that a complete and comprehensive review of the QMS will occur over the course of several scheduled "mini-reviews." A monthly interval seems to be appropriate for most organizations. The total QMS should be discussed, with emphasis on the more critical elements. Once a review interval is established, management can evaluate its effectiveness, and alter the schedule if necessary. Under no circumstances would a review cycle of longer than three months be adequate; an interval less frequent than quarterly will send the message that reviews, and perhaps QMS performance, are unimportant, in which case the focus on quality issues may be lost.

QMS management reviews *must* include the following information categories:[8]

- Results of audits
- Customer feedback
- Process performance and product conformity
- Status of preventive and corrective actions
- Follow-up actions from previous management reviews
- Planned changes that could affect the QMS
- Recommendations for improvement

QMS management reviews *should* also include the following:[9]

- Status and results of quality objectives and improvement activities
- Status of management review action items
- Feedback on the satisfaction of interested parties other than customers
- Market-related factors such as technology, research and development, and competitor performance

- Results from benchmarking activities
- Performance of suppliers
- New opportunities for improvement (not necessarily recommendations)
- Control of process and product nonconformities
- Marketplace evaluation and strategies
- Status of strategic partnership activities
- Financial effects of quality-related activities
- Other factors which may impact the organization

The intent of the management reviews is that the QMS, and all of its components, will be continually improved to ensure that performance is also continually improved. Therefore, the management review process and procedures should ensure that the continual improvement process is always a major element of the reviews. Continual improvement is the subject of Chapter 8.

CASE STUDY

ISO 9000 in Action

In looking back at his last meeting with AMI's executive managers, Jake Butler knew he had made significant progress. The purpose of the meeting from Butler's perspective had been twofold. First, he wanted to make sure that AMI's executive managers understood their responsibilities in building a quality management system. Second, he wanted to have AMI's executives develop a quality policy. Both of these goals had been accomplished, but not without some head-butting.

Butler had gotten into trouble right off when Arthur Polk, AMI's CEO, had suggested that he—Butler—draft a quality policy for the company. Butler knew that this seemed like a reasonable request to executive managers. After all, it was how each of them normally worked. Turn over the developmental work to a subordinate, and save the higher-level work for themselves. The problem, as Butler saw it, was that developing a quality policy is higher level work that must be done by executive managers. The quality policy must be their policy. Butler knew that AMI's success in implementing TQM and achieving ISO 9000 registration could turn on this one point.

In an attempt to convince Arthur Polk and the other executives that they should develop AMI's quality management system and quality policy, Butler had used the analogy of staying fit. Butler had said, "If you want to stay in shape, you must work out and diet. These things cannot be delegated, or they won't have the desired effect. Developing a quality system is like working out and dieting. Nobody can diet for you."

As a fitness buff, Arthur Polk had understood the analogy and responded to it. When it became apparent that Polk was ready to take responsibility for developing AMI's quality management system, the other executives began to warm to the idea. After about twenty minutes of direct and pointed discussion, AMI's executives had agreed to take on the project of building the company's QMS, and to begin by developing a quality policy.

Wanting to strike while the iron was hot, Butler had distributed copies of several different quality policies he had collected as examples. Arthur Polk had asked the executives to review the example policies and extract quality concepts that appealed to them. Butler served as a facilitator and recorded the various concepts extracted by the executives. The concepts he had recorded were as follows:

- Teamwork
- Employee involvement
- Continual improvement
- Customer satisfaction
- Customer retention
- Control of processes
- Input from all stakeholders

Using these key concepts as the starting point, AMI's executives had drafted a quality policy on the spot, and, to Butler's delight, it was an excellent policy.

Once the quality policy was in place, Butler had facilitated a discussion of the structure of the quality management system, making sure that continuous improvement be built into the system, and that documentation of the system and system evaluation be stressed. Butler had left the meeting knowing that in its effort to build a QMS, AMI's executives had laid down a solid foundation.

SUMMARY

1. The quality management system is comprised of all the organization's policies, procedures, plans, resources, and processes, and its delineation of responsibility and authority, all deliberately aimed at achieving product or service quality levels consistent with customer satisfaction and the organization's objectives. When these policies, procedures, plans, and so on are taken together, they define how the organization works, and how quality is managed.
2. The senior executive manager is responsible for committing the organization to quality and continual improvement of quality, and for developing the organization's quality policy. Responsibility for developing and deploying a quality policy cannot be delegated.
3. Once an organization has developed a quality policy and quality objectives, it is ready to develop the rest of its quality management system. This involves providing a supportive organizational structure, the necessary processes, procedures, documentation, and resources, a management representative, communication to all employees, QMS performance reviews, commitment, and an environment in which continual improvement can occur. All of the various tasks involved in developing the quality management system fall into one of the following broad categories of responsibility: management responsibilities; resource management; design and

production operations (product realization); test, measurement, and analysis operations; and continual improvement.

4. The organization must commit itself to continual improvement. The insistence of ISO 9000:2000 on this commitment is spelled out over and over again in numerous clauses beginning with the Introduction clause 0.2 that states "*Continual improvement of the organization's overall performance should be a permanent objective of the organization.*" This is one of the most positive aspects of ISO 9000:2000 when compared with the earlier version.
5. The documentation required by ISO 9000:2000 has four distinct levels as follows: Level 1 includes the quality management system manual and the quality policy; Level 2 includes procedures; Level 3 includes practices; and Level 4 includes the organization's records or "proofs."
6. For registered organizations two levels of audits are used to verify the integrity of the quality management system: external audits performed by the registrar and internal audits performed by the organization's employees. The program and procedures for an internal audit should nominally cover the following: (1) the organization's processes, functions, and departments to be considered in the audits; (2) the frequency of audits; (3) the responsibilities and requirements for planning and conducting audits and reporting the results; (4) the communication of audit results; (5) auditor competence requirement and (6) audit methods.
7. Reviews of the quality management system *must* include the following: results of audits, customer feedback, process performance and product conformity, status of preventive and corrective actions, follow-up actions from previous management reviews, planned changes that could affect the quality management system, and recommendations for improvement. There are other factors that *should* be included, but those just listed are considered mandatory.

KEY TERMS AND CONCEPTS

Audit report
Competence and awareness training
Configuration management
Continual improvement
Customer focus
External audits
Guiding principles
Internal audits
Internal communication system
Management commitment
Management of resources
Management representative
Management responsibility
Measurement, analysis and improvement
Mission statement
Operational procedures
Policy level
Practice level
Procedure level
Product realization
Proof level
QMS documentation hierarchy
QMS Management review
QMS structure

Quality management system
Quality objectives
Quality policy
Record keeping
Vision statement

REVIEW QUESTIONS

1. What is a quality management system?
2. Explain management's responsibilities with regard to the quality management system.
3. What are the key elements of the quality policy?
4. What are the various tasks in the implementation and operation (product realization) component of the ISO 9000 QMS model?
5. Explain what is meant by the term *customer focus.*
6. Give four reasons for having quality objectives.
7. List the six procedures 9000:2000 explicitly requires be documented.
8. What are the five components of good documentation?
9. Explain the responsibilities of the management representative required by ISO 9000:2000.
10. Explain the term *continual improvement.*
11. Briefly summarize all of the tasks associated with the following components of a process-based quality management system: management responsibilities, resource management, design and production operations (product realization), and test, measurement, and analysis operations.
12. Explain each element in the "Quality Management System Structure."
13. Describe the various elements in the QMS documentation hierarchy.
14. Explain the concept of verifying the integrity of the QMS, and how this is done.
15. QMS management reviews *must* include information from seven prescribed categories. What are they?

APPLICATION ACTIVITIES

1. Develop (on paper) a fictitious manufacturing or service company. Know well the product or service your company will provide.
2. Write a quality policy for your company.
3. Develop a set of quality objectives for your company.
4. Make a list of all the various documents that will be necessary in order to fully document your company's quality management system.
5. Describe how your company will conduct internal audits.

ENDNOTES

1 ISO 9000:2000, clause 3.2.7.

2. ISO 9004:2000, clause 5.1.1.

3. Peter R. Scholtes, *The Team Handbook* (Madison, WI: Joiner Associates, 1992), pp. 1–11.

4. ISO 9001:2000, clause 0.2

5. David L. Goetsch and Stanley B. Davis, *Quality Management: Introduction to Total Quality Management for Production, Processing, and Services,* 3rd ed. (Upper Saddle River, NJ: Prentice-Hall, 2000), p. 604.

6. Allan J. Sayle, *Meeting ISO 9000 in a TQM World,* 2d ed. (London: AJSL, 1994), p. 141.

7. ISO 9004:2000, Clause 8.2.1.3

8. ISO 9001:2000, Clause 5.6.2

9. ISO 9004:2000, Clause 5.6.2

CHAPTER SIX

QMS Documentation

MAJOR TOPICS

- Documentation: Definition
- ISO 9000 Documentation Requirements
- Documentation Format Requirements
- Reformatting Preexisting Documentation
- Combining Required Documentation
- Cross-Referencing Required Documentation
- Structure of the Documentation System
- Electronic Documentation

DOCUMENTATION: DEFINITION

According to *Webster's New World Dictionary, document* means "1. Anything written, printed, etc., relied upon to record or prove something." *Documentation,* according to the same source, means, "1. The supplying of documents or supporting references. 2. The documents or references thus supplied." For our purposes we can consider ISO 9000 documentation to include written and pictorial information used for defining, specifying, describing, or recording the activities and processes that lead to the design, development and production of products and provision of services. This documentation may be in hard copy or electronic media. In a practical sense it may include the following elements, as shown in Figure 6-1:

- Written policies
- Written procedures
- Written instructions
- Written quality objectives
- Written records
- Written plans

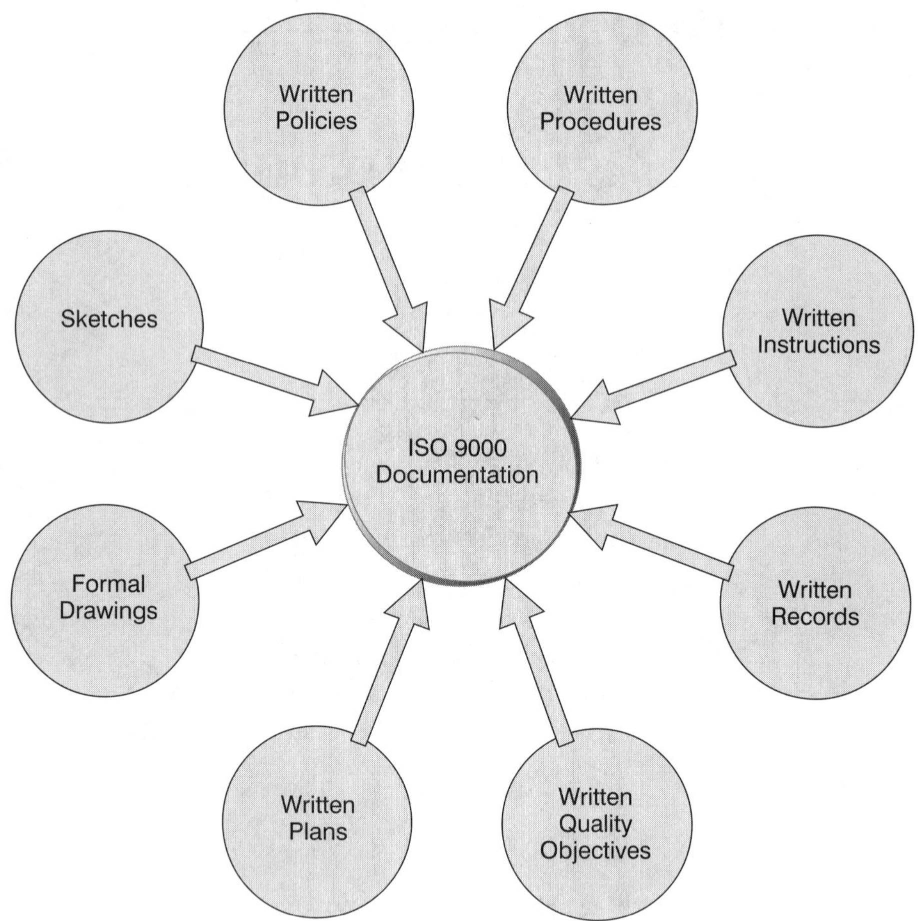

Figure 6–1
Documentation Required by ISO 9000

- Formal drawings
- Sketches

For ISO 9000's purposes, documentation includes all the specific items addressed in the next section.

ISO 9000 DOCUMENTATION REQUIREMENTS

In this section we list the documentation explicitly required by ISO 9001, as well as that for which a requirement is inferred. Whether the requirement is explicit or implicit is determined by the wording of the standard. For example in ISO 9001, clause 8.2.2

requires every registered organization to conduct internal audits at planned intervals. The clause further states that, "The responsibilities and requirements for planning and conducting audits, and for reporting results and maintaining records shall be defined in a documented procedure." This is an explicit requirement for a documented procedure. For all such requirements the organization must have a documented procedure to conform with ISO 9000. On the other hand, Clause 7.6 requires every organization to ". . . establish processes to ensure that monitoring and measurement can be carried out and are carried out in a manner that is consistent with the monitoring and measurement requirements." Yet clause 7.6 carries no stated requirement for a documented procedure. In order to satisfy clause 7.6, most organizations will have to:

- Calibrate the monitoring and measurement equipment at specified intervals
- Adjust or readjust the equipment
- Provide identification and calibration data for specific instruments
- Safeguard against adjustments that would invalidate calibration
- Protect the equipment appropriately
- Maintain calibration and verification records
- Verify the results of previous measurement records when the equipment is found not to conform to calibration requirements
- Confirm the ability of computer software used in the monitoring and measurement processes

If the organization is confident that it can accomplish all of these actions consistently, then ISO 9001 does not require a documented procedure. However, for all but the smallest organizations requiring little process or product monitoring and measuring, the complexity and potential for deviation of the requirement makes it necessary to have a documented procedure. This is an implicit requirement under clause 4.2.1 that requires documented procedures wherever their absence could render planning, operation, and control of processes ineffective.

In the list that follows, explicitly required documents are followed by "Exp" and they are highlighted. Implicitly required documents are followed by "Imp (#)." This indicates that our explanation for the implicit requirement is included in the correspondingly numbered notes that follow the document list. We have listed the requirements in the order they are presented in ISO 9001.

Document Required	Requiring Clause	Exp/Imp
Quality manual	4.2.2, 4.2.1b	Exp
Documented procedure for the control of QMS documents	4.2.3	Exp
Documented procedure for the control of QMS records	4.2.4	Exp

Document Required	Requiring Clause	Exp/Imp
Quality policy	5.3, 4.2.1a	Exp
Quality objectives	5.4.1, 4.2.1a	Exp
QMS planning	5.4.2	Imp (1)
Responsibilities and authorities and their interrelations	5.5.1	Imp (2)
Appointment of management representative	5.5.2	Imp (3)
Procedure for management review	5.6	Imp (4)
Records of management reviews	5.6.1, 4.2.4	Exp
Determination of needed resources (provision of resources)	6.1	Imp (5)
Determination of needed employee competence	6.2.2	Imp (6)
Procedure for assessing training needs, conducting training, and evaluating training effectiveness	6.2.2	Imp (7)
Records of competence and awareness training	6.2.2, 4.2.4	Exp
Determination of needed infrastructure	6.3	Imp (8)
Determination of the work environment	6.4	Imp (9)
Procedure for planning of product realization	7.1	Imp (10)
Objectives for the product—planning of product realization	7.1a, 4.2.1a	Exp
Product requirements—planning for product realization	7.1a	Imp (11)
Processes documentation—planning of product realization	7.1b	Imp (12)
Planning of product realization—verification, validation, monitoring, inspection, and test requirements	7.1c	Imp (13)
Records providing evidence that processes and product fulfill requirements	7.1d, 4.2.4	Exp
Determination of product requirements	7.2.1	Imp (14)
Procedure for determining and reviewing product requirements	7.2.1, 7.2.2	Imp (15)
Records of product requirements review	7.2.2, 4.2.4	Exp
Procedure for effective customer communications	7.2.3	Imp (16)
Procedure(s) for design and development	7.3	Imp (17)
Design and development plans	7.3.1, 4.2.1d	Imp (18)

Document Required	Requiring Clause	Exp/Imp
Records of design and development review	7.3.4, 4.2.4	Exp
Records of design and development verification	7.3.5, 4.2.4	Exp
Records of design and development validation	7.3.6, 4.2.4	Exp
Records of design and development changes	7.3.7, 4.2.4	Exp
Procedure for purchasing	7.4	Imp (19)
Records of supplier evaluation	7.4.1, 4.2.4	Exp
Procedure(s) for production and service provision	7.5	Imp (20)
Plans for production and service provision	7.5	Imp (21)
Records of the validation of production and service provision processes	7.5.2, 4.2.4	Exp
Records of identification and traceability (if required)	7.5.3, 4.2.4	Exp
Records of lost or damaged customer property	7.5.4, 4.2.4	Exp
Procedure for control of monitoring and measuring devices	7.6	Imp (22)
Records of control of monitoring and measuring devices	7.6, 4.2.4	Exp
Plans for monitoring, measurement, analysis and improvement processes needed	8.1	Imp (23)
Procedure for monitoring customer satisfaction	8.2.1	Imp (24)
Procedure for planning and conducting internal audits	8.2.2	Exp
Records of internal audits	8.2.2, 4.2.4	Exp
Records of verification of actions from internal audits	8.2.2, 8.5.2	Exp
Procedure for monitoring and measuring QMS processes	8.2.3	Imp (25)
Procedure for monitoring and measurement of product	8.2.4	Imp (26)
Records of monitoring and measurement of product (conformity and releasing authority)	8.2.4, 4.2.4	Exp
Procedure for control of nonconforming product	8.3	Exp
Records related to nonconforming product	8.3, 4.2.4	Exp
Procedure for analysis of data	8.4	Imp (27)
Procedure for continual improvement	8.5.1	Imp (28)

Document Required	Requiring Clause	Exp/Imp
Procedure for corrective actions	8.5.2	Exp
Records of corrective actions and reviews thereof	8.5.2, 4.2.4	Exp
Procedure for preventive actions	8.5.3	Exp
Records of preventive actions and reviews thereof	8.5.3, 4.2.4	Exp

Notes:

1. *QMS planning (4.1).* Clause 4.1 states that QMS planning include: identifying processes needed and their application; determining the sequence and interactions of the processes; determining how to control the processes; making resources and information available to support operation of the processes; monitoring, measuring, and analyzing the processes; and implementing actions necessary for achieving planned results and continual improvement of the processes. All of these require that the process planning be documented.
2. *Responsibilities, authorities, and their interrelations (5.5.1).* These must be defined and communicated within the organization. To be effective in any but the smallest, simplest organization, this has to be done by published organizational charts, job descriptions, etc.
3. *Management representative appointment (5.5.2).* This will require, as a minimum, inclusion on the documented organization chart. Related to note 2 above.
4. *Procedure for management reviews (5.6).* These are to be held at planned intervals. Clause 5.6.2 lists seven distinct input categories to be reviewed. Clause 5.6.3 requires management review decisions and actions on at least three categories. Each review will have multiple participants from across the organization. Conducting the reviews without a documented procedure will not be effective.
5. *Determination of needed resources (provision of resources) (6.1).* The standard specifies that resources needed to implement, maintain, and continually improve the QMS are to be determined. It would make no sense to determine the resource requirements without documenting them. The intent is clear.
6. *Determination of needed employee competence (6.2.2).* This is done so that only employees or candidates with the necessary competency will be placed in relevant assignments, and to indicate the training required for employees to achieve competency. Since this is not a one-time thing, it has to be documented to be of continued value.
7. *Procedure for assessing training needs, conducting training, and evaluating training effectiveness (6.2.2).* Since training will be a continuing function over time, the organization should have a procedure explaining how it determines needs for competence and awareness training, how it conducts the training, and how it evaluates training effectiveness. Only when there is a documented procedure can these tasks

be carried out consistently and equitably. We don't think training is a function that should be operated ad hoc.

8. *Determination of needed infrastructure (6.3).* One requirement of this clause is for the organization to determine the needed infrastructure. That would include machines, buildings, environmental controls, software, and so on. In many organizations this just seems to develop as first one piece of equipment, then another is added, new software is thrown in, the infrastructure "just growing" like Topsy. The odds are that it is never coherently documented. Although not required by ISO 9000, smart organizations will document the required infrastructure for each process.
9. *Determination of the work environment (6.4).* Same comment as item 8.
10. *Procedure for planning of product realization (7.1).* ISO 9001 does not require a procedure for planning of product realization. However, it is a very complicated process involving several subprocesses. If the organization cannot guarantee that product realization planning can be effectively and consistently carried out without a documented procedure, then a documented procedure is required. (See clause 4.2.1 d.)
11. *Product requirements—planning for product realization (7.1a).* Product requirements will be used in the development, production, and validation of the product. Therefore it is necessary that product requirements be documented.
12. *Processes documentation—planning for product realization (7.1b).* The link to 4.2.1d makes process documentation necessary unless the organization is certain its absence will not cause deviations. We see this as an implicit requirement for all but the simplest processes.
13. *Planning of product realization—required verification, validation, monitoring, inspection and test activities (7.1c).* It would make no sense for the organization to develop these requirements along with the product acceptance criteria without documenting them.
14. *Determination of product requirements (7.2.1).* Product requirements will be used in the development, production, and validation of the product. Therefore it is necessary that product requirements, whether from customers or internal, be documented.
15. *Procedure for determining and reviewing product requirements (7.2.1, 7.2.2).* ISO 9001 does not specify such a procedure, but the organization should have one. There are too many variables and paths. A step-by-step procedure can keep the organization out of trouble, and the process consistent.
16. *Procedure for effective customer communications (7.2.3).* Although there is no stated requirement for a procedure to operate these effective customer communications once they are determined, they are required to cover a wide range of categories, and will, in many organizations, involve many employees. We think a customer communications procedure is necessary for consistency and effectiveness.
17. *Design and development procedure.* Clause 7.3, Design and Development, has seven subclauses covering requirements for design and development planning (7.3.1),

inputs (7.3.2), outputs (7.3.3), review (7.3.4), verification (7.3.5), validation (7.3.6), and change control (7.3.7). In these seven subclauses there are 22 *shalls*. It is clear that ISO is very specific in its requirements for design and development. There is simply no way to do what ISO requires in the absence of documented procedures—either a single procedure covering all seven phases, or seven separate procedures.

18. *Plans for design and development (7.3.1).* Plans for designing and developing new products are far too complex to leave undocumented. In addition, clause 4.2.1d seems to place further emphasis on the requirement for planning documents, although addressing planning for processes.
19. *Procedure for purchasing (7.4).* ISO is very specific in its requirements for the organization's purchasing process. There are three subclauses and nine *shalls*. For such a sophisticated purchasing process to be consistently applied, there has to be a documented procedure. In this case one procedure covering all the aspects of purchasing is quite feasible.
20. *Procedure(s) for production and service provision (7.5).* ISO applies tight controls to product production and service provision. There are five subclauses covering control of production and service provision (7.5.1), validation processes for production and service provision (7.5.2), identification and traceability of product (7.5.3), handling and care of customer property (7.5.4), and preservation of finished product (7.5.5). These subclauses contain 13 *shalls*. Subclause 7.5.1 requires the *availability of work instructions* (a form of documented procedure) and 7.5.2 requires the *use of specific methods and procedures.* That is as close as the standard comes to requiring a documented procedure(s), but it is implicit throughout the subclauses. Only the smallest of organizations could hope to conform to the production and service provision requirements without documented procedures.
21. *Plans for production and service provision (7.5).* The organization is to plan for its production or service provision. Such plans are too complex to be left undocumented except in the smallest, simplest organizations. Most organizations will have to document them. (See clause 4.2.1 d.)
22. *Procedure for control of monitoring and measuring devices.* This is a critical section of the ISO 9000 requirements, since without tightly controlled monitoring and measuring equipment and software, it may not be known whether products will fulfill customer requirements. ISO has incorporated nine specific requirements in clause 7.6, and although there is no explicit requirement for documented procedure(s), it will be impossible to satisfy the clause without one or more.
23. *Plans for monitoring, measurement, analysis and improvement processes needed (8.1).* As with the plans discussed above, these will be too complex and too critical to the effectiveness of the QMS to leave undocumented. These plans must be documented for consistency and validity in demonstrating product and QMS conformance, and for continual improvement of the QMS.
24. *Procedure for monitoring customer satisfaction (8.2.1).* The requirement is for the organization to monitor customer perceptions regarding whether customer requirements have been fulfilled by the product or service, and for the organization

to figure out how to do this. Once determined, the process and its procedure should be documented so that everyone involved in the present and in the future will know what they are to do.

25. *Procedure for monitoring and measuring QMS processes (8.2.3).* The organization is to *apply suitable methods* for monitoring and measuring processes that fall within the QMS's jurisdiction. In most organizations there will be many such processes, each requiring different methods and equipment. Such organizations need documented procedures to assure consistency and legitimacy of monitoring and measuring.
26. *Procedure for monitoring and measurement of product (8.2.4).* Note 25 is concerned with monitoring and measurement of the processes that produce the product. This item deals with monitoring and measurement of the product itself to verify that product requirements are fulfilled. This is usually a complex process, and only becomes more so with more products and production processes. Most organizations will require documented procedures to assure that it is done right and consistently.
27. *Procedure for the analysis of data (8.4).* Clause 8.4 requires the organization to *determine, collect and analyze appropriate data* for the purpose of demonstrating that the QMS is suitably implemented and operated, and to mine opportunities for continual improvement. The data are to come from a variety of sources. This is no simple thing, and as such cannot be effectively accomplished in the absence of a documented procedure.
28. *Procedure for continual improvement (8.5.1).* The organization is required to continually improve the effectiveness of the QMS. In the process of doing that several individual elements are employed, including the quality policy, quality objectives, audit results, analysis of data, corrective and preventive actions, and management review. To bring all this together and provide focus for continual improvement, the organization will need a documented procedure.

In the list just concluded you will note that of all the procedures required, only six are explicitly required to be documented by ISO 9001:2000. Common sense, along with the caution of clause 4.2.1, which requires that documentation be used if, in its absence, the organization cannot *ensure the effective planning, operation and control of its processes,* suggests that all plans and procedures related to ISO 9000 be documented. Beyond the plans and procedures, all other elements of the QMS documentation are straightforward. Policies must be documented. Practices (i.e., the work instructions) also must be documented if their absence might lead to deviation. Finally, proof, or the records, must be documented.

ISO INFO

Regarding the need for documented procedures, Joseph Cascio says, "Word-of-mouth information rarely is communicated consistently. Only written information—clearly written—is constant."[1]

ISO 9000 requires that quality policies be understood, and that procedures be consistent with ISO 9000 requirements and the organization's quality policy. ISO 9000 documentation must meet the following criteria:

- **Clarity.** Documentation must be sufficiently clear to be capable of being understood by employees and other interested parties. Clause 5.3 d states, "Top management shall ensure that the quality policy is communicated and *understood* within the organization. . . ." (Emphasis added.) This is specifically related to the quality policy, but it is implicitly applicable to all QMS documentation. Users of the documentation may find comprehension difficult if it uses jargon and acronyms that are unfamiliar to employees and other interested parties, or if it uses a writing style not attuned to them. When it is necessary to use jargon and acronyms, definitions must be provided. Documentation should be written in the simplest terms possible consistent with the subject matter. Auditors will seek to confirm that employees understand the documentation. If they find that employees do not understand the relevant documentation, the documentation, not the users of the documentation, will be declared deficient.
- **Effectiveness.** The emphasis of ISO 9000 is on the effective implementation and operation of the QMS. Clause 4.2.1 d states, "The quality management system documentation shall include documents needed by the organization to ensure the *effective* planning, operation and control of its processes. . . ." (Emphasis added.) The implicit requirement is that QMS documentation be effective, although one does not find the word used with every document requirement. This means that plans and procedures implemented in order to consistently produce some quantitative or qualitative result must, in fact, produce that result. Procedures that do not consistently produce the desired results are unacceptable. Beyond ISO 9000 considerations, ineffective plans and procedures nearly always result in suboptimal financial performance, and reduced customer satisfaction.

DOCUMENTATION FORMAT REQUIREMENTS

ISO 9001 does not specify the form or format for the various documents and records required. It generally describes what must be documented, but allows the individual organization to determine the format. This is appropriate given the fact that ISO 9000 organizations range from small businesses employing just a few people to huge multinational corporations. Simple, low-cost, low-maintenance documentation may suffice for a small, simply structured business, but it will be inadequate for the more numerous and complex functions of a larger organization. The key is that the documentation should not be more elaborate than necessary; the simplest documentation that meets the organization's needs and, at the same time, is responsive to ISO 9001 requirements is best.

Although ISO 9000 does not dictate a particular format, the QMS documentation must use appropriate, consistent formatting for its policies, procedures, process work instructions, reports, drawings, charts, and other documentation. A number of elements should always be part of the written documentation for policies, procedures, and work instructions, as follows:

Element	Purpose
Organization name or logo	Identify organizational ownership
Title	Name the document
Related to	Cross-reference ISO 9001 clause or other requirement
Document number	Identify and control document
Revision status	Identify and control revisions of document
Date of issue/effectiveness	Support document control and applicability
Review/approval signatures	Signify authority and control
Purpose	Describe intent of document
Body	Impart information or content of document

If the organization already has a documentation system under ISO 14000 or Total Quality Management, the formats used may suffice. If the organization is creating an initial documentation system, it must develop formats for each document type (drawing, policy, procedure, report, work instruction) that standardizes terminology, logos, provision for revision number and date, approval authority, and the other elements just listed. To be most effective, a procedure should be immediately identifiable as a current, authorized procedure; a policy as a current, authorized policy; a work instruction as a current, authorized work instruction; and so on.

Several commercial software packages are available that are designed to help organizations develop documentation that is acceptable to ISO. One advantage of these packages is the format discipline imposed by the software. If several people are involved in writing the documentation, the documents still will have a common format that promotes identification, understanding, and document control.

An example of a suitable format for a quality policy was shown in Figure 5-2. For a typical procedure format incorporating the elements listed above, see Figure 6-2. It is important to remember that while the formats chosen must promote identification, understanding, and document control, they must also work for the organization. It is possible to become so fascinated with a new documentation system that the reason for the documentation is lost. The objective is to help the organization implement, maintain, and continually improve a QMS that consistently results in the performance required by ISO 9000 and the organization's quality policy. Try to find the simplest documentation system that will accomplish this objective.

REFORMATTING PREEXISTING DOCUMENTATION

If the organization already has most of the documentation required, but in a format incompatible with ISO 9000 requirements, the task is to reformat. This usually means copying the text to the proper formats. Since the writing is already done, this is usually

Aero Dynamics, Inc. Procedure	**EP 1011** Procedure No.
Title: **DESIGN REVIEW**	**B** Revision
Related to: ISO 9001, Clause 7.3.4	

Purpose: To ensure that engineering designs are systematically reviewed at appropriate stages to:

1. evaluate the ability of the design to fulfill requirements, and
2. identify any problems and determine actions.

Procedure:

(Body of procedure. Use additional pages as required.)

Approved by M. Alexander
VP ENGINEERING

Approved by Sam South
PRESIDENT

Date of issue 4/18/2001

Page 1 of 1

Figure 6–2
Typical Procedure Format

ISO ISSUE

What Would You Do?

Marilyn Goldstein has two issues to work out. She is the Quality Director for Tram Tech, Inc. and has been asked to coordinate the company's preparations for ISO 9001 certification. Her first issue is to decide what elements will be included in Tram Tech's documentation system. Her second issue involves choosing a uniform documentation format for Tram Tech. ISO 9000 is new to Goldstein. If she asked for your help, what advice would you give her?

simple, although time consuming. The computer programs mentioned earlier might make the task easier. Keep in mind that the format of new documentation must promote identification, understanding, and document control, and it must work for the organization.

COMBINING REQUIRED DOCUMENTATION

Combining documentation is a task that is separate from formatting or reformatting. This can be illustrated best by example. Consider ISO 9001 clause 7.3, Design and Development. The organization might develop (or have) at least seven separate procedures to satisfy this clause, as follows:

1. Procedure for design and development planning (7.3.1)
2. Procedure for design and development inputs (7.3.2)
3. Procedure for design and development outputs (7.3.3)
4. Procedure for design and development review (7.3.4)
5. Procedure for design and development verification (7.3.5)
6. Procedure for design and development validation (7.3.6)
7. Procedure for control of design and development changes (7.3.7)

The organization could handle design and development by applying the seven procedures, or it could decide to have a single design and development procedure that covers all seven areas. Additionally, closely related elements of the seven procedures could be combined, resulting in perhaps three or four procedures. For example, the organization might combine the first, second, and third procedures (design and development planning, inputs, and outputs) into one, or it could combine procedures 4, 5 and 6 (design and development review, verification, and validation). Similar opportunities for combining can be found throughout the ISO 9001 clauses.

The organization may even combine procedures that have their corresponding requirements in different clauses. For example, clauses 5.5.3, Internal Communication, and 7.2.3, Customer Communication, offer such an opportunity since they are both concerned with communications.

It may also be possible to combine documents across different systems. For example, documentation could be developed for use with both the ISO 9000 quality management system and the ISO 14000 environmental management system, or even other systems used by the organization.

The individual organization must determine the best approach—a large number of narrowly focused documents or a smaller number of broadly focused documents—depending on its unique operations, environment, and culture. Items to consider in deciding whether to combine or in determining the degree of combining include the following:

- How will the documents work best for the people who have to use them?
- Will combining documentation increase or decrease the organization's workload in:
  - Establishing the documents?
  - Using the documents?
  - Controlling the documents?
  - Updating the documents?
- Which philosophy is likely to cost less in the long run?

The conclusions of one organization may be different from those of another that has different circumstances. Therefore, organizations should review each documentation element individually, and then combine them or keep them separate based on the necessities and constraints of the individual situation. Organizations should exercise caution, however, because too much combining can create problems when it comes to keeping documents current and using them. Carried too far, combining can make the system less efficient, especially for users of the documents.

CROSS-REFERENCING REQUIRED DOCUMENTATION

Keep in mind that ISO 9000 allows organizations to refer to some documentation instead of including it in particular documents. For example, a quality manual may refer to an organization's procedures rather than actually including them in the manual itself. Specific process procedures (work instructions) may be referenced in the same way within the quality manual or within the related procedures. Reference to legal, regulatory, and other requirements is also essential. Supporting documentation such as the quality objectives may also be included in the quality manual by reference.

Cross-referencing should occur in two directions. For instance, the quality manual can contain a reference to a particular procedure and its location; similarly, the procedure itself can contain a reference (page, paragraph, clause of the quality manual) and the ISO 9001 clause to which it is directed. Most ISO 9000 organizations use cross-referencing in their documentation systems to preserve order and simplicity.

The four levels of documentation required by ISO 9000 may rely on external supporting documentation and elements of successive levels in order to attain or verify their intended function. Cross-referencing between the levels and supporting documentation is necessary if the QMS is to function efficiently. Refer to Figure 6-3, QMS Documentation Cross Referencing.

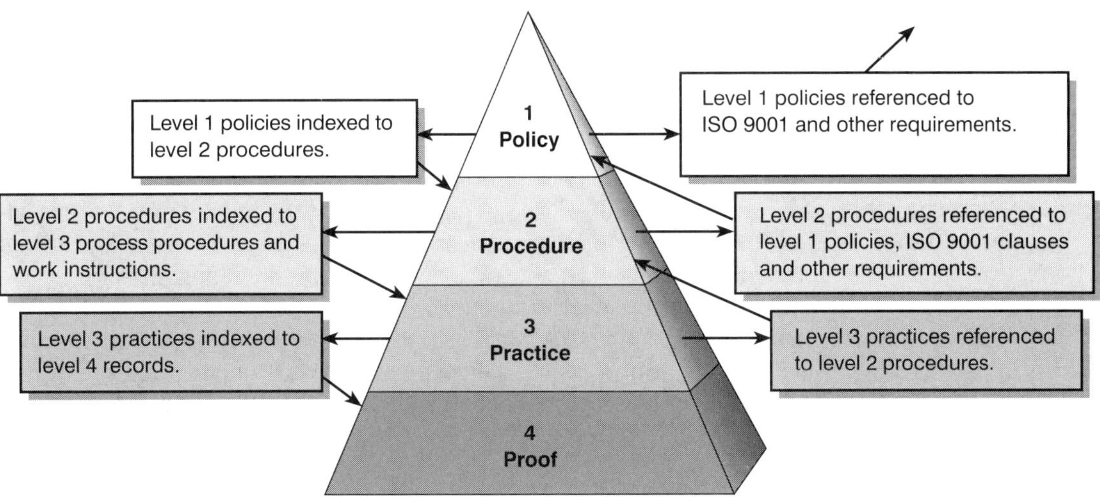

Figure 6–3
QMS Documentation Cross-Referencing

STRUCTURE OF THE DOCUMENTATION SYSTEM

The structure of the ISO 9000 documentation system was discussed in Chapter 5 in the section Documenting the Quality Management System. The structure of the ISO 9000 documentation system is often depicted as a pyramid as shown in Figures 4-4, 5-8, and 6-3. This is the typical illustration for the ISO 9000 documentation system, but not the only one.

The organization should design its system for compatibility with its business, its everyday needs, with ISO 9001 and the QMS, and any applicable regulatory requirements. The organization must have a quality policy. It must have procedures addressing the requirements of ISO 9001. Further, it must keep records. But beyond the guidance found in ISO 9001 and 9004, organizations have a great deal of latitude in organizing their documentation systems.

ISO 9000 even will allow the QMS documentation to be integrated with other documentation systems. For example, such systems may be integrated with an ISO 14000 system or another documentation system already employed. Regardless of how the organization structures its QMS documentation, it is important to remember two things: (1) users must know what applies to their operations, and (2) it must be easy for employees and auditors to locate the relevant document and its supporting documentation.

ELECTRONIC DOCUMENTATION

ISO 9001, clause 4.2.1, Note 3 states that the QMS documentation "can be in any form or type of medium." ISO 9004, clause 4.2 states that it "may be in any form or media *suitable for the needs of the organization.*" (Emphasis added.) Organizations are free to choose from the available media (paper, electronic, film), although paper and electronic media appear to be the most viable approaches.

An all-electronic documentation system has advantages. For example, an electronic documentation system will make it easier to ensure that documentation is kept up to date and that any document used is the latest approved version; distribution of new releases is simplified; and private collections of documents, which lead to the use of superceded documents, are eliminated. A disadvantage of electronic documentation is availability. Documentation such as work instructions must be readily available to employees who need it, at their work locations. If an employee does not have the equipment necessary to view the documentation at the work location, the procedure is not readily available.

From a practical standpoint, most organizations with electronic documentation systems find it necessary to print copies of the documents for day-to-day use. Even so, the advantages and usefulness, speedy revising, orderliness, and compactness make the idea of an electronic master file worth exploring.

CASE STUDY

ISO 9000 in Action

Jake Butler can tell that his colleagues are beginning to grasp that significant changes have occurred with regard to past and present ISO 9000 standards. This morning's meeting about documentation had gotten off to a rough start, but things got better as the meeting proceeded. At first there had been a lot of grumbling about "more paperwork" and claims that ISO 9000 registration was just a "paper chase." One manager had even commented that ". . . this is going to require enough documentation to sink a ship." But Jake Butler had persevered in making three critical points. First, he explained to participants that ISO 9000 registration required nothing more than the types of documentation any well-managed company should maintain. Second, he convinced participants that most of the documentation that would be needed already existed in one form or another. Finally, Butler explained how combining and cross-referencing could be used to cut down on the amount of documentation actually generated. Consequently, pulling documentation together into a cohesive system and making sure that it is properly formatted would be the real challenge.

Once everyone had gotten past the grumbling phase of the meeting, Butler had distributed a checklist of the documentation required—explicitly and implicitly—by ISO 9001. He had then asked each participant to identify existing forms of documentation that would satisfy any requirement on the checklist. By the time the meeting had ended, Butler had identified 90 percent of the documentation that would be needed.

In reviewing what had been accomplished during the meeting, Butler felt like real progress had been made. At the next meeting of the company's management team he would make his final recommendation concerning documentation. Butler planned to ask that he be given a "dedicated" secretary for a period of time to get all documentation scanned onto compact disks and converted into the proper format as specified by ISO. Once this task has been accomplished, all departments and functional units will have their documentation in an electronic file that can be easily updated as needed. In addition, the company will have a central "electronic storage room" containing all required documentation.

SUMMARY

1. For ISO purposes, "documentation" is written and pictorial information used for defining, specifying, describing, or recording the activities and processes that lead to the design, development, and production of products and the provision of services. Documentation may be in hard or electronic media and may include any of the following types of materials: written policies, written procedures, written quality objectives, written records, written plans, formal drawings, and sketches. Two critical requirements relating to documentation are "clarity" and "effectiveness."
2. ISO 9000 does not dictate a specific format for documentation, but it does require that the organization use appropriate, consistent formatting for its policies, procedures, process work instructions, reports, drawings, charts, and other forms of documentation. The following elements must be part of the written documentation of policies, procedures, and work instructions: organization name or logo, title, related to, document number, revision status, date of issue/effectiveness, review/approval signatures, purpose, body.
3. Similar procedures or the same procedures used for different tasks may be combined to simplify documentation and to reduce its size. Organizations may even combine procedures that have their corresponding requirements in different clauses.
4. Cross-referencing may be used as another method for holding down the size and amount of documentation. Cross-referencing involves referring to procedures that already exist in written form rather than including them in their entirety in another documentation package. When this method is used, it should be "two-way" in nature. When the quality manual refers the reader to a piece of documentation located elsewhere, the location should be specified. Likewise, that piece of documentation should also contain a specific reference back to the quality manual. All levels of documentation—policy, procedure, practice, and proof—may rely on cross-referencing, and there may be cross-referencing between the levels.
5. Regardless of how an organization structures its documentation, it is important to remember two things: (1) users must know what applies to their operations, and (2) it must be easy for employees and auditors to locate relevant documentation.
6. Documentation may be in hard copy or electronic format. Both have advantages. Electronic documentation is easier to keep up to date. On the other hand, hard copy documentation may be more readily available to employees responsible for operating specific processes. A recommended approach is to use both. Maintain an electronic master file and print hard copies for employees.

KEY TERMS AND CONCEPTS

Clarity
Combining required documentation
Cross-referencing required documentation
Design and development procedure
Determination of needed employee competence

Determination of needed infrastructure
Determination of needed resources
Determination of product requirements
Determination of the work environment
Effectiveness
Electronic documentation
Formal Drawings
Management representative appointment
Planning for product realization
Plans for design and development
Procedure for effective customer communication
Procedure for management reviews
Procedure for monitoring continual improvement
Procedure for monitoring customer satisfaction
Procedure for production and service provision
Procedure for purchasing
Procedures for assessing training needs
QMS planning
Reformatting preexisting documentation
Sketches
Structure of the documentation system
Written instructions
Written plans
Written policies
Written procedures
Written quality objectives
Written records

REVIEW QUESTIONS

1. Define the term *documentation* for ISO 9000:2000 purposes.
2. List eight elements that might be included in a comprehensive ISO 9000 documentation package.
3. What ISO clauses require the following documentation?
   - Records of management reviews
   - Records of product requirements review
   - Records of supplier evaluation
   - Records of internal audits
   - Procedure for preventive actions
4. Explain the format requirements of ISO 9000:2000.
5. Explain the concept of combining required documentation.
6. Explain the concept of cross-referencing documentation.
7. Describe the advantages and disadvantages of electronic documentation.
8. Explain how an organization can establish a documentation system that makes use of the benefits of electronic documentation while overcoming the disadvantages.

APPLICATION ACTIVITIES

1. Locate a copy of the ISO 9001:2000 standard. Review the standard carefully and develop a list of potential cross-referencing opportunities that might be used to reduce the volume of an organization's documentation package.
2. Using the ISO 9001:2000 standard, look up the documentation requirements for the following:
   - Quality manual
   - Quality objectives
   - Records of competence and awareness training
   - Records of internal audits

ENDNOTE

1. Joseph Cascio, *The ISO 14000 Handbook* (Fairfax, VA: CEEM Information Services, 1996), p. 200.

CHAPTER SEVEN

Registration and the Audit Process

MAJOR TOPICS

- The Registration Process
- Selecting a Registrar
- QMS Audit—Definition
- The Auditors
- Types of QMS Audits
- Objectives of QMS Audits
- Audit Scope
- Audit Process
- Follow-up Activities
  - Registrar Audits
  - Internal Audits

THE REGISTRATION PROCESS

Chapter 4 explained the requirements of ISO 9000, Chapter 5 examined the elements of a sound quality management system, and Chapter 6 described the documentation and documentation system necessary to support the QMS. If the organization has followed through this far, it should be ready to start the registration process. There are eight sequential steps in the registration process, as follows:

1. Decision by the organization to conform to ISO 9000 and to seek registration
2. Internal preparation by the organization to achieve conformance
3. Internal determination that the organization has achieved conformance and that the QMS is functioning

4. Accredited ISO 9000 registrar engaged to certify the organization
5. Preliminary assessment and document review by the registrar
6. Formal QMS audit and certification assessment by the registrar
7. Elimination of nonconformances preventing registration
8. Registration awarded by the registrar

Figure 7-1 illustrates the eight steps of the registration process in a flow diagram. The next several paragraphs provide a guided tour of Figure 7-1.

The first element of the diagram, Decision to Conform, is satisfied when the organization makes a commitment to conform to ISO 9000. This must include a commitment to comply with all applicable requirements and to continual improvement. (ISO 9001, clause 5.3).

In the second element, according to ISO 9001, the organization must prepare its quality policies, plans, procedures, and practices required for its QMS, and must communicate them to appropriate employees. The organization must set up control processes for its QMS operations and documents, including records. Procedures and facilities must be established for product realization, measurement, analysis and improvement, for obtaining customer requirements and satisfaction information, for establishment and fulfillment of quality objectives, for maintenance of QMS records, for internal QMS communications, and for management of resources. Finally, the organization must develop processes for checking and monitoring its QMS operations, for making corrective actions when needed, and for reviewing the adequacy and effectiveness of its QMS.

It is a good idea for the organization to get at least a preliminary feel for the suitability of its QMS and its constituent parts before getting involved with a registrar. The organization should define the employees' quality management system roles, responsibilities, and authority, and conduct the training that it finds necessary, and only then activate the QMS.

In the third element of the diagram, the organization asks, "Is the QMS functioning properly?" The organization should correct obvious malfunctions through its internal preparation loop (return to second element in diagram) before getting a registrar involved. Remember, the registrar cannot help an organization conform with ISO 9000; all that the registrar may do is determine whether or not the organization conforms. A conflict of interest would result if a registrar identified a particular problem and then proceeded to tell the organization how to correct it.

When the organization feels that the QMS is functioning properly, it is time to engage the services of a registrar, as depicted by the fourth element in the flow diagram. (Details on selecting a registrar will be discussed later in the chapter.)

As the fifth element of the diagram illustrates, the registrar first will conduct a preliminary assessment (or initial assessment) and document review. Normally but not always this is accomplished in two stages. First, the registrar may require copies of critical QMS documentation to review before visiting the site. This action allows the registrar to alert the organization about any obvious areas that need attention prior to the visit, thereby helping the organization's preparation efforts. The actual preliminary assessment is nearly always done at the organization's site. Most registrars prefer to assess on-site because of the volume of material to be reviewed and the ease of asking and obtaining

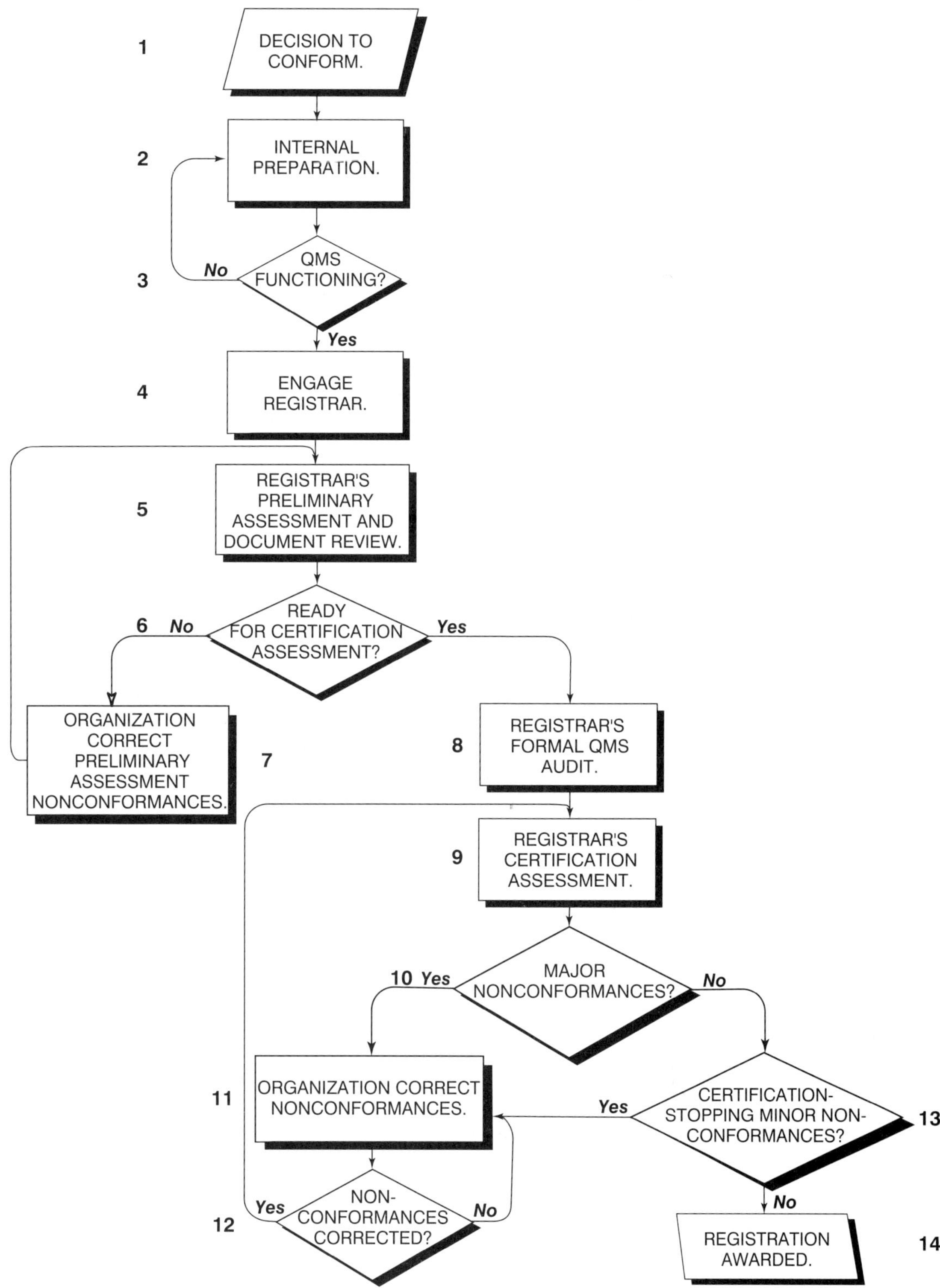

Figure 7–1
ISO 9000 Registration Process

direct responses, as well as to get a sense of the organization's operation. Because different registrars use different methods, it is important for the organization to understand exactly how its registrar will conduct the preliminary and certification assessments before they occur.

During the preliminary assessment and document review, the registrar will, as a minimum, want to review the following:

- QMS manual, including quality policy, policies addressing the ISO 9001 clauses, procedures, organization charts defining roles, responsibilities, and authority, quality objectives, improvement plans, and details and justification for any exclusions. The quality manual must establish the scope of the QMS, and describe the interaction of processes within the QMS. Procedures and other relevant documentation do not have to be in the quality manual if reference to them is provided.
- Training programs, procedures, and records.
- Internal audit reports.
- Records of management review of the QMS.

As the sixth element of Figure 7-1 illustrates, the registrar determines the organization's readiness for certification assessment. If the registrar finds nonconformances, which is usually true in preliminary assessments, the organization will be advised of areas that require additional work. The process proceeds to the seventh element of the diagram, and a nonconformance correction-and-verification loop is developed using the seventh, fifth, and sixth elements. Although the registrar will not (may not) provide substantive assistance, the organization is free to retain a consultant for such purposes if it chooses to do so.

If the registrar finds no significant nonconformances (at the sixth element of the diagram) or if the organization has corrected to the registrar's satisfaction any problems that were noted in an earlier pass, the organization is considered ready for its formal QMS audit.

The eighth element of the diagram reflects that the registrar will conduct a formal audit of the organization's QMS. (The audit process is described later in this chapter.)

Next, at the ninth element, the audit report will be used by the registrar to assess whether certification should be granted.

As the tenth element of the diagram indicates, if there are any major nonconformances, registration cannot be granted until they are eliminated. It would be unlikely, though not impossible, that a major nonconformance would exist at this point because of the preliminary assessment previously completed by the registrar. It would be most embarrassing for the registrar to uncover a major nonconformance at this point. Nonetheless, if any major nonconformances are discovered, the organization implements corrective action (eleventh element of the diagram) and verifies the corrective action (twelfth element). When the organization is satisfied that the nonconformance is corrected, it reverts back to the registrar's certification assessment (ninth element), providing the registrar with evidence of conformance. At this point the registrar usually revisits the site to confirm the elimination of major nonconformances (tenth element).

Once any major nonconformances are eliminated, the certification process advances to the thirteenth element of the diagram. Minor nonconformances do not necessarily

ISO INFO

Major and Minor Nonconformances

*A **major nonconformance** results from (1) a failure to fulfill any requirement of ISO 9001, or (2) multiple minor nonconformances sufficient to lead auditors to conclude that an ISO 9001 requirement is not effectively implemented.*

*A **minor nonconformance** is a single lapse in implementing a QMS requirement. It is seen as an anomaly rather than a systemic issue.*

cause the registrar to withhold registration. However, registration may be withheld if (1) there are minor nonconformances that fall into a pattern which suggests that an element of the QMS is not effectively implemented, or (2) too many minor nonconformances exist, suggesting that the organization has not taken the QMS design or implementation seriously. If registration is withheld for either of these cases, the minor nonconformances must be corrected and verified by the organization (eleventh and twelfth elements of the diagram) and then verified by the registrar through the final verification-and-certification assessment loop (ninth, tenth, and thirteenth elements) *before registration can be issued.* Depending on the nature of the nonconformances, the registrar may schedule another site visit in order to verify correction.

Once no major nonconformances or certification-stopping minor nonconformances are present (tenth and thirteenth elements), the registrar may award registration, as reflected in the fourteenth element.

The ISO 9000 registration process can lead to three possible outcomes:

- **Approval**—The QMS is found to be in substantial conformance with ISO 9001 requirements. In this case a document certifying registration to ISO 9000 is issued to the organization by the registrar.
- **Conditional approval**—All elements of ISO 9001 have been addressed, but either they are not fully documented, not fully implemented, or objective evidence leads the auditors to believe there is a systematic problem. These issues must be corrected by the organization, and reviewed by the registrar before full approval can be granted. The registration certificate will be withheld until the issues are satisfactorily closed.
- **Disapproval**—A least one requirement of ISO 9001 has not been addressed by the organization, or the QMS is demonstrated to be ineffective at meeting policy commitments and/or objectives.

SELECTING A REGISTRAR

While preparing for ISO 9000 certification, one of the most important steps will be the selection of your registrar. This step is important for several reasons, including cost (which can vary widely), recognition (by your customers, both present and potential), familiarity with your industry (it helps if you and the registrar speak the same language), and integrity. The best policy is to consult with registered firms to solicit information about their registrars. Registrars should also provide references that can be checked. Remember, the choice is strictly up to your organization. You will spend a significant amount of money for the registrar's services, to say nothing of the internal cost of developing your ISO 9000 infrastructure. Consequently, you should protect your investment by securing a registrar who is competent, whose services are fairly priced, and whose integrity is beyond question.

This integrity issue is a double-edged sword. You might find a registrar who has the reputation of being "easy," and, on the surface, this might seem like a plus. However, there are two problems with this. First, even if you can secure an "easy" certification by developing a third-rate QMS, the benefit will be only short term and superficial. Remember, no matter what the registrar says, your customers are the final arbiters in determining the quality of your products or services. While other organizations are becoming more competitive by genuinely improving their quality systems, your "just-get-by approach" will put your organization at risk regardless of ISO 9000 certification.

Second, if you are able to identify a registrar as one who has looser certification standards than normal, so can everyone else, and that registrar's reputation is probably already known throughout the world. What this means is that your certification will not carry the weight you need among your customers and competitors. Remember, advertising ISO 9000 certification requires the listing of your registrar's name. This credibility issue with registrars is similar to the one faced by students selecting a college. A degree from a college with a reputation for maintaining high academic standards will be more valuable in the long run than one from a fly-by-night diploma mill, even though it will be more difficult to earn. Even worse, if this kind of practice were to become widespread the entire ISO 9000 movement would ultimately fail because its perceived value depends largely on the integrity of the registrars. Firms preparing for registration should choose their registrars based on past ethical performance, familiarity with the industry, reputation for fairness and objectivity, and cost. Schedule availability may also be a factor as the number of firms requiring services from registrars increases. In addition, the organization may also consider professional chemistry, or the likelihood of being able to develop a good working relationship.

QMS AUDIT—DEFINITION

ISO 9000:2000 defines a **quality management system audit** as a

> *Systematic, independent and documented process for obtaining audit evidence and evaluating it objectively to determine the extent to which audit criteria are fulfilled.*[1]

From this definition we can determine that a quality management system audit has the following characteristics:

- They are systematic—not improvised, not casual.
- The audit process is documented and will be performed accordingly.

- The audit team will objectively seek and evaluate information and evidence concerned with QMS activities, events, conditions, management systems, and related information.
- The objective is to determine whether these activities, events, conditions, management systems, and related information conform with the audit criteria.

These characteristics are logical and straightforward. However, of the fourth point one might query, What are the audit criteria? Who establishes them? Again, according to ISO, audit criteria are a:

Set of policies, procedures or requirements used as a reference.[2]

In effect, the QMS audit criteria are the organization's own statements of intent. The policies, practices, and procedures are those developed by the organization. Requirements may originate from standards (such as ISO 9001), customers, the organization itself, and legislative or regulatory agencies. The registrar will plan the audit around the organization's QMS documentation, and will confirm the audit criteria with the organization prior to an audit.

In its QMS—with the policies, procedures, practices, records, and legal and regulatory requirements—the organization has told the registrar how it intends to operate and maintain its quality management system activities. Through the QMS audit, the registrar will verify whether the organization is doing what it said it would do (conformity), and whether the results confirm that the QMS is effective.

THE AUDITORS

In this section we explain several topics dealing with the different kinds of QMS auditors, their qualifications, audit teams, and auditor certification. We begin by examining the need for quality management system auditors.

Why Auditors are Needed

ISO 9000 is an international standard developed to assist organizations to implement and operate effective quality management systems. In a perfect world it might be sufficient to publish an environmental standard and expect organizations around the world to conform on their own for the benefit of their customers and themselves. In such a world, certification would be of no concern. Unfortunately, ours is far from being a perfect world. Even in an imperfect world, organizations would not need a certification process if there were a single world authority that could mandate adoption of the standard and enforce adherence. The International Organization for Standardization (ISO) has no authority to mandate or enforce its standards; its job is to develop standards through its international committees. Individual organizations decide whether or not to use the standards. Any organization worldwide is free to adopt or not adopt ISO 9000. The decision is inherently an internal one, although organizations may feel pressure from customers or other interested parties.

Internal and External Auditors

Organizations seeking to use the ISO 9000 QMS model need "independent" observers to verify conformance with the QMS and its constituent elements. These independent observers are the **auditors.** Auditors include employees of the organization and employees of a third-party registrar firm. While it is not normally a problem for third-party auditors to be independent, objective observers, it is an issue for internal auditors. Internal auditors should never audit their own organizational component (i.e., department, function, group). For example, an engineer should never audit the engineering department to which he or she is assigned. Internal auditors must not audit their own work.

Auditor Qualification

ISO 10011-2 details the qualification for QMS auditors. Qualification criteria include the following:[3]

- Education—at least secondary (high school) or equivalent.
- Demonstrated competence—in clearly and fluently expressing concepts and ideas orally and in writing.
- Training—to ensure competence in carrying out and managing audits, including:
  - ISO 9000
  - Assessment techniques
  - Planning, organizing, communicating, directing
- Experience—four years relevant workplace experience, of which at least two years should have been related to quality assurance activities. Before assuming responsibility for performing audits, should have participated in a minimum of four audits, for a total of at least twenty days.
- Personal attributes—must be open minded, mature, possess sound judgment, analytical skills, and tenacity. Must have the ability to perceive situations realistically, understand complex operations, and understand the role of individual units within the overall organization.
- Lead auditors have the same qualification criteria as auditors, but require more experience and demonstrated leadership capability (ISO 10011-2, clause 11).

ISO INFO

The technical committees for ISO 9000 and ISO 14000, TC 176 and TC 207, respectively, are working on a single auditing standard that will be applicable to quality and environmental management systems. It will be ISO 19011, and will replace ISO 10011, and ISO 14010, 14011, and 14012 documents. ISO 19011 should be available in 2002.

In addition to the requirements for demonstrated auditor skill, knowledge, and experience, auditors are now required to demonstrate a knowledge of the eight quality management principles (see Chapter 1) on which ISO 9000:2000 is based.[4]

The auditor qualification criteria just explained were developed by ISO for auditors who work for the registrars. ISO contends that "internal auditors need the same set of competencies, but may not meet in all respects the detailed criteria [listed above] depending upon such factors as the size, nature, complexity . . . of the organization; and the rate of development of the relevant expertise and experience within the organization."[5] This means that the organization's employees may be used as internal auditors without meeting all of the qualification criteria. However, no employee should be allowed to audit a QMS without first receiving training on the fundamentals of quality system auditing. In addition, we stress again that internal auditors must be independent of the function being audited.

QMS Audit Teams

The registrar's audit team will typically be comprised of several certified quality system auditors and a lead auditor. The number will vary according to the size of the organization being audited, and can in rare cases be a single individual. The lead auditor considers the following when selecting members of his or her team:

- Qualifications of potential auditors
- The type of organization, processes, activities, or functions being audited
- The number, language skills, and expertise of the audit team members
- Any potential conflict of interest between the audit team members and the organization being audited
- Requirements of clients, and of certification and accreditation bodies

Most registrars attempt to include subject matter experts on their audit teams. These are people with specific kinds of experience beyond ISO 9000. For example, if a facility that manufactures electronic equipment is to be audited, the audit team might include one or more people who are experts in such an environment. Some registrars, however, believe that a certified auditor, regardless of specific technical experience, can satisfactorily audit the QMS of any kind of organization.

Occasionally audit teams may include technical experts who do not serve as auditors. These experts assist the auditors in areas (processes) outside the realm of the auditors' technical expertise. This is a compromise or middle ground position when the registrar's auditors are not familiar with the specific processes of the organization being audited.

The audit team may occasionally have one or more auditors-in-training. These are auditors who are accumulating QMS audit hours in order to obtain full certification.

Auditor Certification

Auditors working for registrar firms must be certified as meeting the ISO quality system auditor qualification criteria. QMS auditors who work for registrars must demonstrate appropriate experience and qualifications as described earlier and must complete a course and pass a comprehensive examination for ISO 9000 auditors. In the United States, quality system auditor certification is provided by the Registrars Accreditation Board (RAB). Other nations use similar auditor certification methods.

TYPES OF QMS AUDITS

There are three general types of QMS audits:

- Registration audits
- Surveillance audits
- Internal audits

Registration Audit

The **registration audit** is performed by the registrar firm that is hired by the organization seeking registration. The intent of this audit is to verify to the registrar initial conformance with ISO 9000 and the organization's QMS. Registration audits are repeated every three years. If a registration audit is completed satisfactorily, the organization becomes registered.

The initial registration audit has at least two time-phased elements. The first element is a documentation review by the lead auditor (usually) at the registrar's place of business. This involves a detailed review of the organization's QMS documentation; the documentation is compared against the requirements of ISO 9001 and other relevant requirements.

The second element is an optional on-site preliminary assessment of the organization's readiness for registration. The preliminary assessment should be viewed as a trial run for the registration audit. It allows the auditors to gain first-hand knowledge of the organization's operations and its personnel. From the preliminary assessment, significant gaps in the QMS implementation can be identified, so that the organization can close the gaps before the registration audit.

The final phase, the actual registration audit, is performed after the registrar determines that the organization is prepared. The registration audit is conducted at the organization's site. Areas audited include:

- Conformance audit: audit policies, procedures, practices, proof—against ISO 9001
- Conformance audit: audit practices, proof—against the elements of the QMS
- Process audits: audit conformance with QMS practices, work instructions

Figure 7-2 is a graphical representation of a QMS audit.

Should the registrar find nonconformances that cannot be corrected while the audit team is on site, one or more follow-up audits may be required to verify subsequent elimination of the nonconformances.

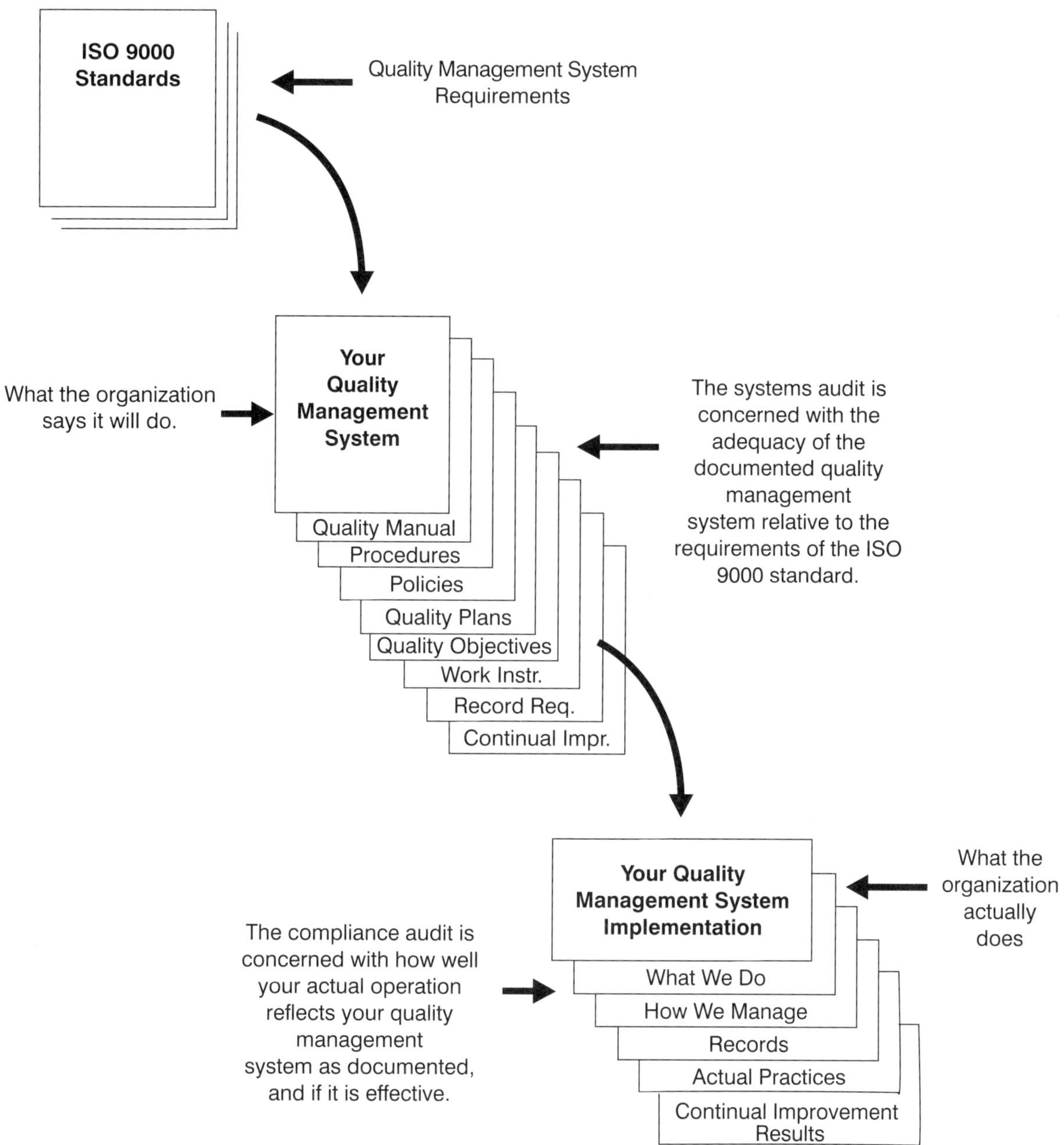

Figure 7–2
QMS Audit by Registrar or Internal Auditors

Surveillance Audit

A **surveillance audit** is also performed by the registrar to verify continued conformance with ISO 9000 and the organization's QMS. These audits, depending on the individual registrars, are conducted at six or twelve month intervals. Satisfactory surveillance findings will result in continued registration of the organization.

Internal QMS Audits

An **internal audit** is conducted by audit teams comprised of employees from the organization. These audits are typically performed between visits by the registrar's auditors in order to verify to management that nonconformances have been corrected, and that the QMS is working effectively. However, the organization may increase the audit frequency in areas where nonconformances have been noted or where extra attention is needed. Internal audits focus on the same areas as the registrar's audits, but the audience is the organization's management team. Satisfactory results in an internal audit should mean that no nonconformances will be noted at the next surveillance audit.

Internal audits are also performed prior to the initial registration audit to confirm an organization's readiness for registration.

OBJECTIVES OF QMS AUDITS

If you ask someone why he or she goes to college, that person is likely to respond, "I'm going to college to earn a degree in ________." However, just earning a degree is not the real objective of four or more years of study. The real objective of going to college may be to enter (and be successful in) a prized career, to be better able to provide for one's family, to have a better chance of achieving a secure future, or to develop personally. The work behind the degree prepares one to pursue the real goal, and is therefore only a means to the desired end.

QMS audits are similar. On the surface, audit objectives would seem to be obvious. For example, most organizations would declare that their objective for a certification audit is registration. Similarly, the objective of surveillance audits may be seen as retaining registration. Both are reasonable, valid objectives. Certainly, though, that is not the end. Surely there is more to it than having the newly earned or updated ISO 9000 registration certificate displayed in the organization's lobby. Like an individual's degree displayed on an office wall, the registration certificate is merely a symbol signifying that development and improvement have occurred. The real importance is that the improvement has allowed the organization to achieve certain objectives, such as:

- Confirmation by an independent, knowledgeable, objective party that the organization's quality management system is effectively implemented and can be expected to aid the organization in meeting customer expectations
- Improvement (or confirmation) of its image and, hence, its business by being able to legitimately claim to customers, employees, and other interested parties that it is a quality oriented organization

A registrar and its audit team have several objectives in performing audits. They include:

- Perform audits objectively, without bias
- Determine conformance with ISO 9000 and the organization's QMS
- Determine that the QMS has been properly implemented and operated (maintained)
- Determine management's ability and willingness to review QMS processes for continued suitability and effectiveness, and for continuous improvement
- Identify improvement opportunities

Registrars should have no objective relating to the organization's certification. Rather, they should ensure that audits are conducted with objectivity, fairness, confidentiality, and integrity. Auditors, in turn, may be indifferent concerning an organization's objective for certification. This may sound harsh, especially since the organization must pay a significant fee to the registrar for the audit. However, if the integrity of ISO 9000 certification is to be maintained, auditors must remain independent and objective, guarding against any possibility of disregarding or compromising standards.

AUDIT SCOPE

The **audit scope** establishes the boundaries of the audit. It specifically establishes:

- Physical locations to be visited
- Organizational activities to be audited
- Resources required for the audit
  - Audit team
  - Organization's audit support requirements
  - Physical resources, such as audit team work space
- Manner of reporting on the audit

The formal audit scope is developed by the lead auditor in consultation with the organization to be audited (the client). Consequently the organization knows beforehand what the auditors will be looking at, how many auditors will participate, how long the auditors intend to be on site, and what resources the organization must provide.

The initial certification audit, as well as the recertification audit every third year, must cover your entire quality system. These audits include an examination of your quality policy, quality manual, policies, plans, quality objectives, procedures, work instructions, continual improvement and records. Collectively, these documents tell the auditor what you plan to do and how you are going to do it. The auditor, then, determines by interview, observation, and through an examination of records, how effective your implementation of the quality management system has been. "Effective" in this context means QMS conformance to ISO 9001 requirements, and the degree to which you are doing what you said you would do, and doing it how you said you would do it. Any activity that may affect the quality of your product or service may be assumed to be of interest to the auditor. Certainly any activity specified in ISO 9000 or in your quality management system documentation will be within the scope of the audit. These audits are usually conducted by a team, and may take several days to a week to complete.

The semiannual registrar surveillance audits are not as far ranging, but may probe any area of interest or concern to the auditor. These audits usually focus on randomly selected target areas together with any nonconforming activities that were noted during the previous audit, or that may have been highlighted by customer input or other sources of feedback. Typically the semiannual surveillance audits are conducted by a small team, or even by one person, and may be completed in as little as one day.

In summary, audits may involve any part of the organization that is specified in ISO 9000 or in your quality management system documentation—or any activity that may in any way affect the quality of your product or service. Assume that the major audits (for certification or recertification) will be comprehensive, and that semiannual surveillance audits may be as comprehensive as the lead auditor thinks necessary and appropriate.

AUDIT PROCESS

Both internal and third-party audits by the organization's registrar are based on the requirements of the ISO 9001 standard, and are governed by the principles and procedures of ISO 19011. The auditor's checklist will contain items from ISO 9001 clauses starting with clause 4. The audit process will attempt to determine the following:

- Whether the quality management system conforms to its requirements, as stated in its policies, procedures, practices, objectives, hierarchy, staffing, and the like; to its planned arrangements; and to the requirements of ISO 9001
- Whether the quality management system is effectively implemented, maintained, and operated

Audits Performed by Registrars

Registrars perform registration audits and surveillance audits. In the section that follows we have concentrated on the registration audit because it is the most pervasive. However, we have included notes to indicate applicability or differences relative to the two audit types.

Pre-Audit Activity

(Note: Applies only to the initial registration audit.) The registrar requires copies of critical QMS documentation in advance of site visits. This helps in planning the registration audit and is less expensive than an on-site review of the materials. In reviewing these copies, the registrar can note areas in the QMS documentation that need attention before the on-site visit.

Preliminary Assessment Visit (Optional)

After having completed a study of the organization's QMS documentation, the registrar may, before the initial registration audit, send a representative, normally the leader of the audit team, to the organization's facility in order to develop a sense of the operation. This step, referred to as the preliminary assessment visit, is useful for the final planning of the audit—plans that specify the composition of the audit team, number of team members, and number of days required for the audit. The plans are also helpful for identifying obvious areas of concern that will be shared immediately.

Preparation for the Audit

(Note: Applies to registration audits, including reregistration.) The audit will be scheduled for a date on which the organization and the lead auditor agree. Before the date, the organization should have completed all preparations for ISO 9000, and all employees should have fully prepared for the audit. Employees preparation should include details of what to expect as well as proper conduct when dealing with the auditors. Employees should be cooperative, open, nondefensive and nonargumentative. In addition, they should answer all questions truthfully. On the other hand, employees should not attempt to answer questions if they do not know the answer. In these instances a simple "I don't know" is the best response. Employees should not disclose information that is not requested. It does neither the organization nor the auditors any good when employees at any level volunteer information. The auditors look for specific facts, and they know the questions to ask in order to obtain these facts. (Note: Employee conduct should be the same for all types of audits.)

Opening Meeting

(Note: Applies to registration and surveillance audits, although surveillance audits will generally be shorter and require less support.) Immediately before the audit begins an **opening meeting** will be held, involving the audit team and key personnel from the organization being audited. During this meeting, which usually lasts one hour, the lead auditor will accomplish the following tasks:

- Emphasize the positive nature of the audit (i.e., the team is present by invitation to enable the organization to be registered to ISO 9000, not to prevent that outcome)
- Introduce members of the audit team to the organization's key staff
- Explain the scope, objectives, plan, and timetable of the audit, making certain that everyone present understands
- Emphasize the importance of timely employee participation
- Confirm the mutually agreed audit criteria (ISO 9001, less any agreed-to exclusions in clause 7, and the QMS and supporting documentation and records)
- Summarize audit procedures and methods
- Review confidentiality issues (i.e., nonattribution for information gained during the interviews, registrar's confidential treatment of all information)
- Discuss the types of findings and observations, and major and minor discrepancies and their significance to the audit's outcome

- Acquaint the organization with the auditing assignments (i.e., who will review specific items or areas)
- Review any relevant safety and emergency procedures that may be appropriate
- Confirm availability and location of required resources and facilities (employees to be interviewed, guides, phones, reproduction machines, team workspace)
- Establish time for end-of-day reviews and for the audit closing meeting
- Respond to any questions

Facility Tour

(Note: May not be required for surveillance audits, depending on the auditors.) It is customary for the organization to conduct a tour of the facility for the auditors. This **facility tour** allows the audit team to familiarize itself with the location of the various activities and the locations of employees to be interviewed. It also enables the auditors to develop questions they had not considered before seeing the activities in operation. Depending on the size of the organization, this tour could require as little as one hour or as long as half a day. If the tour could not be completed in half a day, the audit team should be divided to tour only the specific areas of interest for their work on the audit.

Registration Audit

(Note: The surveillance audit generally seeks to confirm continued conformance and continued effectiveness of the QMS, including any changes introduced through continual improvement, or corrective or preventive action. Therefore it is a more narrowly focused audit.) The purpose of the registration audit is to determine objectively if the organization conforms to the requirements and intent of ISO 9001, its own QMS, and other QMS-related documents. Determining conformance requires the following:

- The QMS document system must be complete, responsive to ISO 9001, and comprehensive.
- The total QMS must have been implemented.
- The QMS implementation must be shown to be effective.

The first requirement in this list may be satisfied by comparing the QMS documents with the requirements of ISO 9001. The second and third requirements must be satisfied through objective evidence, which should be collected by the audit team.

Depending on the size and nature of the organization, the audit lasts up to one week. The audit team will spend approximately 25 percent of its time examining QMS documentation. The remainder of the time will be spent collecting evidence to verify operational conformance with the QMS. Remember, the QMS tells the auditors (as well as the employees) what the organization says it is going to do. The auditors, then, must verify that the organization is actually doing what it said it would do. A major nonconformance will stop the registration, and perhaps the audit itself. Minor nonconformances, unless the numbers reach the lead auditor's threshold, or unless they are grouped in such a way as to indicate that the QMS is not effectively implemented, will neither stop the audit nor cause the registration to be withheld. Often minor discrepancies can be remedied instantly (although corrections will require documentation).

The auditors try to keep the organization informed of their findings through the course of the audit, but inevitably some audit observations or minor nonconformances can be revealed only at the exit meeting after the auditors have concluded their analysis of all the data. In order for the process to continue, the organization and the auditors must agree to a formal schedule for taking corrective action on any open issues. Figure 7-3 is a flow diagram of the corrective action process.

End-of-Day Review

(Note: The end-of-day practice is usually observed for both registration and surveillance audits.) At the end of each audit day, the team holds a short (thirty to sixty minutes) meeting, referred to as the **end-of-day review,** with the organization's key personnel to discuss the information learned during the day. The three purposes of the end-of-day meetings are:

1. Allow the organization to correct nonconformances during the audit
2. Keep the organization informed concerning findings and progress of the audit and to mitigate any administrative difficulties
3. Minimize, or prevent, surprises at the exit meeting

Nonconformances are discussed at the end-of-day meeting. The organization may correct the nonconformances during the audit or, if this is not possible, propose a schedule for correcting them.

Exit Meeting Preparation

(Note: The exit meeting applies to registration and surveillance audits. Techniques used are the same.) Once its investigative tasks have been completed, the audit team assembles to prepare for the **exit meeting.** Collected evidence is discussed, evaluated, and grouped into three categories:

- Conforming—no action required by the organization
- Nonconforming—the organization must address
- Observations and concerns—the organization may or may not take action

The auditors will confirm all input with evidence and reach a consensus regarding all nonconformances and recommendations. Where insufficient evidence is available on a particular point, the audit team may present an observation or concern.

Audit findings, showing areas of conformance or nonconformance, will be summarized, documented, and prepared for presentation to the organization's key staff at the exit meeting. Each nonconformance is numbered and documented on nonconformance report (NCR) forms along with a statement of the requirement not satisfied. A typical NCR is shown in Figure 7-4. Major nonconformances prevent registration; minor nonconformances usually do not.

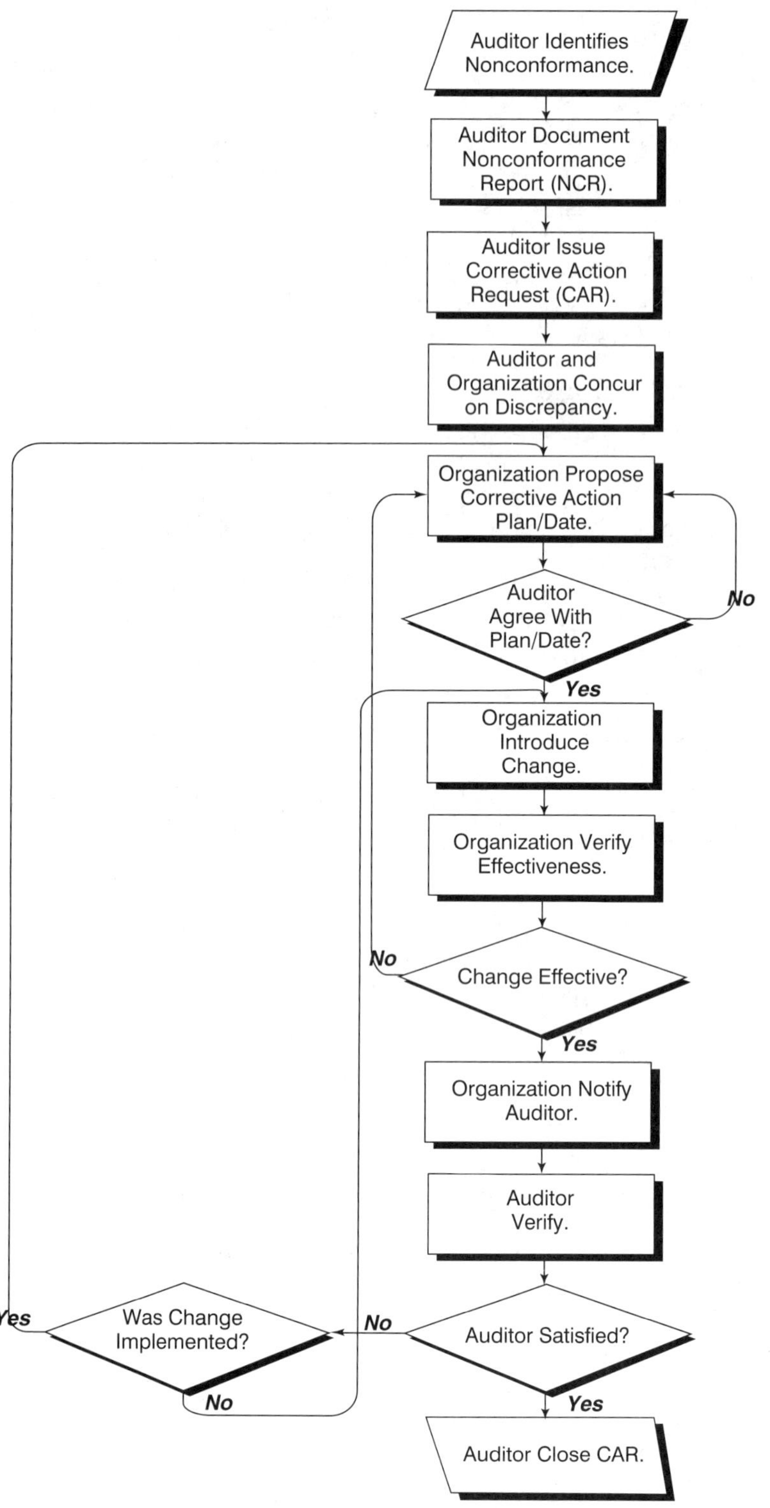

Figure 7–3
Corrective Action Process

QMS Nonconformance Report SDC

AUDIT FOR (FIRM):	NCR#
ACTIVITY:	DATE
ISO 9001 CLAUSE #	REF. DOCUMENT:

NONCONFORMANCE:

☐ MAJOR ☐ MINOR ☐ OBSERVATION

RESPONSIBLE MANAGER — DATE

AUDITOR — DATE

Figure 7–4
Typical QMS Nonconformance Report Form

The nonconformances are also documented on corrective action request (CAR) forms. A typical CAR form is shown in Figure 7-5. CARs are signed by the auditor and the organization's manager in charge of the relevant activity. In addition, the responsible manager must sign off on the date that has been agreed to for completion of corrective action.

Exit Meeting

(Note: The surveillance audit exit meeting is generally the same as for the registration audit, but shorter.) The purpose of the exit meeting, which can last up to four hours, is for the audit team to present its findings. Employees who should attend the exit meeting include key management staff and managers or supervisors of the functions audited. A typical agenda for the exit meeting is as follows:

- Reintroduction of the audit team
- Review of the scope and purpose of the audit
- Brief review of how the audit was conducted
- Review of the audit criteria, namely, ISO 9001, the organization's QMS documentation, and other related documentation
- Review of nonconformance levels
  - Major—failure to meet a requirement of ISO 9001
  - Minor—a single lapse
- Summarize findings of conformance, presented by the lead auditor
- Present nonconformances (including those which were cleared during the audit), presented by the auditors
  - Obtain assurance that findings are understood by the organization
  - Obtain acknowledgment from the organization that findings were based on objective evidence
  - Resolve any disagreements
  - Agree on corrective action schedule
- Disclose the recommendation of the team (for the registrar) in regard to registration, presented by the lead auditor

Copies of the NCRs and CARs are released to the organization, usually with a preliminary copy of the final report containing factual statements of discrepancies and objective evidence in support of the findings. (Note: This assumes that the organization is the registrar's client.)

Awarding or Withholding Registration

Following the registration audit, the registrar's review of the lead auditor's report, and the recommendation of the audit team, the registrar may do the following:

1. *Award registration*—if no major nonconformances were found, if there was no pattern of minor nonconformances to indicate failure in the QMS implementation, and if the number of minor nonconformances was acceptably low

QMS CORRECTIVE ACTION REQUEST

SDC

AUDIT FOR (FIRM) | CAR#

ACTIVITY | DATE

ISO 9001 CLAUSE # | REF. DOCUMENT

AUDITOR | FIRM REP.

DATE | DATE

DESCRIPTION OF PROBLEM: (ADD SHEETS AS REQ'D.)

CORRECTIVE ACTION PLANNED:

PLANNED COMPL. DATE:

FIRM REP SIGNATURE | DATE

AUDITOR'S FOLLOW-UP DETAILS:

SATISFACTORY? ☐ **YES** ☐ **NO** (CHECK ONE)

LEAD AUDITOR SIGNATURE | CAR CLOSEOUT DATE

Figure 7–5
Typical QMS Corrective Action Request Form

2. *Award conditional registration*—pursuant to timely and effective resolution of specified minor nonconformances
3. *Withhold registration*—until the organization demonstrates to the lead auditor's satisfaction that major nonconformances, and/or patterns of minor nonconformances, and/or an excessive number of minor nonconformances no longer exist

(Note: For surveillance audits, the team's recommendation may be to continue or discontinue registration.)

Internal Audits

Internal QMS audits are a requirement of ISO 9001, clause 8.2.2 in support of clause 5.6, which is concerned with management review of the QMS to assure its continuing suitability and effectiveness. The audits must be periodic, performed according to a schedule, rather than on an ad hoc basis. The purpose of the internal audits is exactly the same as audits performed by the registrar, that is, to determine (1) whether the QMS conforms to the requirements of ISO 9001 and other planned arrangements, and (2) if it has been properly implemented and maintained. Results of the internal audits are released to management.

Schedules for internal audits should consider the importance of the activities concerned, from the point of view of the QMS, and results of previous internal and external audits.

The process used is similar to that described earlier for registrar audits, but may not require the formal opening meeting or the end-of-day review. An exit meeting should be held with management and supervisors of the activities audited.

Nonconformances must be documented in a manner similar to the registrar's process. Schedules for corrective action must also be documented and signed off by the manager of the relevant activity.

Obviously, internal audits cannot result in certification or withholding of registration.

FOLLOW-UP ACTIVITIES

Follow-up Activities—Registrar Audits

Note: Follow-up activities will be essentially the same for registration and surveillance audits.

Audit Reports

A formal audit report is written by the lead auditor after the audit has been completed. This report will contain the elements presented and provided to the organization in draft form, at the exit meeting. There should be no surprises; anything included in the report should have been presented during the exit meeting.

The audit report is compatible with the agreed audit plan. If during the course of the audit an auditor went beyond the scope of the audit plan, any resulting findings should not appear in the report. On the other hand, if a finding resulting from efforts beyond the scope of the audit plan is relevant, it should not be ignored. In such cases it is best to inform the organization's management through an informal channel. At the very least, the lead auditor should be certain to include the issue in the scope of the next audit.

Through coordination between the lead auditor and the client, distribution of the audit report is established in the audit plan. In most cases the client and the organization are one and the same. Occasionally the client will be a second party firm, possibly an attorney. The audit report is normally distributed to the client, whether the client is the organization, its legal representative, or some other entity that contracted for the audit. To illustrate the latter, a corporate headquarters in Cleveland could contract a registrar to audit one of its divisions in Baltimore. The lead auditor will distribute the report in accordance with the client's instructions, regardless of the client's identity.

Confidentiality regarding the audit report must be maintained on two levels. First, the lead auditor must never distribute the report to any third party, unless directed to do so by the client. Second, statements used as evidence in the audit process must not be attributed to individuals in the organization.

Corrective Action

Open nonconformance reports and corrective action requests must always be completed; otherwise there is nothing gained by auditing nonconformances. Refer to Figure 7-3 to review the **corrective action** process. Note that no NCR/CAR resulting from a registrar's audit is considered closed until the lead auditor signs off, indicating a satisfactory close-out. In some instances it will be necessary for the auditor to make a special visit to the site to verify satisfactory closure.

The lead auditor tracks the outstanding CARs against their scheduled completion dates. When dates are missed, the lead auditor *may* initiate action to determine the reason. Another lead auditor (or registrar) may not do so, however, believing that an organization wanting registration, or wanting to retain it, will take care of corrective actions without prodding from the registrar. This is not an unreasonable attitude; the registrar is not in the business of managing the activities of the organization. For this reason the organization should always be proactive in taking the initiative to keep the lead auditor informed of progress and problems associated with completion of CARs. Every reasonable attempt should be made to meet the completion dates agreed to, but when it becomes apparent that a date will be missed, it is prudent to inform the lead auditor or the registrar immediately with an explanation. As long as the organization is working in good faith to clear CARs, and there is no pattern of inattention or disinterest, negative consequences are not likely.

Preparation for the Next Registrar Audit

(Note: This applies to both registration and surveillance audits.) The lead auditor will ensure that any areas of concern from the just-completed audit are targeted for follow-up at the next audit. Areas of concern include the following:

- Any activity that resulted in a reported unfavorable observation
- Any area that had nonconformance reports and corrective action requests, even though implemented

- Any area that resulted in an informal negative report (as in the earlier example)
- The effectiveness of all completed corrective actions

The organization should be aware that the next audit will inevitably emphasize areas that demonstrated weaknesses in the QMS on the last audit, so that the organization can prepare appropriately.

Follow-up Activities—Internal Audits

Follow-up activities from internal audits generally follow the same scheme outlined for registrar audits, with the exception that internal audits are not concerned with awarding registration.

Audit Reports—Internal Audits

At the conclusion of an internal audit, information on the results of the audit must be provided to the organization's management. Information from internal audits forms one element of management's review of the QMS's continuing suitability, adequacy, and effectiveness.

Corrective Action—Internal Audits

The internal corrective action process is the same as that involving the registrar's auditors. In Figure 7-3 the word *activity* can be substituted wherever *organization* is used. This creates a flow diagram that can be used to deal with internal NCRs and CARs. *Auditor* in the flow diagram becomes *internal auditor.*

Nonconformances discovered during internal audits must be documented in the same manner as those of registrar audits, which is through a system of NCRs and CARs. The major difference is that completion of corrective action will be signed off by the internal lead auditor rather than the registrar's lead auditor. The forms used should be similar to those shown in Figures 7-4 and 7-5 but tailored for internal use.

Corrective action requirements stemming from internal audits should carry the same weight as those from an audit performed by the registrar. Internally generated NCRs and CARs become part of the permanent QMS record system, and are reviewed by the registrar's auditors. If corrective action is incomplete or ineffective, the organization will be faced with a nonconformance by the registrar.

The internal lead auditor should track completion of corrective actions against the agreed schedule. The internal lead auditor should be actively engaged in pushing for on-time completion, and advising both the activity and the organization's management of insufficient attention to corrective action requirements, as well as schedule slippage.

Preparation for the Next Internal Audit

Like registrar audits, the next internal audit should be influenced by past performance. If an activity has demonstrated weakness in the QMS or its implementation, that activity should be targeted in the next audit. In discussing internal audits, ISO 9001, clause 8.2.2 says in part,

> An audit program shall be planned, taking into consideration the status and importance of the processes and areas to be audited, as well as the results of previous audits.

Planning for internal audits should always include information from registrar audits. For example, corrective action resulting from a registrar's audit should be internally audited for proper implementation and effectiveness before the next registrar audit. By doing so any lingering difficulties may be eliminated before the registrar's next visit.

CASE STUDY

ISO 9000 in Action

Jake Butler felt as if he had dodged a bullet, and, from a quality perspective, he had. The issue of selecting an ISO 9000 registrar for AMI should have—in Butler's opinion—been a simple enough task. Consequently, he had approached it as such.

In preparation for his meeting with AMI's executive management team, Butler had obtained a directory of registrars, identified those that could be available during the time frame that suited AMI, and made some calls to determine which registrars could provide team members with knowledge relating specifically to companies such as AMI. Having conducted this preliminary research, Butler had been able to present the *short list* of registrars for consideration. This had been where things began to unravel.

AMI's Vice-President for Marketing had gotten the discussion of registrars off to a bad start by asking Butler, "Which of these companies is the easiest when it comes to the certification standards?" Two other executives had picked up on where the Marketing VP seemed to be heading. Before he could apply the brakes, the question of the best registrar had been turned into a question of the *easiest* registrar, and the company's executives were caught up in this approach.

Rather than interject his opinion concerning the easy route, Butler had remained silent waiting instead for someone else to step forward. However, when it had become apparent that selecting the easiest registrar was gaining support as a viable option, Butler had finally joined in the discussion. His rebuttal had been straightforward and to the point. "What good will it do to select the easiest registrar? We should be asking which one is the toughest."

AMI's executives had been temporarily taken aback, and looked at Butler as if they weren't sure if he was serious. The Marketing VP had said, "What do you mean we should be looking for the toughest? The point here is to get registered as soon as possible, is it not?"

Butler had responded, "No sir. The point is to get *competitive* and registered. If we select a registrar that will let us pay the fee and get by in spite of shortcomings, we might be registered but we won't be competitive. We'll be like a kid who has a high school diploma, but can't read. The diploma looks good hanging on the wall, but it doesn't change the fact that the kid can't read. If our customers don't see a discernible difference as a result of ISO 9000, the registration will be meaningless."

Discussion had stopped dead, and the room had fallen silent until Arthur Polk, AMI's CEO, had said, "He's got a point, you know. Why don't we find ourselves the most credible, most reputable registrar on Butler's list?" There had been nods of approval from all participants, except the Marketing VP, John Summerhill, who apparently was still not satisfied. Polk, sensing his hesitation, had said, "Say what's on your mind, John." Summerhill had responded, "Look, we need to be registered fast—yesterday would be ideal. We are being clobbered by the competition simply because we can't wave the ISO 9000 flag. I'm against anything that is going to slow the process down."

Arthur Polk had nodded in agreement. To Butler he had said, "Here is what you do. Get on the phone and find us a registrar with a good reputation and plenty of credibility that understands the urgency of the situation. Tell them we will pull out all stops at our end if they can move the process as fast as possible." Butler had agreed to take care of finding such a registrar right away, and the meeting had broken up.

SUMMARY

1. There are eight sequential steps in the registration process: (1) decision by the organization to conform to ISO 9000 and seek registration; (2) internal preparation by the organization to achieve conformance; (3) internal determination that the organization has achieved conformance and that the QMS is functioning; (4) engagement of an accredited ISO 9000 registrar to certify the organization; (5) preliminary assessment and document review by the registrar; (6) formal QMS audit and certification assessment by the registrar; (7) elimination of nonconformance preventing registration; and (8) registration.
2. Selecting the registrar is one of the most important steps in the registration process. It is important to select a registrar who is familiar with the industry sector in question, charges appropriate and affordable fees, has credibility in the eyes of the industry in question and the ISO 9000 world, and has a proven track record.
3. The QMS audit is defined as the following: A systematic, independent and documented process for obtaining audit evidence and evaluating it objectively to determine the extent to which audit criteria are fulfilled.
4. Organizations seeking to use the ISO 9000 QMS model need independent observers to verify conformance with the QMS and its constituent elements. These independent observers are the auditors. There are internal and external auditors. Internal auditors are employees of the organization to be audited. Their independence is critical. Employees who are internal auditors should never audit their own department or functional area. Auditor qualifications are based on education, demonstrated competence, training, experience, and personal attributes.
5. There are three types of QMS audits: registration, surveillance, and internal audits. Registration audits are performed by the registrar hired by the organization seeking registration. If such an audit is successful, the organization is awarded ISO 9000 registration. Surveillance audits come after registration has been awarded. They are performed by the registrar to verify continued conformance. Internal

audits are conducted by employees of the organization in question to prepare it for a successful registration or surveillance audit.

6. Objectives of an organization when pursuing ISO 9000 registration are: (1) confirmation by a credible third party that an effective QMS has been implemented and can be expected to help the organization satisfy customer expectations; and (2) improvement (or confirmation) of the corporate image and, hence, its competitiveness by being able to legitimately claim to be a quality-oriented organization.
7. The audit scope establishes the boundaries of the audit. It specifically establishes the following: (1) physical locations to be visited; (2) organizational activities to be audited; (3) resources required for the audit (audit team, organization's audit support requirements, physical resources such as work space for the audit team); and (5) manner of reporting on the audit.
8. Audits performed by registrars involve the following activities: (1) pre-audit activities; (2) a preliminary assessment visit (optional); (3) preparation for the audit; (4) opening meeting; (5) facility tour, the registration audit itself; (6) end-of-day review; (7) exit meeting preparation; (8) exit meeting; and (9) awarding or withholding of registration.
9. Once a registration audit has been completed, there are necessary follow-up activities. These include audit reports, corrective action, and preparation for the next registrar audit.
10. Once an internal audit is completed, there are necessary follow-up activities. These also include audit reports, corrective actions, and preparation for the next internal audit.

KEY TERMS AND CONCEPTS

Approval
Audit process
Audit scope
Auditor certification
Auditor qualification
Auditors
Conditional approval
Corrective action
Disapproval
End-of-day review
Exit meeting
Exit meeting preparation
Facility tour
Internal audits
Internal preparation
Major nonconformance
Minor nonconformance
Objectives of QMS audits
Opening meeting
Pre-audit activity
Preliminary assessment and document review
Preliminary assessment visit
Preparation for the audit
QMS audit
QMS audit teams
Registrar audits
Registrar's certification assessment

Registration audits
Registration process
Selecting a registrar
Surveillance audits

REVIEW QUESTIONS

1. List the various steps in the registration process.
2. Make a case for selecting a "tough" registrar as opposed to an "easy" registrar.
3. Define the term *QMS audit.*
4. Compare and contrast internal and external auditors.
5. Briefly describe each of the following types of audits:
   - Registration audit
   - Surveillance audit
   - Internal audit
6. Describe the objectives of an audit from the perspective of the organization seeking registration.
7. Describe the objectives of an audit from the perspective of the third-party auditor.
8. What are the various factors established by the audit scope?
9. Describe each step in the audit process for an audit performed by the registrar. What follow-up activities go with such an audit?
10. Describe what takes place in an internal audit. What follow-up activities go with this type of audit?

APPLICATION ACTIVITIES

1. Secure a list of certified ISO 9000 registrars. Contact several registrars to determine the cost of a registration audit for an electronics manufacturing company with 700 employees. Also determine how long the audit process is likely to take provided there are no major discrepancies.
2. Identify a company that has been certified and awarded ISO 9000:2000 registration. What were the most difficult challenges? What advice would they give other companies seeking registration? What would they do differently?

ENDNOTES

1. ISO 9000:2000, Clause 3.9.1.
2. ISO 9000:2000, Clause 3.9.3.
3. ISO 10011-2.
4. ISO, *Communique: Results of the IAF—ISO/TC 176—ISO/CASCO Joint Session on Transition Planning for the Year 2000 ISO 9000 Standards,* 9/27/1999.
5. ANSI/ISO 14012-1996, *Guidelines for Environmental Auditing—Qualification Criteria for Environmental Auditors,* Introduction, p. v.

CHAPTER EIGHT

Continual Improvement

MAJOR TOPICS

- Improvement Versus Maintenance—A Definition
- Continual Improvement from the Perspective of the Customer
- Continual Improvement from the Perspective of the Organization
- Continual Improvement Requirements of ISO 9000
- Use of Statistical Techniques in Continual Improvement
- What World-Class Organizations Do

The concept of **quality improvement** refers to greater suitability, improved performance, increased longevity, enhanced reliability, and even better appearance of a product. At least as important, but often overlooked, is improvement to the process that produces the product. Process improvements are undertaken for the purpose of making the product better in some way, making it easier to produce, eliminating the possibility for errors, ensuring consistency, or reducing costs. The concept of process improvement applies equally to both products and services.

Improvement of quality on a continual basis is one of the foundational tenets of Total Quality Management. Suppliers of goods or services can no longer get by on just being good enough. Organizations that rest on their laurels will fall behind and lose customers as their competitors pass them by. Consumers once believed, as did manufacturers, that better quality meant higher prices. While this view prevailed, consumers grudgingly accepted that the lower priced goods they purchased would have some defects, and would wear out faster than the more expensive brands and models. The Japanese, through continual quality improvement, proved this concept wrong. Today one will find just as few flaws in a bottom of the line Toyota as in a top of the line Lexus. Of course, the Lexus will offer more features and luxuries than the Toyota, but this is what commands the higher price, not a difference in quality. The same kind of comparison can be made across the entire spectrum of manufactured goods. Low cost does not have to mean low quality (Figure 8-1).

Conversely, high quality does not have to mean higher costs for consumers or manufacturers or service providers. The fact is, by setting up the processes—people, tools, procedures—for improved quality, real costs can be reduced through the avoidance of waste and defective goods or services. Higher quality with lower cost; it sounds almost too

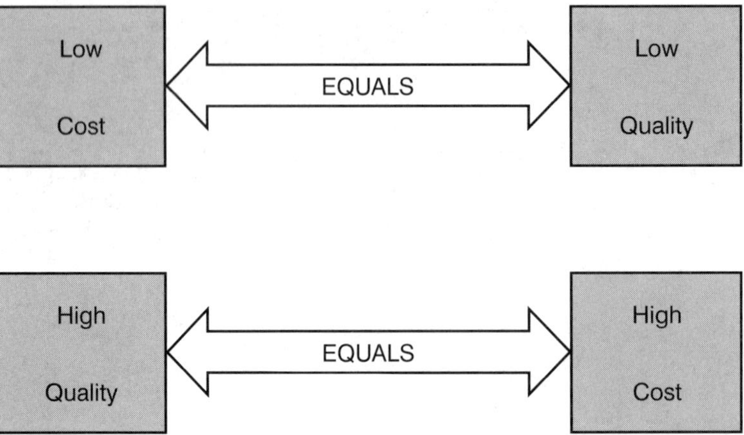

The Traditional Quality/Cost Paradigm

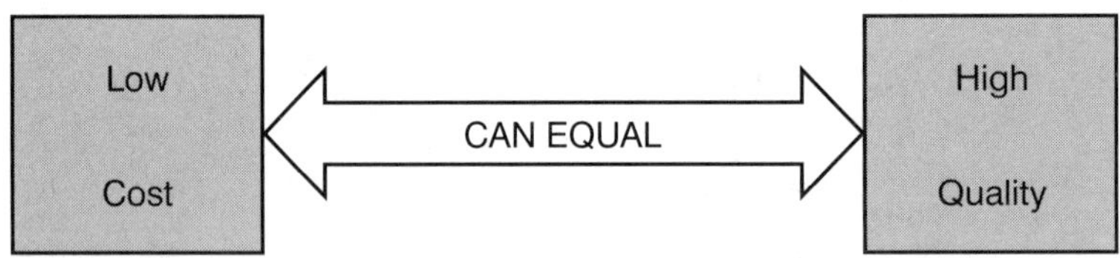

The New Quality/Cost Paradigm

Figure 8–1
Quality/Cost Paradigms

> **ISO INFO**
>
> *Chain reaction. Improve quality, what happens? Your costs go down. . . . That is one of the main lessons that the Japanese learned and that American management doesn't even know about and couldn't care less about. . . .*[1]
>
> W. Edwards Deming

good to be true. Firms that learn this lesson and exploit it effectively can take market share from those that do not, leading ultimately to the demise of the firms that fail to understand and adopt the new paradigm. This is why companies can no longer be satisfied with their current state of product or service quality. Good enough today will not be good enough tomorrow. Continual improvement is the key to longevity in today's market place.

IMPROVEMENT VERSUS MAINTENANCE—A DEFINITION

It is important to have a common understanding of the terms *improvement* and *continual improvement* before reading this chapter. *Improvement* is often misinterpreted and misused, as shown by the following scenario.

A metal shop uses a chemical plating process. Eighteen months ago a spill occurred. The process was stopped quickly to prevent more of the chemical from overflowing. A team was activated to find the cause of the spill. It determined that an automatic valve had jammed in the open position, hence failing to cut off flow to the plating tank when the proper plating bath level had been achieved. A new valve was installed and tested with satisfactory results. The process was restarted, and has experienced no spills to date. Has improvement taken place?

In another case a manufacturing process had historically produced electronic devices at an acceptance rate of 99.4 percent—six parts defective for every thousand produced. Without any warning, the process's acceptance rate fell abruptly to 73 percent, or 270 parts defective of a thousand produced. It was determined that a temperature sensor was out of calibration. When the sensor was replaced with a new, calibrated unit, the acceptance rate went from 73 percent to 99.4 percent, as illustrated in Figure 8-2. Is that an improvement?

The answer to both questions is, *No, no improvement occurred.* In both cases the processes in question were only returned to their typical performance levels. This is called **maintenance.** Both of these processes can expect to experience the same problems in the future, since nothing was done to eliminate the root causes of the problems.

Suppose the electronics plant learned of a new technology that could enable it to increase the process's acceptance rate to 99.7 percent. If the plant incorporated the new technology, and it produced the anticipated results, would that represent an improvement? Yes, this would be an improvement, even though a small one. See Figure 8-3.

In the electronic device example, the acceptance rate increased all the way from 73 percent to 99.4 percent. This was called maintenance, not improvement. Then, when the acceptance rate only changed from 99.4 percent to 99.7 percent, this was called improvement. Why? Three points need to be clearly understood.

1. When a process's performance deteriorates and is then restored to its historic performance level, no improvement has occurred in the process's capability. The process has merely been returned to its normal performance (maintenance). Maintenance, however, is very important and is an essential element of any management system.

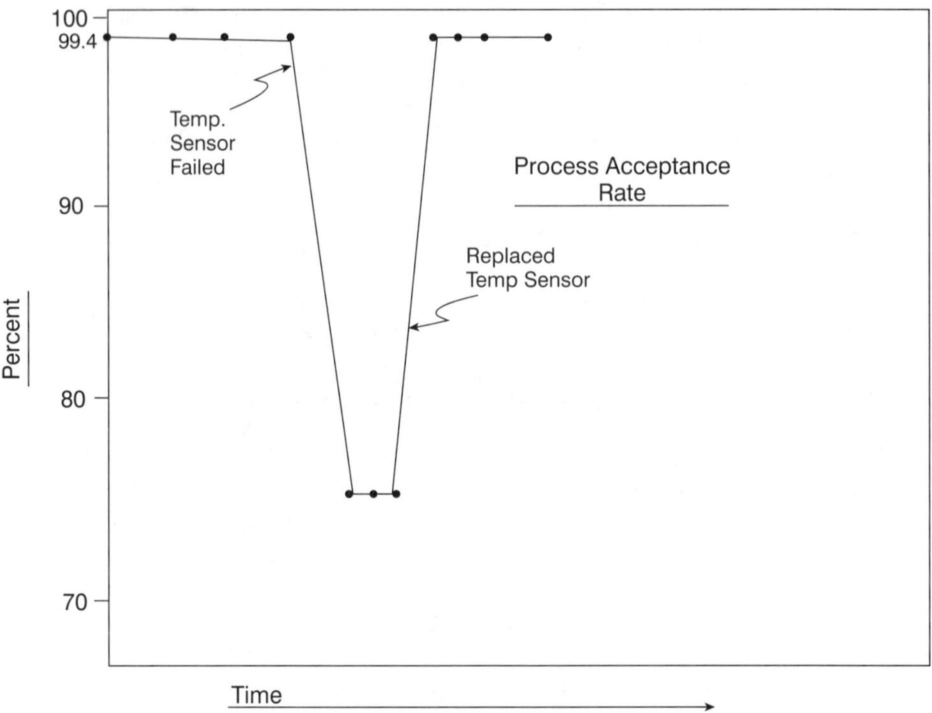

Figure 8–2
Restoring Historical Performance—Maintenance

2. Although major, breakthrough improvements are wanted, any change for the better from a process's historic capability represents improvement—no matter how small. Most improvements fall into the "small" category. When the management system is targeting continual improvement, small, incremental improvements result in significant improvement over time. That is the power of continual improvement. See Figure 8-4.
3. It is possible to achieve process improvement without changing the absolute performance of the process. If a process can be made more reliable or consistent, and therefore, less likely to fail, that is an improvement. Typically this kind of improve-

> **ISO INFO**
>
> *Putting out fires is not improvement. Finding a point out of control, finding the special cause and removing it, is only putting the process back where it was in the first place. It is not improvement of the process.*
>
> Attributed to Dr. Joseph Juran by W. Edwards Deming[2]

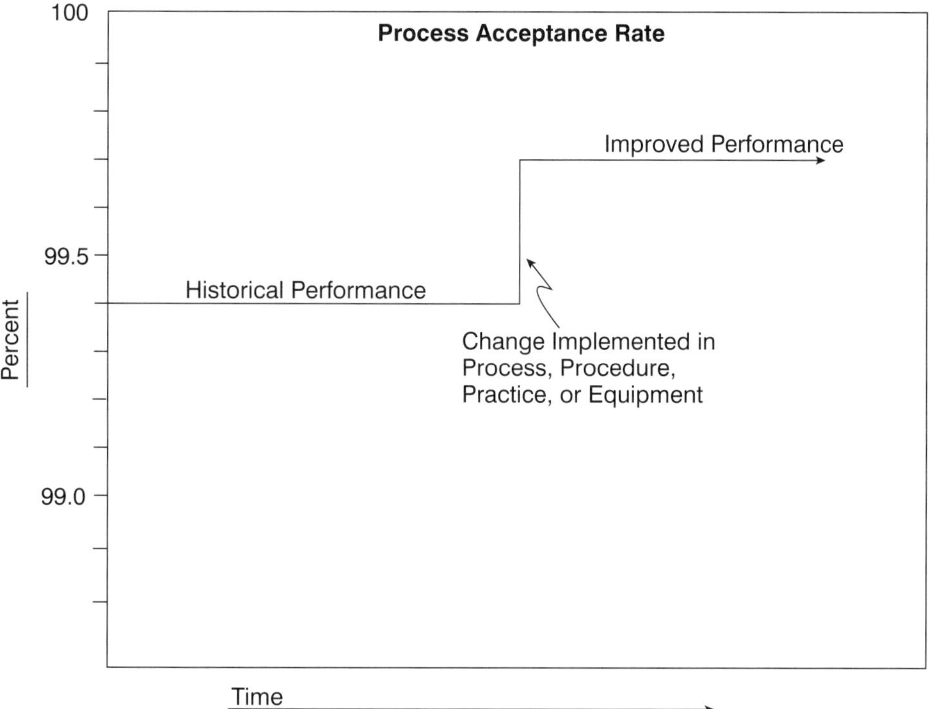

Figure 8–3
Achieving Better Performance—Improvement

ment is the result of locating and eliminating root causes, not just symptoms, of process variability or failure modes. Also, if a process can be made more difficult to be operated incorrectly—the Japanese call it *poka-yoke*[3], foolproofing the process—that is an improvement. See Figure 8-5.

Improvement, therefore, takes the process to a new, higher level of performance, or renders the process more reliable, more consistent, or less likely to permit operator induced errors. **Continual improvement** is simply the relentless pursuit of process or product/service improvements on a continuing basis, never becoming satisfied with the current state. No matter how well a process performs or how reliable and consistent it is, it falls short of perfection, the real objective.

CONTINUAL IMPROVEMENT FROM THE PERSPECTIVE OF THE CUSTOMER

In the mid-1950s Volkswagen advertised that every car it produced was subjected to more than a thousand quality control checks before leaving the plant. Those were the days when the Volkswagen Beetle could be purchased new for around $1,700. If Volkswagen

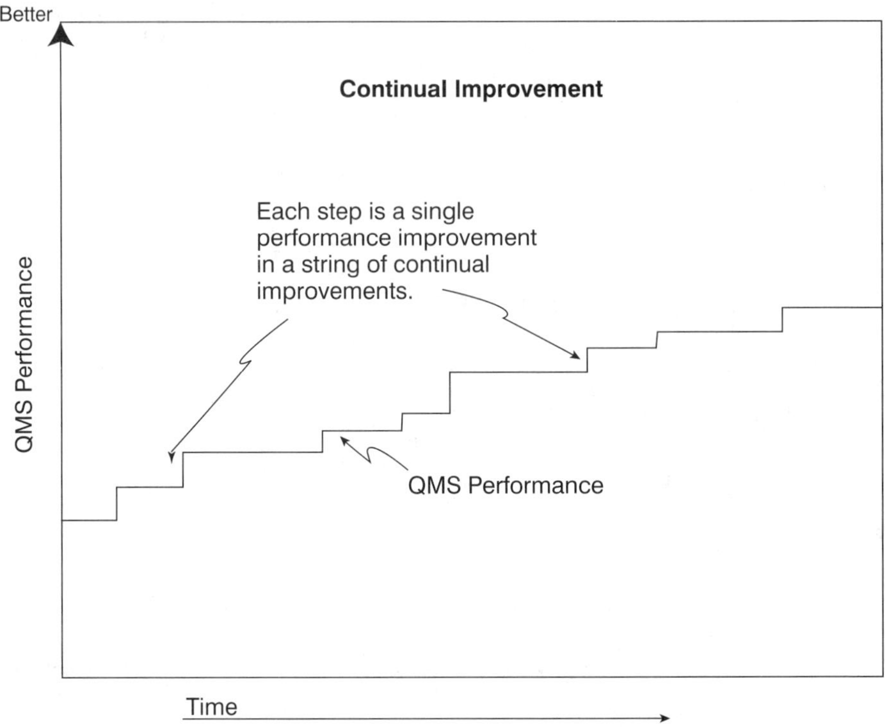
Better
Continual Improvement
Each step is a single performance improvement in a string of continual improvements.
QMS Performance
QMS Performance
Time

Figure 8–4
Performance Improvement Through Continual Improvement

made a thousand inspections on its little car, imagine how many General Motors must have made on its Cadillac, which was thousands of dollars more expensive. However, one can only imagine that, since no automobile maker but Volkswagen considered quality important enough to be part of an advertisement. Volkswagen had a reputation in those days for being a stickler for quality, and its cars reflected this attitude. One had to poll a lot of Volkswagen owners to find any dissatisfied with their Beetle.

VW's New Beetle is a completely different car, which has also become popular, but what was there about the original Beetle that inspired such delight in owners? Actually there were several things. First, the car's fit and finish were impeccable. It was steadfastly reliable, always doing what it was supposed to do (provide reliable, inexpensive transportation with reasonable comfort, considering its diminutive size). It was the very embodiment of the current ISO 9000 phrase, *suitability for purpose*. Was it perfect? Far from it. In terms of comfort, the Beetle with four adults in it was fully loaded and was cramped for space. Its handling, especially the early versions with rear swing axles, could be tricky. While fuel consumption was quite good (in the high twenties), its engine usually required major work at about 75,000 miles. And although it would traverse snow-covered roads better than any other passenger car of its time, on cold days its heating system left much to be desired. Finally, there was the car's appearance. In a word, the old Volkswagen Beetle was ugly; and it has stayed ugly throughout its long production life. The original Beetles

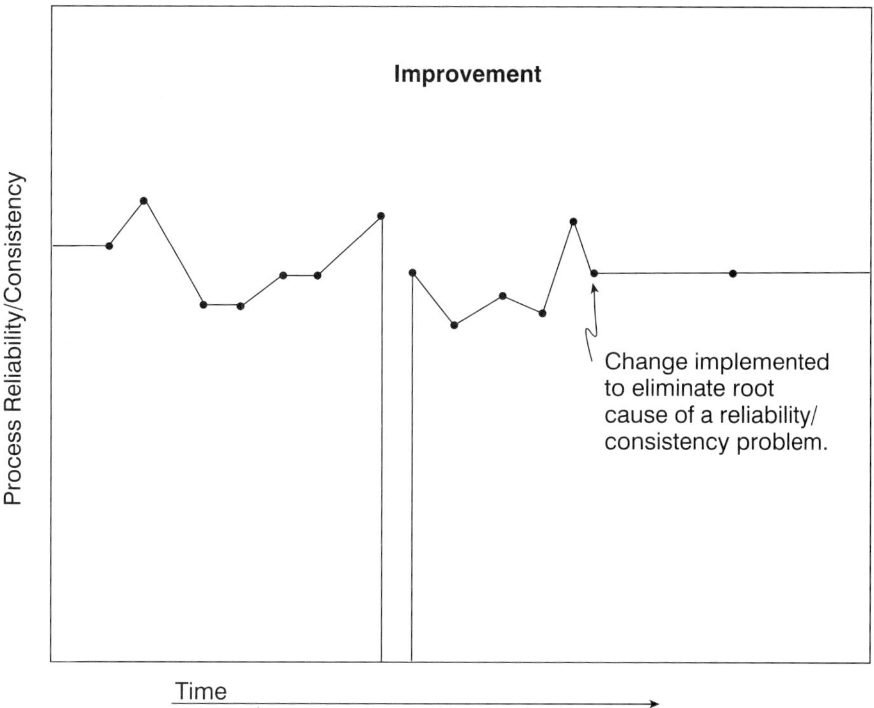

Figure 8–5
Improvement in Process Reliability and/or Consistency

are still manufactured in Mexico, although they have not been sold in the United States for many years. In spite of all these negatives, Volkswagen Beetle owners may have been the most delighted customers on the planet. The car did what they wanted it to do, and it was affordable. In fact, in an ironic twist, its ugly appearance became one of its better selling points, and is the reason the New Beetle is a success.

Over the period from the 1950s to the present, the Japanese found the secret of manufacturing high-quality low-cost automobiles, while Volkswagen drifted into that sphere of car companies, run mostly by accountants, that produced so-so products. As Japanese automobile manufacturers began to emerge as quality leaders, their European and American counterparts were slow to respond and, as a result, lost substantial market share.

There is a lesson here about customer satisfaction. The lesson is just this: Customer satisfaction is an on-the-spot, instant response that can change in a blink of the eye. Today's delighted customer can quickly turn into tomorrow's dissatisfied customer. This can happen even though nothing in the product or service has changed. It can be the result of increased expectations in customers. In Volkswagen's case, the Beetle did not get worse over the years. On the contrary, it got better in those areas that could be altered within the basic concept of the car. Volkswagen's problems began when the price of the Beetle was driven higher and higher by the method used to achieve the expected level of quality; a method known as **inspecting quality in.** Rather than improve their processes so that every component was produced right the first time, Volkswagen fell into a pattern of inspect-reject-rework that is too costly to justify in a competitive marketplace (Figure 8-6). Consequently, Volkswagen could not keep pace with the speed with which Japanese cars were being improved in all aspects. Rather than adopt the old inspect-reject-rework approach of automakers in the United States and Europe, the Japanese applied the concept of continual improvement to all of their operations, designing and building cars that

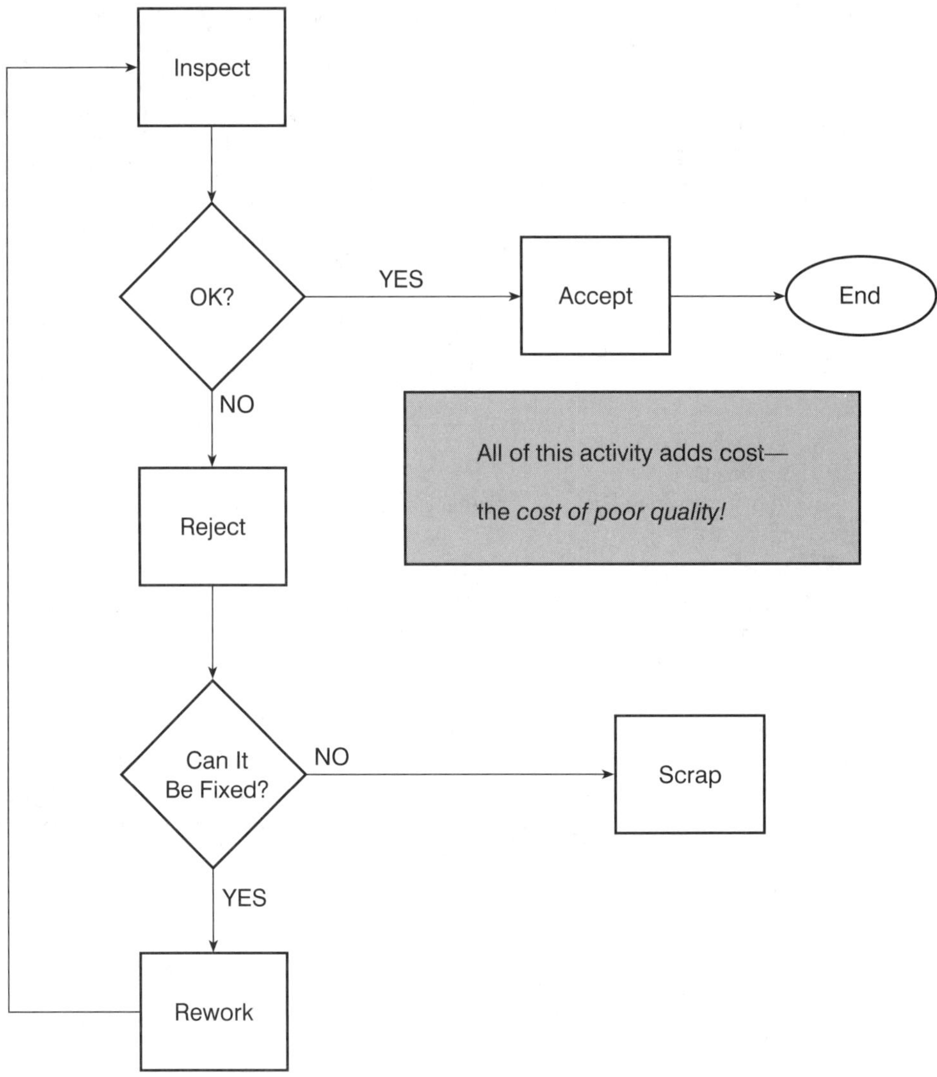

Figure 8–6
The Traditional "Inspect-the-Quality-in" Approach

could do more, and do it better and cheaper than those of their competitors. In the process, Japanese automakers raised the expectations of consumers, including the formerly satisfied Beetle owners, all around the world. The Japanese emerged as market leaders, and they did it using the concept of continual quality improvement.

Quality improvement in any product or service will result in more-satisfied customers and, indirectly, in heightened expectations throughout the marketplace. This can mean just one thing for organizations that do not respond to the market demands for ever better quality and value—loss of market share. Ultimately this will lead to one of two outcomes. Either an organization will adopt the continual quality improvement approach and survive, or it will fail. In today's marketplace failure is almost assured for any company that refuses to heed the customer's demand for ever improving quality.

CONTINUAL IMPROVEMENT FROM THE PERSPECTIVE OF THE ORGANIZATION

From the perspective of the manufacturer or the provider of services, continual improvement is not just a way to increase market share, but is also a method of reducing costs. As paradoxical as it might seem, real quality improvement results in lower costs to the producer. However, many people still cannot grasp this concept. Some of the best known organizations have not yet accepted the concept as being plausible. Nevertheless, a large and growing list of organizations around the world has proven conclusively that improving quality does, in fact, lower costs.

Any organization wants to offer goods or services at prices that will attract customers, but still produce a profit. They want their goods or services to be as useful and serviceable to potential customers as possible so that customers will prefer theirs to those of the competition. Continual improvement as now required by ISO 9000 and promoted by Total Quality Management offers organizations the chance to accomplish both competitive pricing and competitive quality.

CONTINUAL IMPROVEMENT REQUIREMENTS OF ISO 9000

Since the first version of ISO 9000, continuing until the year 2000 release, continual improvement has not been a requirement. The 1994 versions of ISO 9001, 9002, and 9003 made implicit references to quality improvement—without requiring that it be continual. See Figure 8-7.

Continual improvement is an especially significant element of the ISO 9000:2000 QMS, which is illustrated by the sheer number of references to it in the standard, as explained next.

ISO 9001

- Introduction, clause 0.2, discussing the process approach, "When used within a quality management system, such an approach emphasizes the importance of . . . d) *continual improvement* of processes based on objective measurement."

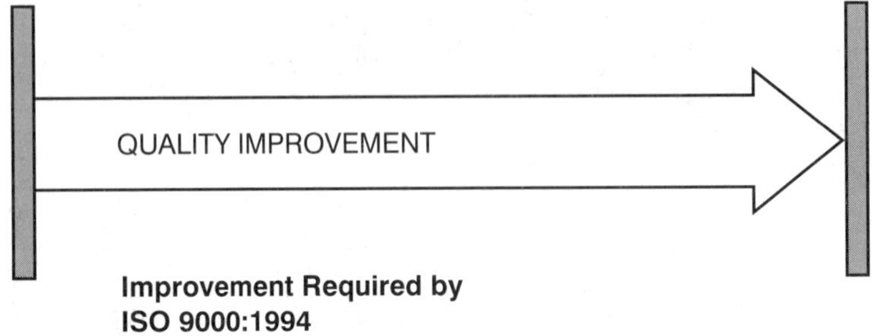

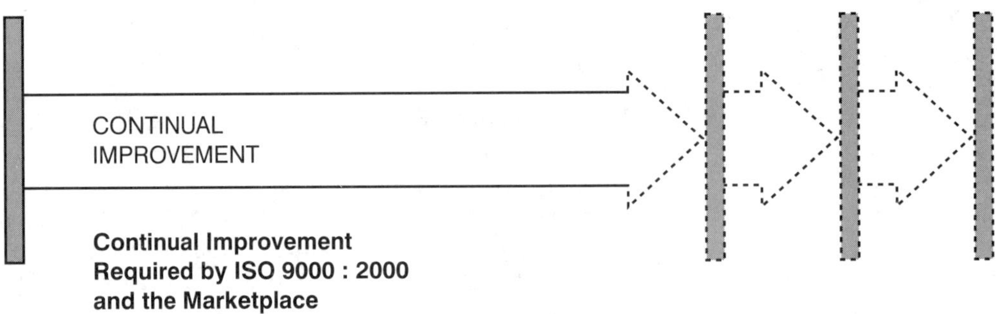

Figure 8–7
Comparison of Continual Improvement Requirements

- Introduction, clause 0.2, introduces the "Plan-Do-Check-Act (PDCA) Cycle to "*continually improve* process performance."
- Introduction, Figure 1 illustrates a major output of the process approach QMS is "*Continual improvement* of the quality management system."
- Introduction, clause 0.3, points out that ISO 9004's guidance covers a wider range of QMS objectives "than does ISO 9001, particularly for the *continual improvement* of an organization's overall performance and efficiency, as well as its effectiveness."
- Clause 1.1: "This international standard specifies requirements for a quality management system where an organization . . . b) aims to enhance customer satisfaction through the effective application of the system, including processes for *continual improvement* of the system . . ."
- Clause 4.1: Requires the organization to *continually improve* the effectiveness of its QMS.
- Clause 4.1 f): Requires the organization to "implement actions necessary to achieve planned results and *continual improvement* of these processes."
- Clause 5.1: Requires top management to provide evidence of its commitment to "*continually improving*" the effectiveness of the QMS.

- Clause 5.3: Requires top management to ensure that the quality policy includes a commitment to "**continually improve** the effectiveness of the quality management system."
- Clause 6.1: Requires the organization to determine and provide resources needed to *continually improve* QMS effectiveness.
- Clause 8.1: Requires the organization to "plan and implement the monitoring, measurement, analysis and improvement processes needed . . . c) to *continually improve* the effectiveness of the quality management system."
- Clause 8.4: Requires the organization to determine, collect and analyze data "to evaluate where *continual improvement* of the effectiveness of the quality management system can be made."
- Clause 8.5.1: Requires the organization to "*continually improve* the effectiveness of the quality management system through the use of the quality policy, quality objectives, audit results, analysis of data, corrective and preventive actions and management review."

In addition, there are many other references to *improvement* in ISO 9001, and they generally have reference to, or support, continual improvement.

ISO 9004

Through the main body of ISO 9004 the references to continual improvement follow those of ISO 9001. However, ISO 9004 adds two important sections:

- Clause 8.5.4: This clause offers guidance for the *continual improvement* of the organization itself.
- Annex B is a detailed discourse on the *continual improvement* process.

Given the emphasis placed on continual improvement in ISO 9000:2000, it is clear that the organization must incorporate the continual improvement philosophy as a key element of its QMS. Continual improvement in a QMS setting requires a relentless, never-ending cycle of improving the processes and products or services, as well as improving the policies, procedures, and practices of the QMS. When problems arise, it is not enough to fix the problem; the relevant employees must constantly seek ways to improve processes, procedures, and practices to achieve improved QMS performance and effectiveness, and to enhance process and product reliability.

Internal and registrar audits will have to consider the results of continual improvement. Records, therefore, must be kept as proof of improvement. Continual improvement results are measured in terms of improved organizational performance and effectiveness, product or service acceptance, and through other relevant data.

An organization accustomed to the demands of ISO 9000:1994 will have to be doubly certain to incorporate continual improvement into its ISO 9000:2000 quality management system. Continual improvement is now a cornerstone requirement of ISO 9000.

USE OF STATISTICAL TECHNIQUES IN CONTINUAL IMPROVEMENT

Total Quality Management has always promoted the use of the seven **statistical tools:**

- Pareto charts
- Cause and effect diagrams
- Check sheets
- Histograms
- Scatter diagrams
- Run charts and control charts (SPC)
- Stratification

Admittedly some are more "statistical" than others, but all qualify as tools to support continual improvement.

ISO does not require the use of statistical techniques, but in ISO 9001, clause 8.1, Measurement, Analysis and Improvement, General, the organization is required to determine "applicable methods, including *statistical techniques,* and the extent of their use." In other words, it is left to the organization to determine what methods, including statistical techniques, are appropriate, and whether or not to use them. The organization should seriously consider the employment of statistical techniques, not because of an ISO requirement, but for the potential benefit to the organization's effectiveness.

WHAT WORLD-CLASS ORGANIZATIONS DO

World-class organizations always strive to improve their processes, procedures, products, and people. One of the techniques through which organizations achieve this is continual improvement. As an integral part of their continual improvement process, they employ the PDCA Cycle and the statistical tools just listed. Everyone in the organization—from top management to hands-on employees—is involved in searching for opportunities for improvement and ways, however small or grand, to accomplish it.

World-class organizations believe that "good enough" is never good enough. They constantly seek to raise the performance and quality bars for their organizations and products or services.

CASE STUDY

ISO 9000 in Action

The debate had focused on the issue of continual improvement. All of AMI's executives had claimed to understand that with ISO 9000:2000 continual improvement is not a suggestion, but a requirement. The problem Jake Butler had run into was how seriously the executives planned to take the requirement. There seemed to be a "just-do-enough-to-get-by" mentality among AMI's executive managers. The problem, as Butler had explained it, was that doing "just enough" was the antithesis of the continual improvement philosophy.

Butler had shown AMI's executives that world-class organizations—something AMI's executives claim they want the company to be—adopt the philosophy that good enough is never good enough. He had driven this point home with all the determination he could muster. Butler had said, "If we aren't going to buy into continual improvement philosophy completely, we are wasting our time with ISO 9000. It would be like building a house without first laying the foundation." Several of AMI's executives had been taken aback by the strength of Butler's statement. In fact, Butler thought he had overstepped his bounds until Arthur Polk, AMI's CEO, smiled at him and gave a slight nod.

With this subtle encouragement from the top, Butler had continued making his case. Rather than focus on the fact that ISO 9000 registration requires continual improvement, he gave all the market and customer-oriented reasons for continual improvement. He closed by saying, "In the long run it won't be ISO 9000 registration that will make our company globally competitive. It will be our commitment to take ISO's requirements to heart and truly commit ourselves to continually improving everything we do forever." There had been more discussion, but Butler's viewpoint had eventually prevailed.

SUMMARY

1. Just solving a problem and returning operations to normal is maintenance, not improvement. Improvement occurs when performance indicators go up. Continual improvement occurs when performance indicators get better continually and incrementally over time. Continual improvement means never being satisfied.
2. Quality improvement as a concept refers to greater suitability, improved performance, increased longevity, enhanced reliability, and/or better appearance. It applies to products, services, and processes. Improvement of quality on a continual basis is a foundational tenet of Total Quality Management.
3. Traditionally, organizations and consumers have associated high quality with high cost. The new paradigm is that higher quality can be linked to lower cost.
4. Organizations that are still trying to "inspect quality in" will have difficulty surviving in the global marketplace. This traditional approach leads to waste and rework, both of which increase the cost of the product in question. By continually improving the processes through which products are produced and services rendered, organizations can get it right the first time and for less.
5. Continual improvement is a basic principle of ISO 9000:2000 and is spoken to in numerous clauses in ISO 9001. ISO 9004 provides guidance for the continual improvement of the organization itself. Annex B is a detailed explanation of the continual improvement process.
6. Tools that can help organizations continually improve are Pareto charts, cause and effect diagrams, check sheets, histograms, scatter diagrams, SPC, and stratification.
7. World-class organizations believe that "good enough" is never good enough.

KEY TERMS AND CONCEPTS

Continual improvement
Good enough
Inspecting quality in
Maintenance
Quality improvement
Statistical techniques
Statistical tools
World-class organizations

REVIEW QUESTIONS

1. What is meant by the term *quality improvement?*
2. What is the rationale for continual improvement from the perspective of organizations?
3. Contrast the concepts of "maintenance" and "continual improvement." Develop a hypothetical example that illustrates the difference.
4. Explain the rationale for continual improvement from the perspective of the customer.
5. What does ISO 9004 add to ISO's requirements for continual improvement?

APPLICATION ACTIVITIES

1. Identify a company in your community that is registered to the ISO 9000:1994 standard. Interview the quality manager for this company and get his or her views on what the emphasis on continual improvement in ISO 9000:2000 is going to mean to the company. What changes will the company have to make in order to maintain its ISO registration?
2. Identify a company in your community that uses statistical techniques. Visit the company and see how these techniques are used. Write a report on your findings.

ENDNOTES

1. Mary Walton, *The Deming Management Method* (New York: Putnam Publishing Group, 1986), p. 25.
2. Walton, *The Deming Management Method,* p. 67.
3. Kiyoshi Suzaki, *The New Manufacturing Challenge—Techniques for Continuous Improvement* (New York: The Free Press, A Division of Macmillan, Inc. 1987), p. 98.

CHAPTER NINE

Implementing ISO 9000: The Steps to Registration

MAJOR TOPICS

- Organizational Decision to Implement ISO 9000
- Common Registration Problems
- Minimizing Registration Costs
- Fifteen Steps to Registration
- Follow-up to Registration

ORGANIZATIONAL DECISION TO IMPLEMENT ISO 9000

The number of ISO 9000 registered firms in North America is approaching 50,000 as this is written. The number of registered firms world-wide is 350,000. One of the principal factors driving registration is customer pressure. As larger firms are registered they in turn pressure their suppliers to become certified. An example of this is the auto industry in the United States where Ford, DaimlerChrysler, and General Motors insist that their suppliers hold both ISO 9000 and QS 9000 certificates. (See Chapter 11 for information on QS 9000.) Even large governmental agencies such as the Department of Defense and NASA now recognize ISO 9000 in lieu of the many traditional government standards when dealing with suppliers. As these supplier organizations become certified, they in turn apply pressure to their suppliers to seek ISO 9000 registration, and the pressure cascades down through the entire supplier chain. Not all of the pressure is coming from customers. Increasingly organizations have begun to understand that a sound quality management system, such as that required by ISO 9000 or Total Quality Management, is a viable approach to becoming and remaining competitive. The marketplace is rapidly approaching the point when a recognized quality management system will become the "admission ticket" to doing business.

Having said this, the decision to become an ISO 9000 certified firm remains a difficult one for many organizations. The dilemma stems partly from a lack of understanding about ISO registration and the benefits it can bring to all organizations. But even with the potential benefits, the decision is still difficult because of the anticipated cost and the amount of work required of a large part of the work force. Most companies have to struggle daily to realize a profit—indeed many struggle to just keep their heads above water financially. Few employers have excess human resources available any more. Consequently, committing the money and the personnel to perform the extra work involved in preparing for certification is no small problem for many of the firms that would benefit most from ISO 9000. When the issue was about just having the ISO certificate hanging on the wall, the decision to seek certification was relatively easy to decline. But now that ISO 9000 is seen as a management tool with the potential for the continuous enhancement of competitiveness, ISO 9000 is difficult to ignore.

A survey of 1,679 firms from the United States and Canada registered to ISO 9000:1994 conducted by *Quality System Update* and the management consulting firm Deloitte and Touche, found that the investment necessary for ISO 9000 certification is typically repaid within three years. Few investments in today's market pay back in so short a time. But as good as the payback considerations are, there is more to be gained from ISO 9000 than just a quick return on investment. Respondents to the survey listed the following as major internal benefits of ISO 9000 registration (Figure 9-1):

- Better documentation
- Positive cultural change
- Greater quality awareness
- Higher perceived quality (listed as the most significant benefit)

All of these factors are beneficial to any firm. Is it possible to achieve these results without ISO 9000? Of course. An organization can implement Total Quality Management and achieve all of these benefits and more. Would this approach cost less and involve less work? TQM may cost less to get to the same point as ISO 9000 since a registrar is not

ISO INFO

Economic Necessity and ISO 9000

As more and more firms are certified by ISO 9000 auditors, the economic pressure on noncertified firms can only increase. What began as a way to ensure a specified level and consistency of quality, may—over time—become a necessary admission ticket to the game of global business.

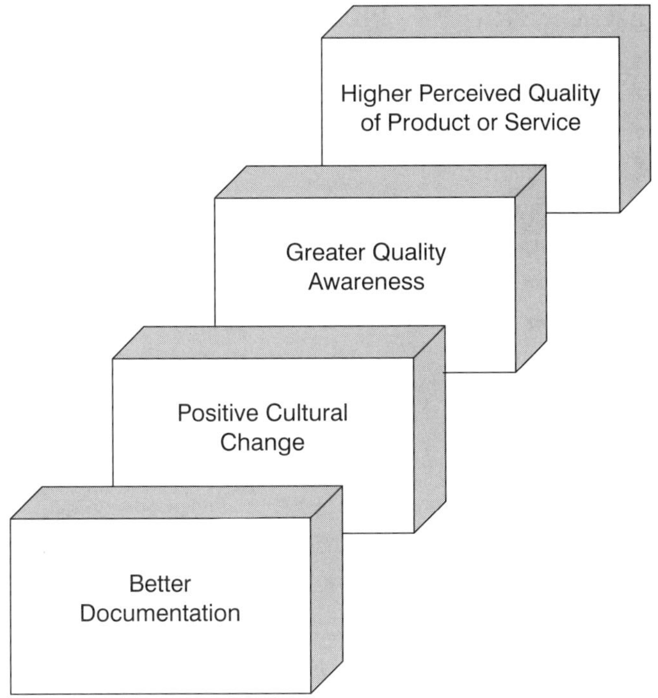

Figure 9–1
Internal Benefits of ISO 9000 Registration

needed, but the amount of work to get to the same place would be identical. TQM alone, on the other hand, does not provide a universally recognized accreditation such as the ISO 9000 certificate. For many firms such a certificate is a business necessity already. For others it may become so in the near future.

Consequently, organizations are better off making the ISO 9000 investment sooner rather than later. Our recommendation to management is to take a serious look at ISO 9000, and if it appears beneficial now, or appears that it will be, make the commitment to become certified. We have used here the key term relating to TQM, ISO 9000, process improvement, cultural change, and organizational development. That word is **commitment.** In our books on Total Quality we have stressed that the first and main thing required for success is the absolute and unwavering commitment of top management. It is no different with ISO 9000, because the same factors are involved; money, labor, training, and corporate culture. Only the highest levels of management can make the necessary commitments on these matters, and only the person at the very top can enlist the full support and cooperation of all employees to work toward a common corporate objective (e.g., ISO registration). If a genuine commitment from the top is not made, it is pointless to proceed with ISO 9000, TQM, or any other quality initiative.

COMMON REGISTRATION PROBLEMS

Approximately 15 percent of firms seeking ISO 9000 registration fail to pass the certification audit. According to Robert Bakker of Entela, Inc., 80 percent of those failures result from just the following ten elements[1]:

1. Failure to provide adequate control over documents and the data in them.
   - Do not allow handwritten and initialed changes. (Unless the formal update is assured and fast).
   - Update cycle is too long—should be few days, not weeks or months.
2. Calibration of tools and gages.
   - If you use gages and tools that require calibration (such as torque wrenches, micrometers, etc.) in your processes, you must be able to show that they provide true and consistent readings.
   - All gages and tools must be labeled and tracked for calibration.
   - You must maintain calibration histories for all applicable tools and gages.
   - You must conduct gage reliability and repeatability (R&R) studies, and keep records of those studies and of resulting recalibrations.
   - Maintain your record system—it is your only proof.
3. Poor receiving practices for incoming materials.
   - You must use accepted practices for receiving and inspecting materials. Such practices include the following:

     Inspection by your inspectors

     Approved suppliers

     Vendor Statistical Process Control data

     Accredited testing labs
4. Failure to adhere to stated process control procedures.
   - If the procedure is valid, it *must* be followed.
   - If the procedure is not valid (impossible or impractical to follow), then the procedure *must* be changed.
5. Supplier control is deficient.
   - Must maintain an up-to-date list of approved suppliers.
   - Must track their performance.
   - Must keep records of supplier performance on key items such as the following: quality, process capability, on-time delivery, shipping costs, and response to quality deficiency reports.
6. Inadequate control over nonconforming product.
   - Nonconforming products at any level must be identified and segregated from conforming products.

   - System must prevent the possibility of nonconforming products going to customers.
7. Training/credentials.
   - Staff credentials must match the job descriptions that are part of the quality system.
   - New employees must receive the training necessary for their assignments.
   - Effectiveness of the training provided must be evaluated through performance.
8. Out-of-date documentation.
   - When design changes are made to processes or products, the corresponding documentation must reflect the changes.
9. Periodic management reviews of the quality system elements are either not held, or records of the reviews are inadequate.
   - Periodic management review of the quality system is a requirement.
   - Minutes of the review meetings are your proof that the reviews were held. They need to reflect what was discussed and by whom.
10. Contract reviews are not conducted, or changes to contracts are not reflected in the paperwork.
    - Review of all contractual requirements vs. capabilities is mandatory. Records must be kept.
    - When customers change contract requirements by any means, you must update your contract paperwork accordingly.

Of all of an organization's functional areas, Yehuda Dror, General Manager of DNV Certification, Inc., finds the following the most likely to be deficient[2]:

1. *Purchasing* (new clause 7.4). Frequent problem areas include the following:
   - Evaluation of subcontractors (vendors).
   - Lack of feedback concerning nonconforming material.
2. *Document control* (new clause 4.2.3). Frequent problem areas include the following:
   - Controlling only upper level documentation, (e.g., the quality manual, policies).
   - Failing to have lower level documents such as procedures and work instructions under control.
   - Having a document control process that is so cumbersome that it is unworkable, and as a result, not followed.
3. *Management responsibility* (new clause 5). Common problems include the following:
   - Failure to define responsibility and authority for personnel.
   - Assigning responsibility without granting the necessary authority.
   - Failure to carry out management reviews of the quality system to ensure the effectiveness of the system.

ISO INFO

Failure to Follow Documented Procedures

One of the most commonly reported problems by ISO 9000 auditors is failure to follow documented procedures. Such a failure is a sign of haphazard management. When documented procedures are not followed, one of two things is happening, both bad. Procedures that have been proven to represent best practices—and, therefore, have been documented for the purpose of standardization—are being ignored. When best practices are ignored, both quality and productivity suffer. The other possibility is that some employees have discovered better procedures than those in the documentation and are using them to improve quality and productivity. If this is the case, the new procedures should be documented, standardized, and used by all employees.

Young Kim, Senior Associate with Coopers & Lybrand lists the following four major roadblocks to ISO 9000 certification[3]:

1. *Misinterpretation of ISO 9000's requirements.* This is certainly a potential problem, and must be guarded against or both time and money will be wasted. Registrars, seminars, consultants, trade journals, and books (including this one) are heavily involved in interpretation of the standard's requirements. Thus an organization can find help in this regard.
2. *Over-development of the quality system.* The word that best describes this problem is "overkill." Some organizations go into overkill mode when setting up their ISO 9000 quality management systems. Such organizations apply so much unneeded control that the system becomes a burden rather than a benefit. Too much control can result in a loss of flexibility, and may even inhibit improvement initiatives by making change too cumbersome. Overkill can defeat the purpose of ISO 9000, and must, therefore, be avoided. Remember, you are designing *your* quality system. It has to work for you. Even if your plan clearly suffers from overkill, the auditor will hold you to it. Make it unreasonably tough on yourself, and you will have problems. There will always be those who insist that procedures be checked, rechecked, and checked again, and who disapprove of any change without dozens of signatures, especially their own. Don't involve these types in the design of your quality system, or it is likely to contain unnecessary controls and encumbrances. Our suggestion is, take the minimalist approach: *Control only what is necessary, and only to the degree necessary.* You can always increase controls later if necessary.
3. *Documentation development and control getting out of hand.* Clearly, ISO 9000 requires that the quality system be adequately documented. However, as Kim so succinctly puts it, ". . . the effectiveness of a quality system should not be represented

by the quantity of the documentation." There are some things you can do to guard against over-development of documentation. First, be absolutely certain that documentation applies to a process or activity that affects the quality of your product or service. We have said that this can include almost anything, but, on the other hand, it does not have to include *everything.* Remember, if you insert some peripheral procedure into your quality management system documentation, it will be subject to the same auditing attention as the more critical procedures. Perhaps a bigger problem is the level of control detail put into the procedures and work instructions. Be careful not to go overboard controlling every process or activity to an inappropriate degree if doing so will not make a difference in the level of quality. Again, it is a good idea to start with the minimalist approach.

4. *Comprehensive estimation of what you need to do, and allocation of resources to do it.* Early in the ISO 9000 process, you need to assess exactly where your organization stands in reference to the standard's requirements. This will indicate what must be done in preparation for certification. This can be done only by first understanding what ISO 9000 requires, and comparing the requirements with the system you currently use. The differences will be the "gaps" that have to be closed. For example, ISO 9000 requires that relevant processes be documented if the organization cannot otherwise "ensure the effective planning, operation and control of its processes . . ." (clause 4.2.1d). If you currently operate without process documentation or work instructions, they may have to be developed. Unfortunately, the situation is seldom straightforward. Most organizations already have procedures in some form and at some level of adequacy. If this is the case, you must determine whether they can be upgraded for ISO 9000, if they should be discarded and developed over from scratch, or if they may be used as currently written. Before making a decision, you must completely understand the processes in question. This will require you to obtain information from the people who operate the processes, and who provide inputs to it or take outputs from it. It is very important to identify the gaps before work begins on developing your documentation. Getting ahead of yourself here can lead to wasted time and money.

Once you identify the gaps, the resources necessary to close them must be allocated. There are a number of ways to approach this, including the use of teams under the direction of a steering committee, or a general allocation of responsibility to the functional department that "owns" the gap. We discuss various approaches later in this chapter in the section entitled Fifteen Steps to Registration. For the moment suffice it to say that: (a) the process owners must be included in the documentation process, and (b) resist using consultants to develop your procedures.

What should result from the gap analysis and the allocation of resources for closing the gaps is a plan. This plan must make sense in terms of workload responsibilities and schedule. If it does not, you can find yourself reacting to problems without ever getting ahead of them. Note: ISO 9004:2000, ANNEX A is a GUIDELINE FOR SELF-ASSESSMENT. Our Appendix E provides an ISO 9000 checklist. Both will be helpful.

MINIMIZING REGISTRATION COSTS

When considering the cost of certification, look at the whole. Your internal costs for developing the quality system, its documentation, and training must be combined with external costs (consultants, the registrar, etc.) to determine the true cost of certification. In fact, it does not stop there. Every time the registrar conducts a follow-up or a surveillance audit, there will be additional costs. We recommend that you combine all costs to get an accurate picture of the true cost of ISO 9000 certification and operation. Most organizations will want to minimize these costs while realizing the maximum benefit from operating as an ISO 9000 firm.

Possibilities to consider for cost minimization include the following (Figure 9-2):

- Consultants—do we need one or more?
- Develop the quality system in-house or out-of-house?
- Develop the documentation in-house or out?
- Which registrar to go with?
- What is our timetable?

Consultants

Many firms achieve ISO 9000 certification without the help of outside consultants. More could do so if they tried. We believe that the best consultant is someone who knows your products or services and is familiar with your processes, culture, organization, and the people within it. Such a person almost has to be an employee. Our philosophy is that the

Characteristics of Total Quality	
✓	Are consultants needed?
✓	Should the quality system be developed in-house or out?
✓	Should the documentation be developed in-house or out?
✓	Which registrar should be used?
✓	What is the necessary timetable?
Comments: ______	

Figure 9–2
ISO 9000 Cost Minimization Checklist

right employee trained in ISO 9000 will be far more effective than the ISO 9000 expert who does not know your organization. If you pay the consultant long enough to learn what he/she needs to know about your organization, certification costs will skyrocket. Alternately, where do you find this employee who is also trained in ISO 9000? You select one and send her or him to one of the RAB certified ISO 9000 Lead Assessor courses. The course is a week long, and currently costs about $1500 plus any transportation and subsistence expense. With this training, the employee will be able to perform as your in-house consultant in preparing for certification. Realize, however, that time spent in this capacity still represents a cost of certification, but a cost that is well below the equivalent cost of an external consultant.

Developing the Quality System In-House or Out-of-House

You can hire a consultant to develop your quality management system, but the results may not meet your needs because no outsider can fully understand your organization, products or services, vision, culture, and processes. If a consultant stays on-site long enough to learn about these things, it will cost you more than the alternative of doing the job in-house. Actually, the in-house versus out decision should be based on just one criterion: What is the best way? It is always better to have the people who use the quality management system develop it. That means the organization's top managers. Nobody knows the organization, its products/services, and its culture better. Nobody knows better where the organization is trying to go or what it wants to become. This is critical information in the development of a viable quality management system. In-house employees may not have an adequate understanding of ISO 9000 requirements, but these requirements can be learned. We recommend that the quality management system be developed in-house regardless of the cost factors. In the long run this approach will cost less anyway—and it will be right for the organization.

Development of the Documentation In-House or Out?

The answer to this question is similar to that presented in the previous section, except the documentation should involve the process owners. At this level it will be necessary to understand the processes, the gaps vis-à-vis ISO 9000 and what must be done to close those gaps, and any current documentation. These are excellent tasks for teams of six to eight people including those directly involved with the processes. Again, regardless of cost considerations, the organization should develop its documentation in-house. Any other way is likely to produce unsatisfactory results. The in-house approach will be less expensive in the long run.

Which Registrar Should We Use?

There are currently 65 accredited registrars in North America—54 in the United States. One of the considerations for selection of the organization's registrar should be cost. After the field is narrowed to those familiar with the relevant industry, who routinely work with

ISO ISSUE

What Would You Do?

The Steering Committee is getting nervous about the preparation costs for ISO 9000 certification. Mike Randall, Quality Manager for Automation International, knows he is in a bind. He is convinced that the company needs a consultant to assist with making preparations for the big certification audit. But consultants can be expensive, and the Steering Committee is already concerned about costs. Randall knows he will have to find ways to cut costs. What cost cutting recommendations would you make in this situation?

the organization's size of business, and who enjoy a good reputation, bids should be solicited from at least three. Information should be obtained about the price not only of the certification audit, but also of a follow-up audit (in case it is required), of a pre-assessment visit, and for the surveillance audits.

The Certification Timetable

How long should it take to secure certification? What is the organization's preferred deadline? There is a general rule that says the longer something takes, the more it will cost. This rule may not apply to ISO 9000 certification—at least not if the time frame is within reason. Our recommendation is that you do as much as possible in-house without the assistance of consultants. When you take this approach, remember that your employees still have their normal duties to perform, and cannot, as a result, spend full time on ISO 9000 preparation. This tends to stretch the timetable out. If you can accept the longer schedule, say 18 months rather than the one year you might desire, proceed with the in-house approach. If not, you will have to resort to outside help and the added costs that come with it. Also, remember that every hour your employees spend on ISO 9000 work is a cost of certification. However, it is a cost that can be recouped through better performance as employees come to understand their processes more fully.

FIFTEEN STEPS TO REGISTRATION

There are certain steps that must be taken in preparation for ISO 9000 certification, and for several of the steps there is an order that should be observed. For example, the leaders who see the need for certification would be foolish to start the preparation before securing the backing of the organization's top managers. ISO 9000 certification is not an undertaking a group of enthusiasts can do without the knowledge and support of top management since there will be significant costs involved and a major investment of employees' time. Similarly, even with the support of top management, it would not be prudent to start changing procedures and publishing a quality manual before determining the organization's current posture vis-à-vis the requirements of ISO 9001. The following steps, taken together as presented, should be considered a model that may be adapted to the organization's particular circumstances and needs. Refer to Figure 9-3.

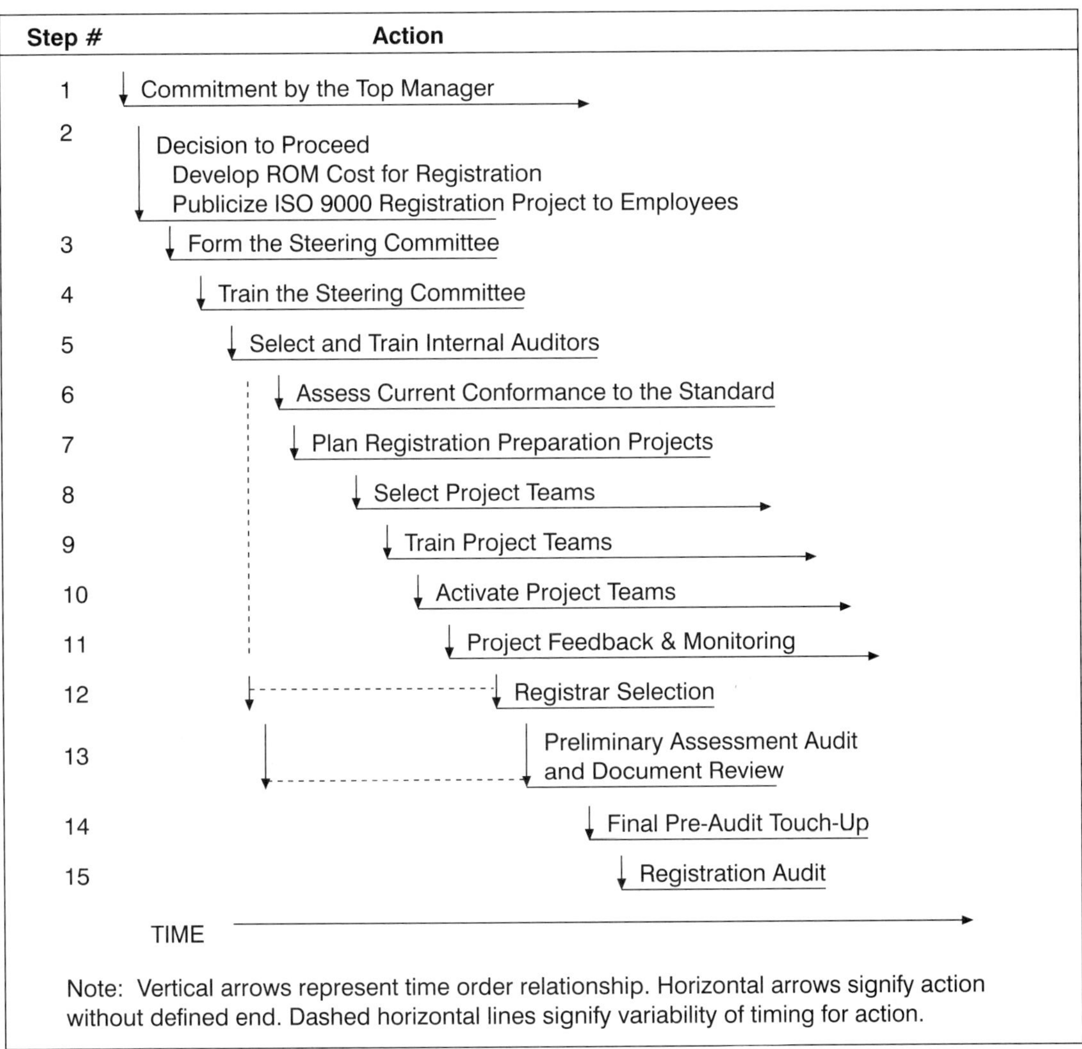

Figure 9–3
Fifteen Steps to ISO 9000 Registration

1. Commitment at the Top

Commitment to achieving registration from the highest level of management in the organization is an essential prerequisite to starting the ISO 9000 journey. There are several reasons for this, including the following:

- The need for *resources* that only top management can authorize. These resources include money and employee time, both in significant amounts.
- Inevitably there will be individuals who *prefer the status quo*. When they hold senior positions in the organization, it takes intervention by top management to overcome their resistance.
- The most important aspect of *leadership* is setting a positive example for employees to follow. In order for the ISO 9000 quality system to gain acceptance in an organization, top management has to demonstrate its commitment by being actively and visibly involved in the preparation process and it must be supportive of the quality philosophy, including the eight quality management principles, continual improvement, and the process approach to the QMS. Only then will employees commit themselves.

No matter how enthusiastic middle or even senior managers are, they cannot overcome a lack of commitment from the top.

2. Decision to Proceed

Having secured the necessary commitment from the top, the decision to proceed follows naturally. Even so, it helps to view it as a separate step in which certain important tasks are carried out. These include the following:

- Developing a rough order of magnitude (ROM) estimate of the effort in terms of costs for external services and time for internal tasks. At this point the estimate cannot be very precise because there are many unknowns (primarily the scope of the internal work required), but a range can be developed within which the actual costs will likely fall. By estimating in this way, top management can gain a sufficient feel for the total cost to make an informed decision to proceed or not.
- Commensurate with the decision to proceed, top management should begin publicizing the ISO 9000 effort to all levels of employees. When communicating with employees, managers should let them know that their input and support will be required, and should attempt to enhance their quality and ISO 9000 awareness. Employees need to know what is going on.

3. Steering Committee

Having confirmed commitment from the top, having decided that the potential investment for pursuing registration is not a roadblock, and having begun acquainting employees with the project, the next step is to choose an approach for managing and distributing the work of preparing for registration. There are a number of approaches to managing the registration project. One frequently used, and seemingly promoted by ISO 9000, is to designate a management representative who leads and coordinates the establishment, implementation, and maintenance of the quality system. Quoting from clause 5.5.2:

> Top management shall appoint a member of management who, irrespective of other responsibilities, shall have responsibility and authority that includes
>
> a) ensuring that processes needed for the quality management system are established, implemented and maintained,
>
> b) reporting to top management on the performance of the quality management system and any need for improvement, and
>
> c) ensuring the promotion of awareness of customer requirements throughout the organization.
>
> NOTE The responsibility of a management representative can include liaison with external parties on matters relating to the quality management system.

Except for the Note, this is, in our opinion, a flawed concept. We do not recommend delegating authority and responsibility to any individual for the establishment (meaning development) or implementation of the quality system. We much prefer the approach of using a **steering committee** and **project teams.** It is our belief, acquired from long and sometimes painful experience with Total Quality Management implementations, that the senior manager must retain the responsibility for developing and implementing any quality system. This is fundamentally a function of leadership, and cannot be delegated.

However, CEOs are not in a position to do it themselves. Consequently, the best course is to establish the senior management team as a steering committee chaired by the CEO. The function of the steering committee is to provide the leadership in establishing and implementing the quality system and in monitoring its performance. This includes all of the work necessary for registration. This does not mean, however, that new procedures must be written by the steering committee itself. Rather, the steering committee's role is to determine what is needed, secure the resources to satisfy those needs, and manage the activities of those given assignments. In the case of procedure writing, the steering committee would assign employee teams to accomplish the tasks and then monitor their progress.

Will ISO 9000 allow this approach? Yes, it will. Although there may be a registrar out there who would try to hold you to a literal interpretation of the cited clause, most would not. In the worst case, the CEO could be designated as the "management representative." This would satisfy even the most literal interpretation of clause 5.5.2 by the registrar.

4. Steering Committee Training

We assume at this point that some members of the senior management staff (the steering committee) are uninformed concerning ISO 9000. Therefore, it will be necessary to provide training that covers at least the following topics: familiarity with the documentation of the ISO 9000 series, the quality philosophy including the eight quality management principles, the ISO 9000 requirements, and the work likely to be necessary to satisfy the ISO 9000 requirements. The steering committee members must also understand the rationale for undertaking the registration project, that the work does not end with registration and that ISO 9000 will be a normal part of doing business forever. It is also advisable to provide the steering committee with **teamwork training** since members will have to work effectively as a team. Further, under the quality philosophy, all work tends to be a function of teams rather than of individuals. If teamwork is new to your organization, include it in the steering committee's training.

With the steering committee approach, all senior managers automatically assume roles in the ISO 9000 registration process. This may be the most powerful argument in favor of the steering committee approach. Any other approach will involve designating as registration project leader (management representative) an employee without the authority to coordinate and direct activities throughout the organization. When these activities require the time of employees, as they will, the project leader will have a difficult time with department heads. The steering committee approach eliminates this problem because department heads and their employees understand that the issue in question is important to top management, and, as a result, are more likely to get the job done.

5. Internal Auditor Selection and Training

It is too early at this point to put internal auditors to work checking the performance of the quality system. But it is not too early to begin their training. The sooner they are trained, the sooner their knowledge of ISO 9000 will begin to pay off. This is why we suggest that they be selected and trained in this step. The number of internal auditors needed will depend on the size and complexity of your organization. The required minimum number should be interpreted as being no less than two, since an auditor should always be independent of the activity being audited (clause 8.2.2). With just one auditor, there will always be one function in the organization that cannot be audited due to the independence rule. The actual number of internal auditors will have to be determined by the organization. In most cases, the auditing function is not a full time job. Rather, it consists of specific tasks that must be performed periodically. It is a good idea to have enough internal auditors to allow audits to be carried out according to the required schedule, unencumbered by workload fluctuations. The restraining factor is the cost of training. To ensure effectiveness, the internal auditors should complete an RAB accredited Lead Auditor course. The course will cost approximately $1500 per person. If travel is required in order to attend, travel and per diem costs must also be factored in.

By designating the internal auditors at this point, and getting them trained, the organization can gain the benefit of their ISO 9000 knowledge throughout the registration preparation process. Having internal auditors during the preparation process is like having your own in-house consultants—only less expensive.

Note: You can see from Figure 9-3 that we placed the selection of the registrar at Step 12, which is the latest possible point for this task. It could occur as early as between Steps 5 and 6 since by this point the steering committee should be capable of making an informed selection. If you intend to use outside help in your preparation work, an early selection of the registrar could be advantageous. For example, you could have the registrar perform a preliminary assessment audit and document review to determine where preparation efforts should be focused. This would take the place of Step 6. Some organizations, on the other hand, will want to become better prepared before they host a visit by a registrar. Such organizations will select their registrar at Step 12 and proceed through the implementation steps as presented herein.

6. Assessment of Current Conformance

Whether this step is performed by an outside party or by an in-house team, it must be completed before serious preparation work is started. Failure to observe this recommendation can result in wasted effort and cost. An effective assessment of your current level of conformance can be accomplished using in-house personnel. Certainly your internal auditors, if they have been trained, have the necessary expertise for this task.

The task involves taking an objective look at all elements of the organization, with an emphasis on your present quality system (if any), and whatever processes you employ that relate in any way to the quality of your products or services. The goal is to determine what needs to be done to satisfy the requirements of ISO 9001. To assist you with this task, we have included an ISO 9000 checklist as Appendix E. This instrument is easy to use. Its questions are linked directly to the requirements of ISO 9001. ISO has also included a Guideline for Self-Assessment as Annex A to ISO 9004:2000.

By completing an assessment that compares your current situation with the requirements of the standard, you will learn where to focus your preparation efforts. The results of this initial assessment should not be assumed because ultimately you will have to conform to the satisfaction of the registrar. There is no point in proceeding without a thorough understanding of your current strengths and weaknesses relative to the standard.

If you do not have the in-house expertise to carry out a valid assessment, employ a consultant or have the registrar do it for you. In either case, the assessment must be done before proceeding.

7. Plan Preparation Projects

Using the results of the initial assessment, develop a list of tasks that must be performed to bring the organization into conformance. We suggest that the list be converted into a timeline chart since you will need some tasks completed before starting others. For example, if one task on the list is to develop a quality policy, this task will have to be completed before other parts of the quality manual can be developed. Likewise, process flow diagrams may be required before process instructions or work instructions can be documented. Another reason for the timeline chart is that the steering committee should have a schedule in mind for the registration project. The registration date may be dictated by customer demand (e.g., terms of a contract), or it may be a target the steering committee believes is reasonable. Either way, the timeline chart of tasks should be set up to accommodate the completion date for the registration project. The timeline developed by the steering committee represents the plan for what must be accomplished and the points in time when the specific tasks must be completed. Never start a project of this magnitude without having a plan. To do so will virtually guarantee failure.

8. Project Team Selection

Once the plan has been developed, the next step is for the steering committee to determine the composition of the teams that will be assigned specific tasks. There is more to this step than meets the eye. For example, if a task is to develop an understanding of a particular process and then document the process, there will be some "givens." First, always put the process owner(s) on the team. They are the ones who live with the process and its idiosyncrasies on a daily basis; therefore, they are better able than anyone else to document how it really works, and what they have to do to make it do the intended job. Second, be sure to include on the team the supplier and customer of the process. They will also have invaluable insights concerning the process. Third, ensure that the team is cross-functional. It will need someone with unbiased views—an outsider—to ask questions or make suggestions that those more familiar with the process might overlook or might not ask. Other functions that have direct input into the process should also be included.

Beyond these types of members, the steering committee is well advised to select team members who have the kind of attitude that will support the activity (i.e., positive toward both team activities and change). The work will be difficult enough without negative, contrary team members.

9. Team Training

Employees of many organizations beginning the ISO 9000 journey will be untrained in some essential subjects. This does not apply to organizations already operating according to the Total Quality philosophy, but for other organizations it will be necessary to provide new teams with training in the following areas before activating projects.

ISO 9000 and Quality

(This item also applies to Total Quality organizations.) Team members need to understand what ISO 9000 is all about and that it is ultimately tied to quality. They should also learn why top management has concluded that ISO 9000 certification is vital for the

ISO INFO

Selecting Project Team Members

The attitudes of employees selected as members of the project teams are an important consideration. Imagine a football team consisting of players who not only don't care about winning, but actually want their team to lose. This is the type of situation that can result when project team members are selected without careful thought concerning their attitude. When selecting employees to serve on preparation projects, choose the most flexible, positive employees available—those who are open to new ideas and change.

organization. The need for all team members to have this understanding is just this: without it, the team cannot achieve the focus necessary to carry out the tasks to be completed. If the internal auditors have been trained, they will be able to conduct this training, along with a senior manager who can speak to the rationale for seeking ISO 9000 certification. This session can be covered in approximately two hours.

Teamwork Training

In most organizations it will be necessary to provide training on the subject of how individuals can work effectively together as a team. Remember, this may be a new experience for many employees, even those who think they are good team players. There are a number of books on this subject, including Peter Sholtes' *The Team Handbook,*[4] and our own *Introduction to Quality Management for Production, Processing, and Services*[5] and *Total Quality Handbook.*[6] You may find it necessary to bring in an outside consultant to conduct this training if no one on your staff is qualified to do so. As an alternative, one employee might be sent to one of the many seminars available to develop the necessary expertise. Having an in-house trainer is advantageous since the training will have to be given over and over as new teams are formed. Using an outside consultant can be expensive. This session can be completed in approximately four hours.

The Quality Tools

Before a new team can be given the go-ahead on a project, it must be equipped with the tools necessary to accomplish the task. In the case of ISO 9000 projects, the necessary tools are the same as those used in a Total Quality Management environment. TQM uses nine fundamental problem solving/analysis tools. They are the following:

- *Pareto charts:* To separate the important from the less important.
- *Fishbone charts:* To identify and isolate root causes of problems.
- *Stratification:* To group data by common elements to facilitate interpretation.
- *Check sheets:* To facilitate collection and interpretation of data.
- *Histograms:* To depict frequency of occurrence within a process.
- *Scatter diagrams:* To determine the correlation between two variables.
- *Run charts and control charts:* To record the output of a process over time, and to separate *special* and *common* causes of variation, respectively.
- *Flow charts:* To describe inputs, steps, functions, and outflows so that a process can be understood and analyzed.
- *Surveys:* To obtain relevant information from sources that might not otherwise be heard from.

All of these tools will be useful to members of your ISO 9000 project teams, but a core set of tools should be taught to all of your new teams. That core set should include flow diagramming, Pareto charts, fishbone charts, surveys, and stratification. For teams tasked with determining the effectiveness of processes, the additional tools of run charts and control charts, histograms, check sheets, and scatter diagrams should be added. Assuming that the teams will have a facilitator available who is competent in tool application (probably the same person who does the tools training), the core set can be taught in four hours. Then the facilitator can assist team members one-on-one.

Timing of the Training

Training is not retained unless it is applied. The sooner it is applied, the better the retention. This is the rationale for our recommendation that training be given on a just-in-time basis. This means that teams are trained just as they are activated. One can make a case for training larger groups all at once. Initial direct training costs are certainly lower with this approach, but the true cost of training must also take into account the long-term effectiveness of the training. If all employees are trained in one large session, but the new information is lost because not all employees can immediately apply what they have learned, the effectiveness of the training will be lost. Ineffective training, no matter how low the cost, can be more costly in the long run than effective training, even though the initial costs are more.

10. Project Team Activation

The steering committee has identified a preparation project and identified employees to serve on the team to which this project will be assigned. The team has received the training it needs to accomplish the task. The temptation at this point is to tell the team to get started. However, our experience with implementing Total Quality Management indicates that team activation should be a formal, structured, interactive process in which the team is given a written charter that documents exactly what is expected of it. In the absence of such a charter, the team members can be unclear about how far they are expected to go, and how much authority they have. The team activation meeting is used by the steering committee to do several things:

- First, the team should be given an overview of the project, an explanation of what led to the development of the project, and a description of any problems or issues perceived by the steering committee to be relevant to the project. The team should also be informed as to why its members were selected and what the steering committee expects of them.
- Second, the schedule of meetings between the team and the steering committee should be set. Through this vehicle, team members will know how often they should report, to whom, and in what format. This schedule will ensure a formal closed-loop system for two-way feedback and interaction between the team and the steering committee.
- Third, the steering committee should give the team a proposed schedule for completion of individual tasks and the overall project. In Step 7, the steering committee planned the overall registration preparation project and placed the individual tasks on a timeline. This information must now be imparted to the team.

- Fourth, the steering committee must clearly assign the team responsibility for accomplishing the project in question, define the authority of the team, and ensure that team members understand both their responsibility and their authority. With these things understood, team members will know what they can do on their own and when they must ask the steering committee for approval or assistance. In the absence of a clear understanding of responsibility and authority, teams will flounder, especially when confronted with cross-functional considerations. In addition, the team should be asked to select a team leader. In some cases the team leader may be designated by the steering committee, but it is usually better to have the team make its own selection. The team leader's responsibilities are to conduct meetings, assign action items to team members, and communicate as necessary with the steering committee between formally scheduled meetings. The leader is also responsible for having a team member record minutes of meetings, including action item assignments. The leader should publish the agenda for all team meetings, distributing it to team members a day or two before meetings.
- Fifth, the steering committee should make a trained facilitator available to the team. The facilitator should attend all meetings. His or her job is to keep meetings focused and moving in the right direction. The facilitator also provides assistance in the use of the appropriate tools (discussed earlier in Step 9), and keeps discussions moving while at the same time preventing domination by individual team members. He or she also sums up, brings discussions to a close when they appear to have run their course, and helps maintain the integrity of the agenda in terms of both time and content. As experience is gained, facilitation functions may be assigned to a team member, but outside team facilitation can be very helpful in the early stages of teamwork.

In addition to being thoroughly discussed at the team activation meeting, all of the items explained in this step should be included in the written team charter.

11. Project Feedback and Monitoring

This phase of the registration preparation process begins with the activation of project teams in Step 10. For the duration of the team projects, the steering committee will receive feedback from all teams in accordance with the schedule and format agreed to during the activation meeting (Step 10). The steering committee uses this information to monitor progress and to provide new instructions for the team when that is deemed appropriate. Thus, a closed loop exists between the steering committee and the teams.

12. Registrar Selection

Typically, companies select their registrars six to eighteen months prior to the target date for the registration audit. Your organization should have a similar timeline. As we pointed out earlier, selection of the registrar could occur as early as between Steps 5 and 6. The choice is yours based on whether or not you want to use the registrar to assist with some of the intervening steps. This step is the latest point at which registrar selection

ISO INFO

Long-Term Value of Registration

Ten years from now will ISO 9000 registration have any meaning? The answer to that question depends on how all the players in the drama act out their parts. If accreditation bodies in the United States and Europe hold unfailingly to the strictest standards when accrediting auditing firms; if auditing firms maintain the proper arms-length ralationship with the firms they audit, while holding them accountable to both the letter and the spirit of the standards; and if individual firms actively seek out registrars of the highest integrity, then ISO 9000 certification will hold its value. But if there is slippage in even one of these critical areas, the ISO 9000 registration process may become a meaningless exercise.

should occur. Refer to the section Which Registrar Should We Use? in this chapter under the heading Minimizing Registration Costs.

Some things to do when selecting a registrar include the following:

- Check with trade organizations and the accrediting bodies for background information on potential registrars. (In the United States the accrediting body for ISO 9000 is ANSI/RAB.)
- Solicit references on registrars from other companies, preferably those in your own industry.
- Question potential registrars directly. The general tone of the response may be as telling as the answers themselves.
- Make sure that potential registrars can accommodate your schedule for registration.

Critical factors to consider in the registrar selection process are the following:

- Does the registrar have the proper accreditation? Note that registrars in the United States normally are accredited through RAB, but firms accredited by foreign agencies are certainly acceptable. There are a number of European registrars with European accreditation doing business in the United States and Canada.
- Does the registrar have sufficient experience in general, and specific experience in your industry? You may not want to be among a new registrar's first clients. Their learning curve could be expensive for you. However, the more important issue is this: Does the registrar speak your industry's language and understand your processes? Do not get into a situation where time is wasted educating your registrar.
- Does the registrar have the resources to satisfy your requirements? Your requirements mean not only the registration audit, but a preliminary assessment, the follow-up visits, and surveillance audits as well. The registrar should be able to accommodate your schedule without requiring major changes to it.

- Does the registrar use its own full-time auditors, or does it engage independent auditors to perform audits? Using independent auditors is acceptable provided they are accredited by RAB. In fact, with independent auditors it can be easier to put together an audit team that is knowledgeable in your industry. However, there is one problem, and we consider it significant. If the registrar uses independent auditors for its audits, it is likely that you will lose continuity as the members of the audit team change from audit to audit. This is an important consideration because continuity from audit to audit is valuable. An audit team that is familiar with your organization does not have to waste time getting up to speed every time a visit is called for. Rather, they know what to look for based on previous visits. In the long run you will benefit from the consistency of a stable team. In addition, costs should be lower since a stable team can get in and out more rapidly.
- When you estimate the cost of the registrar, be sure to compare all costs from potential auditors (i.e., follow-up and surveillance audits, preliminary assessment visits, and the registration audit itself). Be certain that the registrars know the size of your firm. The bigger the firm, the higher the registrar's costs, since more auditors will be required and for a longer period of time.
- Look for favorable "chemistry" between the registrar and your firm. You will in effect be married to the selected registrar for several years, so it is important to select one with whom you are comfortable.
- Reputation comes last in this list, but not as a reflection of its relative importance. Indeed this is the most important factor in registrar selection. Reputation is left until last so that it will be the one criterion you are sure to remember. The ultimate viability of your registration depends on the reputation of your registrar. *Never* select a registrar because its reputation is one of "being easy!" Organizations that make this mistake never get the full benefit of ISO 9000, because the QMS may not be optimized, and the laxity will eventually be found out in the marketplace. You need—and ISO 9000 needs—a registrar that puts ethics at the top of its agenda, followed closely by competence and the willingness to work with customers to ensure valid, fair, and equitable audits. We urge you to put as much emphasis on reputation and ethics as you put on cost. Selecting a good registrar is like selecting a good college. You may have to work harder and longer, but you will learn more; and that is the point.

13. Preliminary Assessment Audit and Document Review

Although not required, many firms contract for a preliminary assessment audit. They hope that the preliminary assessment audit and document review will help them prepare for the registration audit. The preliminary assessment audit is conducted to identify ISO 9000–related deficiencies that have to be corrected prior to the registration audit. If the preliminary assessment audit is conducted early enough, it can serve the same purpose as the assessment of current conformance in Step 6. It also can provide a high-visibility jump start to get the preparation process under way. Whether it is done at Step 6 or at Step 13, it is an important part of the process. It uncovers any remaining areas of nonconformance so that they can be corrected before the registration audit. By identifying areas of nonconformance and taking the appropriate corrective action, an organization can approach the registration audit with a high degree of confidence and can minimize, if not eliminate, follow-up issues and their cost. The preliminary assessment audit can be a good investment. If an organization feels competent to go through Step 6 on its own, it should schedule the preliminary assessment at Step 13. Otherwise, the organization should select a registrar earlier and have the preliminary assessment done at Step 6.

As stated earlier, the preliminary assessment audit is optional. Many organizations and their registrars skip it if they are confident of conformance. However, all registrars will conduct a review of the organization's ISO 9000–related documentation at Step 13. The document review is usually conducted at the registrar's facility, although site visits in connection with the review are not uncommon.

As a word of caution, the organization cannot count on the registrar providing a lot of help when a nonconformance is found, either through a preliminary assessment audit or through the documentation review. Remember that auditors must reveal any nonconformances, but cannot tell the organization how to correct them. No matter what nonconformances are found by the registrar, the organization has the responsibility for determining how to correct them.

14. Final Pre-Audit Touch-Up

Having gone through the previous thirteen steps, you have developed and implemented the quality management system, and you have the information produced by the preliminary assessment audit and/or documentation review. With time relatively short between this step and the formal certification audit, the organization should proceed with correcting the remaining deficiencies. Use the internal auditors to verify compliance as these remaining action items are completed. With the pre-audit touch-up completed, the certification audit should be uneventful. However, if one or more major discrepancies cannot be corrected in time, notify the registrar at once so that the certification audit can be rescheduled. It is pointless to go into an audit with a known major discrepancy because the registrar cannot grant certification until it is eliminated. It is better to postpone the audit than to pay for an extra audit.

15. Registration Audit

The registration audit is carried out according to a structured procedure that typically proceeds according to the steps presented in Chapter 7.

FOLLOW-UP TO REGISTRATION

Following registration the firm must respond to any CARs that are still outstanding and advise the registrar of actions taken in this regard. The registrar will schedule a follow-up visit to verify the effectiveness of corrective actions. If all actions taken are satisfactory, the registrar will close out the CARs. Now that the firm is registered to ISO 9000, it is up to all employees, but especially management, to keep it registered. This requires careful monitoring of the quality management system for effectiveness, for ways to improve it further, and for scheduled internal audits to verify continued conformance and the effectiveness of the QMS. The registrar's auditors will return in six months to a year to conduct their independent verification.

CASE STUDY

ISO 9000 in Action

On the one hand he was shocked. The preliminary assessment audit of AMI had produced a long list of problems that will have to be corrected before the registration audit. But on the other hand, Jake Butler was relieved. If the preliminary assessment had been the actual registration audit, AMI and Butler would have failed miserably. There were problems; lots of them. But they were problems that could be corrected.

It was well after quitting time, and AMI's offices were deserted except for the janitors as Butler sat at his desk reviewing the discrepancy list. The list contained the following items:

1. The cycle for updating documentation is too long. It will have to be reduced to no more than a few days.
2. Employees in several units do not adhere to established, documented procedures.
3. New employees receive insufficient training for their assignments.
4. Contract paperwork is not kept up to date as changes are made.
5. Insufficient communication exists between production units and the purchasing department concerning nonconforming material.
6. Lower-level documents such as procedures and work instructions are not properly controlled.

As he turned out his office light and prepared to go home, Butler placed the deficiency list in the middle of his desk so it would be the first thing he would see in the morning. There was much to do, and little time in which to do it. "But," thought Jake Butler as he walked to his car, "at least now I know where we stand. That's better than not knowing."

SUMMARY

1. The number of ISO 9000 registered firms in the United States is growing rapidly. One of the principal factors driving the increase is customer pressure.
2. The investment needed to prepare for ISO 9000 registration is typically repaid in three years. Internal benefits of registration include better documentation, positive cultural change, and greater quality awareness. The principal external benefit is higher perceived quality.

3. Failure to pass an ISO 9000 registration audit typically results from one or more of the following problems: inadequate control of documents, insufficient calibration of tools, poor receiving practices, failure to adhere to stated process controls, deficient supplier controls, inadequate control over nonconforming products, ineffective training, out-of-date documentation, insufficient management review of the quality system, and insufficient contract review.
4. Four major roadblocks to ISO 9000 certification are misinterpretation of ISO 9000's requirements, over-development of the quality system, over-development of documentation, and underestimation of work and resources needed for registration.
5. An organization can minimize the costs associated with ISO 9000 registration by doing the following: forgoing the hiring of consultants, developing the quality management system in-house, developing the documentation in-house, carefully selecting a registrar that knows the industry in question, and establishing a reasonable timetable.
6. There are 15 steps to ISO 9000 registration. They are as follows: gain commitment from the top, make a decision to proceed, establish the steering committee, train the steering committee, select and train internal auditors, assess current compliance, plan preparation projects, select project teams, train the teams, activate project teams, monitor projects, select a registrar, conduct the preliminary assessment audit and document review, make the final pre-audit adjustments, and undergo the registration audit.
7. After the registration audit, the firm must respond to outstanding CARs, notify the registrar of action taken, and provide objective evidence of conformance or schedule a follow-up visit with the registrar for verification of corrective action.

KEY TERMS AND CONCEPTS

Assessment of current compliance
Calibration of tools and gages
Commitment at the top
Decision to proceed
Final pre-audit touch-up
Internal auditor selection and training
Misinterpretation of ISO 9000's requirements
Out-of-date documentation
Over-development of the quality system
Poor receiving practices
Preliminary assessment audit and document review
Project feedback and monitoring
Project team activation
Project team selection
Registrar selection
Registration audit
Registration timetable
Steering committee
Steering committee training
Team training

REVIEW QUESTIONS

1. Why is the decision as to whether or not to pursue ISO 9000 registration sometimes difficult for an organization?
2. List three internal benefits of ISO 9000 registration.
3. What is the most significant external benefit of ISO 9000 registration?
4. What are the ten elements that typically cause firms to fail their registration audit?
5. Of all an organization's functional areas, which are most likely to be found deficient during the registration audit?
6. List and briefly explain the four major roadblocks to ISO 9000 registration.
7. Explain how the following factors can affect the cost of preparing for the registration audit:
   - Consultants
   - Developing the quality system in-house or out
   - Selecting a registrar
8. Explain why commitment from the top is the first and most important step in the ISO 9000 registration model.
9. Defend or refute the following statement: Responsibility for implementing the quality system should be delegated to the quality director.
10. Explain why the minimum number of internal auditors is two.
11. What are the key elements of team training for members of project preparation teams?
12. Briefly describe what should take place when activating a project team.
13. List several factors to consider when selecting a registrar.
14. The registration audit answers several questions. What are these questions?
15. During the registration audit, is it wise to volunteer information beyond that asked for by the auditor? Why?

APPLICATION ACTIVITIES

1. Poor receiving practices for incoming materials is a frequently cited problem during registration audits. Secure a copy of the ISO 9001 standard and make a checklist of requirements relating to receiving practices. Contact a local company and receive permission to compare its receiving practices with your requirements checklist.
2. Identify a local company that will work with you on the following project. Using the company and its current state of affairs as a case, apply the "15 Steps to registration" model as a guide in developing a plan for achieving ISO 9000 registration. The plan should be comprehensive and detailed. It should describe everything the company will have to do in order to achieve registration. (If this project is too much to pursue alone, a better approach might be to form a team and divide up the various responsibilities and tasks.)

ENDNOTES

1. Joseph C. Quinlan, "Dodging the Potholes," *Quality in Manufacturing* (Jul.-Aug. 1996): 43.
2. Yehuda Dror, from a presentation to the Second Annual International Conference on ISO 9000, 1995.
3. Young Kim, "ISO 9000—Making Companies Competitive," *Quality in Manufacturing* (Nov.-Dec. 1994): 26.
4. Peter R. Sholtes, *The Team Handbook* (Madison, Wis.: Joiner Associates, Inc., 1992).
5. David L. Goetsch and Stanley B. Davis, *Quality Management: Introduction to Quality Management for Production, Processing, and Services,* 3d ed. (Upper Saddle River, N. J.: Prentice-Hall, 2000), particularly Chapter 10.
6. Goetsch and Davis, *Total Quality Handbook* (Upper Saddle River, N.J.: Prentice-Hall, 2001), particularly Chapter 7.

CHAPTER TEN

ISO 9000 As a Stepping Stone to Total Quality Management

MAJOR TOPICS

- Comparative Scope of ISO 9000 and Total Quality Management (TQM)
- Origins of ISO 9000 and TQM
- Aims of ISO 9000 and TQM
- Management Motivation for ISO 9000 and TQM
- Compatibility of ISO 9000 and TQM

COMPARATIVE SCOPE OF ISO 9000 AND TOTAL QUALITY MANAGEMENT (TQM)

The two principle quality initiatives at work in the world today are **ISO 9000** and **Total Quality Management.** Consequently, it is well to begin by explaining the relationship between the two. The following statements outline the relationship between ISO 9000 and Total Quality Management. Each statement is explained in the sections that follow in this chapter.

- ISO 9000 and Total Quality Management are not interchangeable.
- ISO 9000 is compatible with, and can be viewed as a subset of, TQM.
- ISO 9000 is frequently implemented in a non-TQM environment.
- ISO 9000 can improve operations in a traditional environment.
- ISO 9000 may be redundant in a mature TQM environment.
- ISO 9000 and Total Quality are not in competition.

ISO 9000 and Total Quality Are Not Completely Interchangeable

In spite of a commonly held view to the contrary, ISO 9000 and Total Quality are not the same. By definition, ISO 9000 is concerned only with quality management systems, for

ISO INFO

The primary aim of the "consistent pair" [ISO 9001:2000 and ISO 9004:2000] *is to relate modern quality management to the processes and activities of an organization, including the promotion of continual improvement and achievement of customer satisfaction. Furthermore it is intended that the ISO 9000 standards have global applicability. Therefore, the factors that are driving the* [ISO 9000:2000] *revision process, among others* [includes the] *provision of a natural stepping stone towards Total Quality Management.*[1]

ISO

the *design, development, purchasing production, installation,* and *servicing* of products and services.

On the other hand, Total Quality Management, by definition, encompasses every aspect of the business or organization, not just the systems used to design, produce, and deploy its products and services. This includes all support systems such as human resources, finance, and marketing. Total Quality Management involves every function and level of the organization, from top to bottom.

Total Quality Management also means that management is responsible for developing the organization's vision (its dream), establishing guiding principles (a code of conduct for the organization and all of its employees), and setting the strategy and tactics for achieving the vision within the constraints of the guiding principles. In a Total Quality Management organization the vision is pursued with input from an empowered work force that cooperates and collaborates with management.

Total Quality Management based on the teachings of Deming, Juran, Ishikawa, et al, with criteria defined by Deming's Fourteen Points, Juran's Ten Steps to Quality Improvement, and the Malcolm Baldrige National Quality Award, is more pervasive and demanding—literally requiring the transformation of the entire organization.

Before the advent of the year 2000 release, ISO 9000 was concerned only with the standards upon which an organization could build its own version of a quality management system. ISO 9000:2000 has closed much of the gap that existed between the two. The primary remaining difference between ISO 9000 and TQM is in the degree to which the total organization is involved. Where TQM requires the involvement of all functions and levels of the organization, ISO 9000 does not require the QMS to include functions and levels that do not play a direct role in the management and execution of the product/service realization processes. Functions that are typically not involved under the QMS include human resources, finance (accounting), sales, and marketing.

Figure 10-1 illustrates how close ISO 9000's evolution has brought it to TQM.

Total Quality is defined as an approach to doing business that attempts to maximize the competitiveness of an organization through the continual improvement of the quality of its products, services, people, and environments by emphasizing the characteristics listed in Figure 10-1.

Characteristics of Total Quality Management	ISO 9000:2000	TQM
Customer focus (internal and external)	√	√
Obsession with quality		√
Scientific approach to problem solving	√	√
Long-term commitment	partial	√
Teamwork		√
Continual process and product improvement	√	√
Education and training intensive	√	√
Freedom through control		√
Unity of purpose	√	√
Employee involvement and empowerment	partial	√

Figure 10–1
Total Quality Management Characteristics Compared with ISO 9000

In comparison, the ISO 9000 quality management system is designed to "provide the framework for continual improvement to increase the probability of enhancing customer satisfaction and the satisfaction of other interested parties. It provides confidence to the organization and its customers that it is able to provide products that consistently fulfill requirements."[2] ISO claims that beyond customer satisfaction, cost and risk-management benefits will also accrue to the organization. These benefits translate to improved competitiveness—the same as TQM's objective. ISO claims these benefits result from emphasizing the eight quality management principles upon which the standard is based. See Figure 10-2 for a comparison of ISO's eight quality management principles with Deming's Fourteen Points, and TQM.

ISO 9000 Is Compatible with, and Can Be a Subset of, Total Quality Management

Clearly, Total Quality Management and ISO 9000 are not quite the same thing. However, there is nothing inherent in ISO 9000 that would prevent it from becoming part of a larger Total Quality Management environment. There are many examples today of companies that have successfully included ISO 9000 as part of a larger TQM effort. Organizations that are already at some level of TQM maturity have typically found it easy to implement ISO 9000. This is because a Total Quality Management environment with its infrastructure of documented processes and procedures, continual improvement, obsession with quality, and so on easily supports the requirements of ISO 9000.

ISO 9000	Deming's 14 Points	TQM
1. Customer focus		√
2. Leadership	#1, #2, #7	√
3. Involvement of people		√
4. Process approach		√
5. System approach to management		√
6. Continual improvement	#5	√
7. Factual approach to decision making		√
8. Mutually beneficial supplier relationships	#4	√

Figure 10–2
ISO 9000's Quality Management Principles vs. Deming's Fourteen Points and TQM

ISO 9000 Is Frequently Implemented in a Non-Total Quality Management Environment

Although Total Quality Management is compatible with and may well facilitate an ISO 9000 implementation, it is by no means a prerequisite for implementing ISO 9000. In fact, it is safe to say that the majority of ISO 9000 registered organizations have not fully adopted TQM. At least, not yet.

ISO 9000 Can Improve Operations in a Traditional Environment

By **traditional environment,** we mean an organizational environment that has persisted in companies for decades, until the Total Quality Management movement began to change things. In this book, then, a traditional environment, or traditional organization, is one which still operates according to the old way of doing things, rather than according to the principles of Total Quality Management.

When ISO 9000 is implemented by a traditional organization, the company should be the better for it. We will not go so far as to say it *will* be the better for it, because much depends on the organization's reasons for adopting ISO 9000 and the degree of executive-level commitment to it. Said another way, if ISO 9000 is approached inappropriately and for the wrong reasons, it can become nothing more than a marketing ploy, and the organization's functional departments might develop even more problems than they had before ISO 9000.

ISO INFO

As a marketing advantage, ISO 9000 may be only temporary. Organizations achieve an advantage if they are registered before their competition. But the advantage lasts only until the competition adopts ISO 9000. Gaining a marketing advantage is the wrong motivation for adopting ISO 9000.

Goetsch and Davis

ISO 9000 May Be Redundant in a Mature TQM Environment

Just as ISO 9000 should help traditional organizations, it should also benefit TQM organizations. However, in an organization that has achieved a high level of maturity in its Total Quality Management journey, say in the 400–600 range on the Baldrige scale of 1000 points, all ISO 9000 criteria may already be in place. In such a case, the only compelling reason for registration under ISO 9000 would be for marketing purposes. What would a company such as Toyota gain from ISO 9000 registration? Probably nothing. They already do everything required by ISO 9000. Their products and processes are recognized as worldclass. Consequently, they wouldn't gain even a marketing advantage. However, there are many fine TQM organizations that are not as well known as Toyota. Such organizations, even though they may already meet or exceed the requirements of ISO 9000, may find it necessary to register in order to let potential customers know that their products or services satisfy the international standard.

ISO 9000 and Total Quality Are Not in Competition

This is not a case of one or the other. Organizations can adopt Total Quality Management, ISO 9000, or both. While there may be those who advocate one to the exclusion of the other, in the larger scheme of things the two concepts fit well with each other. Both have worthwhile and similar aims. Our view is that not only are TQM and ISO 9000 compatible, they actually support each other and are complementary. There are good reasons for using both in a single management system.

ORIGINS OF ISO 9000 AND TQM

ISO 9000 and Total Quality Management originated independently of each other, for different reasons, in different parts of the world, and at different times. The ISO 9000 series of standards was originally developed in response to the need to harmonize dozens of national and international standards then existing throughout the world. To that end the **International Organization for Standardization (ISO),** a worldwide federation of national standards organizations, formed Technical Committee 176.

Although sometimes considered to be a European standard (certainly the impetus came from Europe), ISO 9000 was developed by an international team that includes The American National Standards Institute (ANSI), the U.S. member of ISO. ANSI was represented by the American Society of Quality (ASQ), its affiliate responsible for quality management and related standards. The first version of ISO 9000 was released in 1987. By this time, the Total Quality Management movement was more than 35 years old. A revised version of ISO 9000 was released in 1994, and most recently in 2000. As a result of this standard, suppliers of products and services are able to develop and employ a quality management system that is recognized by all their customers, regardless of where on the planet those customers might be. Customers around the world who deal with ISO 9000 registered organizations can expect that their purchases will measure up to a set of standards they recognize.

Total Quality Management, on the other hand, got its start in Japan around 1950 when W. Edwards Deming and Joseph Juran introduced Japanese industrial leaders to the concept of quality. Deming and Juran told the leaders of Japanese industry that by adopting their concepts Japan, a nation still reeling from defeat in World War II, and one known at the time for producing shoddy products, could successfully compete in the international marketplace. The work of Juran and Deming, together with the writings of Walter Shewhart, also of the United States, laid the foundation in Japan. Then several Japanese leaders, including Kaoru Ishikawa, Taiichi Ohno, and Shigeo Shingo, developed and expanded the system which we now call Total Quality Management, as shown in Figure 10-3. The historical record shows how well Total Quality Management has served Japan. By the 1990s TQM was transforming organizations in the United States and around the world, enabling them to be competitive in the global marketplace.

AIMS OF ISO 9000 AND TQM

The aim of ISO 9000 has historically been to assure that the products or services provided by registered organizations are consistently fit for the intended purpose. ISO 9000:2000 has raised the standard's aim to a new level. Customer satisfaction and continual improvement, along with the other six quality management principles, seek to make registered organizations more competitive. This is essentially the same objective as Total Quality Management.

MANAGEMENT MOTIVATION FOR ISO 9000 AND TQM

Management motivation for adopting either ISO 9000 or Total Quality Management can vary widely. There are both appropriate and inappropriate motives. For example, if a company seeks ISO 9000 registration in order to obtain a marketing advantage, its motive is inappropriate. As a result, the organization will likely give mere lip service to adopting the standard. Appropriate motives for adopting ISO 9000 include the following:

- To improve operations by satisfying the ISO 9000 requirements for management responsibility, resource management, product realization, and measurement analysis, and improvement

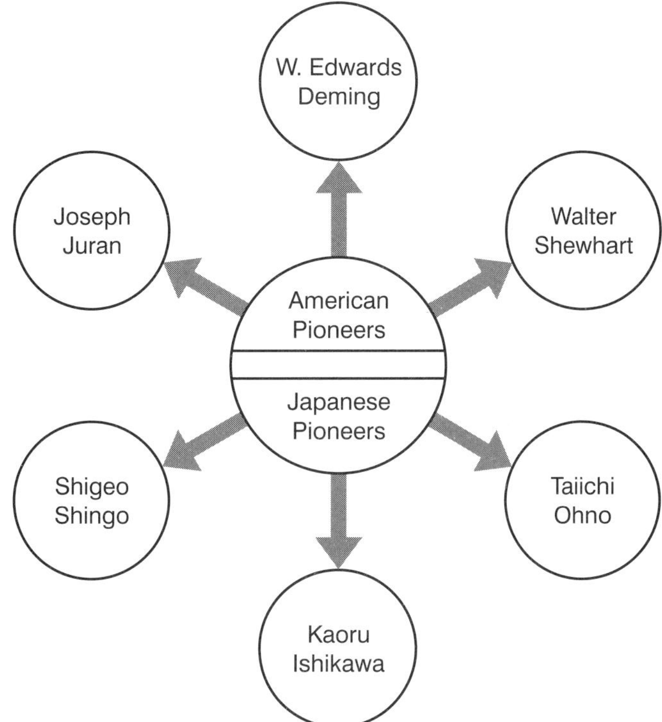

Figure 10–3
Pioneering Leaders Who Led Japan to Total Quality Management

- To create or improve a quality management system that will be recognized by customers worldwide
- To improve product or service quality, or the consistency of quality
- To improve customer satisfaction
- To improve competitive posture
- To conform to the requirements of one or more major customers (although adoption would be better motivated by internal considerations, such as the preceding five)

What we are saying here is that, ideally, management will adopt ISO 9000 as a way to make real improvements in the company's operations, serve its customers in a more responsible way, and, as a result, open new markets. This approach is more likely to assure commitment and participation by top management. Approaching ISO 9000 from a strictly marketing perspective may result in a negative reaction to the amount of work required by the functional departments, and only enough management commitment to do the bare minimum for registration. In other words, if ISO 9000 is viewed as a necessary evil one must adopt in order to compete in certain markets, every dollar and every hour spent on ISO 9000 will be seen as a burden to be endured, rather than an investment in the organization's future. By definition, a burden is a load that is difficult to bear. The connotation is negative. When negative feelings abound among employees, commitment to ISO 9000 will suffer. It may be possible to fool the ISO 9000 registrar's auditor, but we guarantee that customers will not be fooled—at least not for long. Newfound markets will soon wither and disappear. If ISO 9000 is to have a real and permanent effect, it must be approached with a positive attitude and the unwavering commitment of top management.

Management motivation for adopting Total Quality Management can be equally muddled. Too many organizations attempt to implement TQM for the wrong reasons. When this happens, the implementation is nearly always doomed to failure. The appropriate motivation for implementing TQM is that senior management has learned what it can do to improve every aspect of the organization, and they want these improvements to occur. However, we often find examples in which senior managers become enamored of TQM through a seminar or an article in a trade journal, and with this limited amount of knowledge attempt an implementation. This approach hardly ever works.

Interestingly, many organizations turn to Total Quality Management out of desperation. Their business is consistently losing in the marketplace, so TQM is turned to as a last resort for survival. As it turns out, these are often the easiest implementations, because the management team typically gets behind TQM with commitment and enthusiasm rather than see their jobs vanish into thin air. Desperation is a strong motivator. It would be accurate to say that Japan was in a similar situation in 1950, when it originally got the Total Quality movement started. However, organizations with good leadership turn to TQM before a crisis sets in.

COMPATIBILITY OF ISO 9000 AND TQM

We have discussed the fact that ISO 9000 and TQM are different in scope and were developed from different perspectives, but now have similar requirements and objectives. Now, more than ever, the two concepts are compatible. TQM requires everything required by ISO 9000. However, even a mature TQM organization, one that does everything it would do under ISO 9000, and more, will not have the worldwide recognition afforded by ISO 9000 registration. There is no corresponding international certification for TQM. For this reason, even the mature TQM organization may find it necessary to seek ISO 9000 registration as a way to satisfy the demands of its customers. On the other hand, a traditional organization that is registered under ISO 9000 may find that it needs the larger Total Quality Management implementation to become or stay competitive.

CASE STUDY

ISO 9000 In Action

Jake Butler is feeling on top of the world. AMI's executives no longer see ISO 9000 registration as a marketing gimmick or a "necessary evil" required of those companies who want to keep up in the industry. They have finally come to realize that the preparations for the registrar's audit have made AMI a better company. At the weekly executive's meeting this morning, the CEO—Arthur Polk—had announced that AMI was going to expand into the Japanese market, a strategy that would have been unthinkable just twelve short months ago. But first, the company would have to "go the rest of the way" to becoming a Total Quality organization.

Breaking into the Japanese market would be tough. Jake knew this. But, as Arthur Polk had said, "This market represents the big league in our industry, and with the progress we've made since beginning our ISO 9000 journey I am convinced we can play in the big league." Polk had then asked Butler to explain what would be necessary in order to move AMI beyond ISO 9000 registration to full implementation of the TQM philosophy. Polk had begun his presentation with these words: "Ladies and gentlemen, the best way to win in the Japanese marketplace is to beat them at their own game. That game is TQM and we need to play even better than they do."

Butler had then said, "The only real difference between where we are now and where we need to be lies in the following factors: (1) we have to become obsessed with quality; (2) we have to make sure that we are totally committed for the long term; (3) we have to start working in teams; and (4) we have to establish the principle of freedom through control.

Butler described the lengths to which some Japanese firms in AMI's industry go in order to continually improve the quality of their processes, products, and people. "We have made a good start with the continual improvement efforts undertaken in preparation for ISO 9000 registration. But I think you can see that we are not yet *obsessed* with quality," said Butler. "We are also committed now. I've seen our commitment grow from lukewarm to complete in just twelve months. Now we have to solidify our commitment and maintain it for the long term."

Our two biggest challenges will be to implement *teamwork* and *freedom through control* because both will require major cultural change for us and for the employees. We pay on the basis of individual performance, we evaluate on the basis of individual performance, and we promote on the basis of individual performance. If we want our employees to work in teams, we are going to have to make some fundamental structural changes to the way we reward and recognize employees. If we want them to have the freedom to think and innovate within specified parameters, we are going to have to clearly establish those parameters and effectively communicate them to employees. We will also have to train our supervisors to work in this new environment."

The meeting had gone well, but during his presentation Butler could tell that AMI's executives were aware that the company had come to a major fork in the road. There had been a great deal of quiet and deeply personal mental calculation going on during the meeting. Butler knew that some of the people sitting around the table would not be able to successfully make the transition to Total Quality Management, but he also knew that the ones who mattered most would. Butler planned to be right there with them throughout the transition and after. He had never felt better about the future.

SUMMARY

1. The following statements describe the relationship between ISO 9000 and TQM: ISO 9000 and TQM are not completely interchangeable; ISO 9000 is compatible with, and can be a subset of, TQM; ISO 9000 is frequently implemented in a non-TQM environment; ISO 9000 can improve operations in a traditional environment; ISO 9000 may be redundant in a mature TQM environment; and ISO 9000 and TQM are not in competition.

2. The origins of ISO 9000 and TQM are vastly different. ISO 9000 was developed in response to the need to harmonize dozens of national and international standards relating to quality. TQM got its start in Japan around 1950 as a way to help that nation compete in the international marketplace.
3. The aim of TQM is to transform organizations into competitive players in the global marketplace. ISO 9000 now has essentially the same aim.
4. Appropriate motivations for implementing ISO 9000 are as follows: to improve operations, to improve or create a quality management system, to improve the consistency of quality, to conform to the requirements of customers, to improve customer satisfaction and to improve competitiveness. The appropriate motivation for implementing TQM is a desire to continually improve all aspects of an organization.
5. ISO 9000 and TQM are compatible in that ISO 9000 can be a complementary subset of TQM. ISO 9000 can give an organization a head start in implementing TQM.

KEY TERMS AND CONCEPTS

Aims of ISO 9000 and TQM

Compatibility of ISO 9000 and TQM

"Consistent pair"

International Organization for Standardization

ISO 9000

ISO 9000 and Total Quality Management are not completely interchangeable

Origins of ISO 9000 and TQM

TQM

Traditional environment

REVIEW QUESTIONS

1. List several statements that summarize the comparative scope of ISO 9000 and TQM.
2. Explain the origins of ISO 9000 and TQM. How are they different?
3. Compare the aims of ISO 9000 and TQM.
4. List three appropriate reasons for implementing ISO 9000.
5. What is the most appropriate rationale for implementing TQM?
6. Describe how you would use ISO 9000 as an entry into TQM?

APPLICATION ACTIVITIES

1. Undertake a research project to determine how W. Edwards Deming and Joseph Juran came to be the leaders of the Japanese Total Quality movement. Your research should answer the following questions as a minimum:
   - Why did Deming and Juran go to Japan rather than helping manufacturers in the United States?

- Why did Japan need help from Deming and Juran?
- What is the *Deming Prize* in Japan and what are its requirements?

2. Undertake a research project to determine what contributions to TQM were made by Taiichi Ohno and Shigeo Shingo.

ENDNOTES

1. ISO, *The New Year 2000 ISO 9000 Standards: An Executive Summary,* 1999.
2. ISO 9000:2000, clause 2.1 *Rationale for Quality Management Systems.*

CHAPTER ELEVEN

Other ISO Standards, Issues, and Developments

MAJOR TOPICS

- ISO 14000
- Industry-Specific ISO 9000–Based Standards
- Anticipated Issues and Developments

ISO 14000

Product and service quality issues became increasingly important on a world-wide basis beginning in the 1970s. This focus on quality gave rise to the spread of Total Quality Management and the development of ISO 9000. At around the same time, environmental concerns began to emerge to the extent that the developed countries established health, safety, and environmental protection laws that must be obeyed by organizations operating within their respective national boundaries. Public and economic pressure encouraging the developing nations to follow suit is increasing. ISO established **Technical Committee 207 (TC 207)** for the purpose of developing an *environmental management system* that could be universally applied, similar to the ISO 9000 concept for quality. TC 207, with multinational representation, developed the new standard based on the British environmental standard **BS 7750.** TC 207 completed its work on the first generation of **ISO 14000** in mid-1996.

The structure and concept of the ISO 14000 standard are similar to those of ISO 9000. ISO 14000 is the *family* of Environmental System Standards, Figure 11-1. ISO 14001-1996, *Environmental Management Systems—Specifications with Guidance for Use,* corresponds to ISO 9001 in that it is the actual standard used for implementation and registration. ISO 14004-1996, *Environmental Management Systems—General Guidelines on Principles, Systems, and Supporting Techniques,* is intended to assist users in developing and executing their environmental management systems, and coordinating them with other management systems such as ISO 9000. ISO 19011 defines the principles and

ISO 14000 Family Environmental System Standards	
Standard Number	**Application**
ISO 14001-1996	Implementation and registration
ISO 14004-1996	Development of the environmental system and coordination with other systems
ISO 19011	Defines the principles and procedures for internal and external auditing of environmental and quality management systems

Figure 11–1
ISO 14000 Standards Family

procedures for internal and external auditing of the user's environmental and quality management systems.

There are good reasons why organizations that produce hazardous waste either directly or as process by-products should be required to adhere to the strictest environmental standards. The primary reason, or course, is that we cannot justify polluting the planet either willfully or by neglect. This applies to all individuals, businesses, and countries. Another reason is that the playing field that is the global marketplace should be level for all competitors—businesses and countries. The playing field is not level when a country such as the United States requires factories to eliminate toxic discharges into the atmosphere, rivers, and soil, while a similar factory in Central America is under no such requirement. The factory in the United States has made significant investments in expensive technology to clean or eliminate its emissions. These investments must be reflected in the cost of operations, and ultimately in the price of the goods produced. These types of costs do not occur in the unregulated factories of Central America, eastern Europe, Russia, the Caribbean basin, Korea, the South Pacific islands, and dozens of other areas. The result is that the goods produced by polluting factories cost less than those produced by the clean factory. Consumers, as a result, benefit temporarily from lower costs, but lose in the long run from pollution and poison. The playing field must be leveled. There is hope that eventually the clean factories may have a competitive advantage over the polluters. As the technology improves, it may be possible to operate at lower costs through recycling of the problematic chemical compounds, or by finding less expensive, nontoxic ways to do the same job. This is already happening in the electronics industry.

In summary, ISO 14000 is intended to do for an organization's environmental management system what ISO 9000 does for its quality management system. Look for ISO 14000 to evolve from a concept applied by forward looking companies to one that represents the normal way of doing business. The ISO 14000 Standards are available from the American Society for Quality (ASQ) at the following address and telephone number:

ASQ
611 East Wisconsin Avenue
PO Box 3005
Milwaukee, WI 53201-3005
Phone (414) 272-8575 or (414) 272-1734

INDUSTRY-SPECIFIC ISO 9000–BASED STANDARDS

QS 9000

Since 1988 the American Big Three auto makers, DaimlerChrysler, Ford, and General Motors, have been standardizing their requirements for suppliers. In 1988 the Chrysler/Ford/GM Supplier Quality Requirements Task Force standardized reference manuals, reporting formats, and terminology. Before this happened, Chrysler, Ford, and GM each had its own proprietary requirements and expectations for suppliers. If your company was a supplier to Ford only, there was no problem. But if you supplied GM and Chrysler too, the task of keeping requirements straight, using the right formats for reports, and even speaking the right language could be daunting.

In 1990 the task force released a common manual for measurement system analysis, and in 1991, a common approach to statistical process control. These steps simplified operations not only for Chrysler, Ford, and General Motors, but also for their many suppliers. As ISO 9000 became internationally accepted as the world's quality management system, Chrysler, Ford, and GM, working through the Automotive Industry Action Group (AIAG), developed its own version of ISO 9000, calling it initially *Quality System Requirements* (1994 draft). The name was changed to **QS 9000** to acknowledge the link to ISO 9000, which forms the basic structure of the standard. QS 9000 has harmonized Chrysler's *Supplier Quality Assurance Manual,* Ford's *Q-101 Quality System Standard,*

ISO INFO

Why ISO 9000 Is the Foundation for QS 9000

On why ISO 9000 was used as the foundation for QS 9000 by the Chrysler, Ford, GM Supplier Quality Requirements Task Force: one industry leader explained that no one told the task force they had to incorporate ISO 9000. "I told people we can either adopt the ISO standard and build on it, or we would spend the rest of our lives telling people why we didn't. We figured it just made sense."[1]

Rad Smith, Ford Motor Company

and GM's *Targets for Excellence.* While it is based on ISO 9000, using the same words and numbering system, it is expanded to include industry and company-specific quality requirements. One of the areas in which it reached beyond ISO 9000 was in the requirement that the quality system provide for *continual improvement,* an area that ISO had not adequately addressed until the year 2000 release.

All 13,000 first-tier suppliers to DaimlerChrysler, Ford, and General Motors were required to adopt QS 9000. As first-tier suppliers felt the pressure to be certified, they, in turn, pressured the second-tier suppliers and the pressure rippled throughout the supplier chain. Eventually all suppliers to the automotive industry, regardless of their relative level in the chain, will have to be registered to QS 9000 in order to do business.

The suppliers generally see QS 9000 as a mixed blessing. On the positive side, it eliminates the need for multiple—often conflicting—requirements from customers, and should mean just one audit per year instead of three or more. For those companies already certified to ISO 9000, QS 9000 certification is relatively easy. However, on the negative side, those who have put off ISO 9000 certification face the dilemma of leapfrogging ISO 9000 and going directly to QS 9000. This is a large and difficult step akin to going directly from crawling to running races.

QS 9000 consists of five manuals. Copies of the standard are available from the Automotive Industry Action Group, and can be ordered by calling the following telephone number: (810) 358-3003.

The automotive industry is the first to incorporate ISO 9000 into a harmonized, industry-wide/industry-peculiar standard, but there are others.

TE 9000

QS 9000 addresses some of the quality concerns of the automotive industry, including those that deal with materials and components purchased from suppliers. If we believe that products can be no better than the machines and tooling used to produce them, the machines and tooling used in the automotive industry are major factors in the ultimate quality of the industry's products. For this reason, the same task force that developed QS

ISO INFO

QS 9000 Certification

Do not assume that because you have a supplier recognition flag from one of the Big Three that this is the equivalent of QS 9000 registration. Section I of QS 9000 requires you to conform to ISO 9000, something not required before. Section II requires you to conform to the harmonized requirements of the three companies, and these are different from your earlier requirements. Finally, Section III imposes contractual customer-specific requirements for DaimlerChrysler, Ford, or GM. Having one flag, or even all three flags does not translate to QS 9000 certification.

ISO INFO

Although TE 9000 is a supplement to QS 9000, tooling and equipment suppliers register only to TE 9000. Suppliers of parts and materials to the automobile industry are certified to QS 9000, whereas suppliers of the machines and tooling used by the industry seek certification to TE 9000. In either case, conformance to ISO 9000 is built in.

9000 also produced the **TE 9000** Standard. It was released in 1996 as a supplement to QS 9000. TE 9000 is aimed directly at the many suppliers (approximately 50,000) of production equipment and tooling used by the automotive industry. It harmonizes the existing requirements in Chrysler's *Tooling & Equipment Supplier Quality Assurance Requirements,* Ford's *Facilities and Tools Quality Systems Standard,* and the QS 9000 standard itself. It embodies all of ISO 9001 as its basic core structure, and like QS 9000, adds industry-specific and customer-specific requirements for production equipment and tooling.

Like QS 9000, TE 9000 required continual improvement before ISO 9000:2000, and expands briefly on the philosophy and methodologies of continual improvement. It lists, for example, process capability indices, control charts, design of experiments (Taguchi method), equipment effectiveness, value analysis, problem solving (using the Total Quality Tools), and benchmarking. All of this is straight out of Total Quality Management, and goes beyond ISO 9000:2000 requirements; it is a positive trend that should continue as other industries develop standards.

The TE 9000 supplement may be ordered from the Automotive Industry Action Group, phone (810) 358-3003. Information about the supplement may be obtained from the Tooling and Equipment Hotline at (800) 444-2810.

AS 9000

The aerospace industry, led by Boeing, has adopted AS 9000.

TL 9000

The telecommunications industry has TL 9000. Both serve similar functions in their respective sectors as QS 9000 does for the auto industry.

It is likely that still other industries will follow the example of the Big Three American automobile makers by issuing industry-specific standards that embody ISO 9000 as the core quality management system standard. One can envision specialized industries, such as healthcare, air transportation, maritime transport, and others benefiting from similar industry-specific standards.

ANTICIPATED ISSUES AND DEVELOPMENTS

Occupational Health and Safety

Three fundamental issues affect all organizations that must compete in the global market:

- Quality of products and services
- Protection of the environment
- Welfare of employees

ISO has addressed two of the three issues with ISO 9000 and ISO 14000. It seems appropriate, and in fact it has been anticipated, that an international standard for occupational health and safety be developed under ISO and integrated with ISO 9000 and ISO 14000. The designation ISO 18000 has been reserved. However, to date, no technical committee has been established for that purpose. This has been an on-again-off-again proposition with ISO, but the latest information available indicates that it is once again on the table. If it happens, an ISO 18000 Occupational Health and Safety Management Standard will undoubtedly be based on the British standard, BS 8800.

In the United States the Occupational Safety and Health Administration (OSHA) has the statutory authority to develop, proclaim, and enforce occupational safety and health standards. Many other countries have similar organizations, but some do not. The result is that standards for occupational health and safety exist to protect employees in many nations, yet in many others the issue is ignored. An international standard that is based on the management system models provided by ISO 9000 and ISO 14000 and that yields the best practices for occupational health and safety universally would be beneficial to organizations, employees, and consumers around the world. This subject is discussed in depth in the book, *Occupational Safety and Health,* 3d Edition, 1999, by David L. Goetsch.

ISO ISSUE

What Would You Do?

Having just facilitated his company's successful attempt to achieve ISO 9001 registration, Mack Crowner—Quality Director for Magna Tech Limited—was looking forward to a well-deserved break. But it would not happen today. Magna Tech's CEO had just called and said, "Congratulations, Mack! Oh, by the way, How about joining me and the steering committee at our meeting tomorrow morning. After we give you a few well-deserved pats on the back, I'd like you to make a presentation about this new ISO 14000 standard I keep hearing about, and any other new ISO standards that might be coming down the pike." Put yourself in Crowner's place. What should he tell the steering committee about ISO 14000 and the other standards?

Other New Standards on the Horizon[2]

ISO has initiated work on standardizing the **handling of customer complaints.** Australian and British member institutes reportedly have had very positive experience with their national standards. ISO is leaning toward making this a "plug-in" for ISO 9000.

ANSI has proposed that ISO develop standards covering **personal finance planning.** Such a standard would include the certification of personal finance advisers based on elements of education, examination results, experience, and ethical conduct. This standard, if ever developed, will undoubtedly be a stand-alone, since it does not fit into either ISO 9000 or ISO 14000.

Merging Management System Standards

It has been anticipated that eventually there will be three international management system standards—quality, environment, and occupational health and safety—and that they will be merged into a single standard. As we have seen, two of the three International Standards, ISO 9000 and ISO 14000, currently exist and have been accepted worldwide. The third, ISO 18000, remains to be developed. Nevertheless, the concept of merging management system standards is active. The advantages of having the two standards become one are considerable. The possibilities include a single registration taking advantage of common and overlapping procedures and organizational structure. In addition, a single audit could cover both quality and environmental management systems and eliminate duplication in procedures, structure, and methods, and lower overall costs of registration.

On the other hand, merging the two standards poses a problem to the organization that is prepared for registration to one standard but not the other. In this case, registration would be delayed until the organization could satisfy both. For some organizations, this could impose an unacceptable delay or added cost. This may prevent ISO from merging the standards completely. However, many of the advantages of merging may be achieved without a complete merger.

This appears to be the path ISO is following. A Technical Advisory Group (TAG 12) formed to look into merging ISO 9000 and ISO 14000, recommended that the standards not be merged, but be made more compatible. (This is certainly seen in the year 2000 version of ISO 9000.) Specific recommendations for the standards include:[3]

- Relevant terms and definitions should be identical and there should be consistent use of terminology in both families of standards.
- Management system standards in the two families should be compatible and, as far as possible, aligned.
- Auditing standards in the two families should be integrated as far as possible to consist of a common core document with accompanying separate modules on quality and environment.

The product of these recommendations will be found in the single common auditing standard, **ISO 19011.** (See Figure 11-2.) A single auditing standard applicable to quality and environmental management systems enables registrars to use a single surveillance

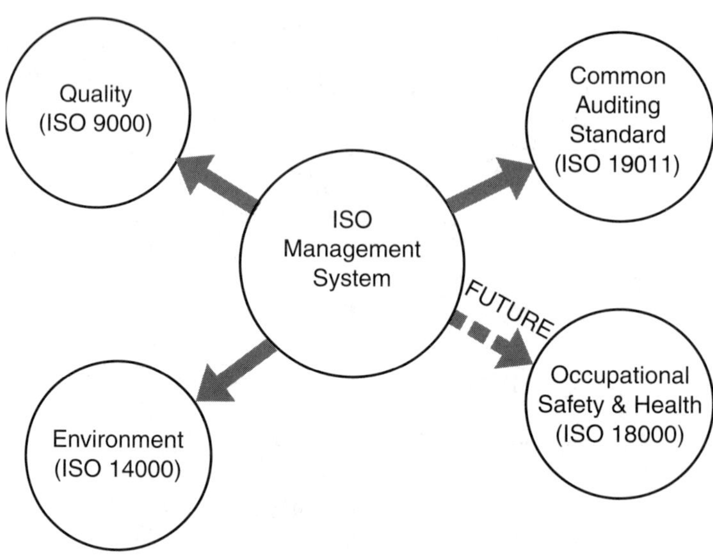

Figure 11–2
ISO Management System Standards

audit, and in certain cases, a single registration audit, to cover the requirements of both standards. For organizations with both management systems this could cut the number of audits by half and would be less expensive. For this to be possible, the periodic revisions for the two standards must occur simultaneously, or one will always be out of step with the other, thereby making joint auditing difficult and possibly adding work for the organization. ISO is working on this issue.

In summary, the merging of ISO 14000 and ISO 9000 will not occur anytime soon, if ever. Nor is an ISO international standard for occupational safety and health likely in the near future. However, ISO 19011, aided by increased compatibility of ISO 9000 and ISO 14000, make joint auditing of quality and environmental management systems possible.

CASE STUDY

ISO 9000 in Action

It had been more than a year since Jake Butler had been given the assignment of guiding AMI through all of the steps necessary to achieve ISO 9000 certification. There had been a lot of hard work and more than a few late nights and weekends at the office. But the hard work and long hours had paid off. It was official as of yesterday. AMI is now an ISO 9001 registered company, and Jake Butler is the relieved but happy recipient of a generous bonus check. Butler plans to take a well deserved vacation during which he will think of anything but ISO, because when he returns to work the whole process begins again.

AMI's CEO, Arthur Polk, has become an ISO advocate. He sees how the company has benefited from the ISO 9001 registration process. Consequently, Polk plans to pursue ISO 14000 registration. In fact, if ISO develops a workplace safety and health standard, Polk wants AMI to pursue it too. For some time to come, Jake Butler is going to be a busy man.

SUMMARY

1. ISO established Technical Committee 207 (TC 207) for the purpose of developing an environmental management system. The committee's efforts resulted in the ISO 14000 family of standards. This family of standards does for environmental management what the ISO 9000 family does for quality management.
2. Working through the Automotive Industry Action Group (AIAG), Chrysler, Ford, and General Motors developed their own version of ISO 9000, calling it QS 9000. QS 9000 is ISO 9000 with additional requirements added that relate specifically to industry and company-specific concerns.
3. The same group that developed QS 9000 (AIAG) also developed TE 9000, a standard aimed directly at suppliers of production equipment and tooling used by the automotive industry.
4. ISO has initiated work on standardizing the handling of customer complaints and has proposed standards covering personal finance planning.

KEY TERMS AND CONCEPTS

BS 7750
Handling of customer complaints
ISO 14000
ISO 14001
ISO 14004
ISO 19011
Personal finance planning
QS 9000
TC 207
TE 9000

REVIEW QUESTIONS

1. List the numbers, titles, and uses of all standards in the ISO 14000 family.
2. What is QS 9000 and how does it relate to ISO 9000?
3. What is TE 9000 and how does it relate to QS 9000?
4. How does TE 9000 differ from ISO 9000?
5. Explain what ISO is doing about merging management system standards.

APPLICATION ACTIVITIES

1. Obtain a copy of ISO 14001. Using the standard as a guide, undertake the following project: Assume that your company has gone through the ISO 9000 certification process and recently received certification. Identify the ways in which the ISO 9000 certification process can short-cut the ISO 14001 process. Also identify all additional work that will need to be done in order to achieve ISO 14001 certification.
2. Obtain a copy of TE 9000. Develop a step-by-step plan for pursuing TE 9000 certification that can be used by any supplier interested in the certification.

ENDNOTES

1. Nanette M. Webb, "Big Three Introduce Harmonized Standards," *Quality in Manufacturing* (Jul.–Aug. 1994).
2. ISO Press Release 772, *ISO addresses climate change, health and safety, complaints handling, personal financial planning,* 21 Feb., 2000.
3. ISO Press Release, *ISO Places "Rush Order" for Joint ISO 9000 and ISO 14000 Audit Standard,* September 16, 1998.

APPENDIX A

Major Differences Between the 1994 and 2000 Versions of ISO 9000

Aim and Mission of ISO 9000:2000

> *To relate modern quality management to the processes and activities of an organization, including the promotion of continual improvement and achievement of customer satisfaction.*[1]

Documents Comprising the Standard

- ISO 9001, 9002 and 9003:1994 have been merged into a single QMS requirements standard, ISO 9001:2000.
- ISO 8402 and part of ISO 9000-1:1994 have been merged into the new standard, ISO 9000:2000.
- ISO 9004-1:1994 has been revised into a new ISO 9004:2000 standard that tracks with and expands upon ISO 9001:2000.
- ISO 10011 (parts 1, 2, and 3) have been merged with ISO 14010, 14011, and 14012, to form a new auditing guidelines standard, ISO 19011.

A Single Standard for Registration

There is no longer a choice of three standards for registration. All registrations will be to ISO 9001:2000. Some tailoring for organization-specific factors is possible through exclusion of clause 7 elements that are not applicable to the particular organization.

Terminology Changes

ISO uses terminology more familiar to the organization. For example, in the 1994 version, *supplier* was the entity seeking registration or registered to ISO 9000. This was always a point of confusion. The new term is *organization*. A supplier is now the entity that supplies materials to the organization.

Adoption of the Eight Quality Management Principles

The year 2000 version of ISO 9000 is based on the following eight quality management principles:[2]

1. **Customer Focus**—Organizations exist because they have customers, and therefore should understand current and future customer needs, should meet customer requirements, and should strive to exceed customer expectations.
2. **Leadership**—The organization's leadership must establish unity of purpose and direction for the organization. Leaders should create and maintain an organizational environment that promotes the full involvement of employees in achieving the organization's objectives.
3. **Involvement of People**—Employee involvement benefits the organization by taking advantage of the abilities of every employee.
4. **Process Approach**—To maximize efficiency of activities and related resources by managing them as processes.
5. **System Approach to Management**—Contributes to the organization's effectiveness and efficiency by identifying, understanding, and managing interrelated processes as a system.
6. **Continual Improvement**—Becomes a permanent objective for the organization's overall performance, and is intended to help the organization respond to the changing needs of its customers, and to improve its competitiveness.
7. **Factual Approach to Decision Making**—Requires that decisions be based on the analysis of data and information rather than impressions and intuition.
8. **Mutually Beneficial Supplier Relationships**—To enhance the ability of both the organization and its suppliers to benefit from working together cooperatively.

Process Oriented QMS Structure and Operation

The QMS is now viewed as a series of processes:

- Management responsibility
- Resource management
- Product realization
- Measurement, analysis and improvement

The result is a sequence that is more logical than just following the twenty requirements clauses of the 1994 version. See Figures 5-1 and 5-6 in Chapter 5.

Incorporation of the Deming/Shewhart Plan-Do-Check-Act (PDCA) Cycle

To be used for the continual improvement of processes and products or services.

Increased Emphasis on the Role of Top Management

Makes top management responsible for commitment, customer focus, the quality policy, quality objectives, planning, QMS structure, communication, and management review.

Customer Satisfaction

The organization is now required to monitor customer satisfaction as a measure of QMS performance.

Required Documentation

There are fewer explicit requirements for documentation, although to avoid documenting a procedure, for example, the organization must demonstrate that it can operate effectively without it. That is, the organization must ensure that deviations cannot occur as a result of the absence of documentation. The net change in documentation required by the typical organization is probably going to be minimal. This invites more interpretation by the organization and registrars.

New Requirements Summary

ISO 9001:2000 carries several new requirements for registered organizations.

- Continual improvement of processes and products
- Increased emphasis on the role of top management
- Consideration of legal and regulatory requirements
- Measurable quality objectives
- Monitoring of customer satisfaction
- Increased attention to resource availability
- Determination of training effectiveness
- Measurements covering the QMS and its processes and products
- Analysis of collected data on QMS performance

ENDNOTES

1. ISO, *The New Year 2000 ISO 9000 Standards: An Executive Summary.*
2. ISO 9000:2000, clause 0.2.

APPENDIX B

ISO Member Bodies

The following membership list was provided by the International Organization for Standardization as of February, 2001. The format lists the full-member nation in bold print, the name of the national standards organization with its abbreviation in parentheses, and finally the organization's address.

Algeria	Institut Algérien de Normalisation (IANOR) 5, rue Abou Hamou Moussa, B.P. 403 - Centre de tri, Alger
Argentina	Instituto Argentino de Normalización (IRAM) Chile 1192, 1098 Buenos Aires
Armenia	Department for Standardization, Metrology and Certification (SARM) Komitas Avenue 42/2, 375051 Yerevan
Australia	Standards Australia (SAA) 1 The Crescent, Homebush - N.S.W. 2140
Austria	Österreichisches Normungsinstitut (ON) Heinestrasse 38, Postfach 130, A-1021 Wien
Bangladesh	Bangladesh Standards and Testing Institution (BSTI) 116/A, Tejgaon Industrial Area, Dhaka- 1208
Barbados	Barbados National Standards Institution (BNSI) "Flodden" Culloden Road, BB- St. Michael
Belarus	Committee for Standardization, Metrology and Certification (BELST) Starovilensky Trakt 93, Minsk 220053
Belgium	Institut belge de normalisation (IBN) Av. De la Brabanconne 29, B-1000 Bruxelles
Bosnia and Herzegovina	Institute for Standardization, Metrology and Patents (BASMP) Hamdije Cemerlica 2, (ENERGOINVEST building), CH-71000 Sarajevo

Botswana	Botswana Bureau of Standards (BOBS) Plot #14391, New Labatse Road, Gabarone West Industrial, BW Gabarone
Brazil	Associacao Brasileira de Normas Técnicas (ABNT) Av. 13 de Maio, no 13, 280 andar, 20003-900—Rio de Janeiro-RJ
Bulgaria	Committee for Standardization and Metrology (BDS) 21, 6th September Str., 1000 Sofia
Canada	Standards Council of Canada (SCC) 45 O'Connor Street, Suite 1200, Ottawa, Ontario K1P 6N7
Chile	Instituto Nacional de Normalización (INN) Matías Cousiño 64-60 piso, Casilla 995—Correo Central, Santiago
China	China State Bureau of Technical Supervision (CSBTS) 4, Zhichun Road, Haidian District, P.O. Box 8010, Beijing 100088
Columbia	Instituto Colombiano de Normas Técnicas y Certificación (ICONTEC) Carrera 37 52-95, Edificio ICONTEC, P.O. Box 14237, Santafé de Bogotá
Costa Rica	Instituto de Normas Técnicas de Costa Rica (INTECO) P.O. Box 6189-1000, San José
Croatia	State Office for Standardization and Metrology (DZNM) Ulica grada Vukovara 78, 10000 Zagreb
Cuba	Oficina Nacional de Normalización (NC) Calle E No. 261 entre 11 y 13, Vedado, La Habana 10400
Cyprus	Cyprus Organization for Standards and Control of Quality (CYS) Ministry of Commerce, Industry and Tourism, Nicosia 1421
Czech Republic	Czech Standards Institute (CSNI) Biskupsky dvur 5, 110 02 Praha 1
Denmark	DANSK STANDARD (DS) Kollegievej 6, DK-2920 Charlottenlund
Ecuador	Instituto Ecuatoriano de Normalización (INEN) P.O. Box 17-01-3999, Quito

Egypt	Egyptian Organization for Standardization and Quality Control (EOS) 2 Latin America Street, Garden City, Cairo
Ethiopia	Ethiopian Authority for Standardization (EAS) P.O Box 2310, Addis Ababa
Finland	Finnish Standards Association (SFS) P.O. Box 116, FIN-00241 Helsinki
France	Association francaise de normalisation (AFNOR) Tour Europe, F-92049 Paris La Défense Cedex
Germany	Deutsches Institut für Normung (DIN) Burggrafenstrasse 6, D-10772 Berlin
Ghana	Ghana Standards Board (GSB) P.O. Box M 245, ACCRA
Greece	Hellenic Organization for Standardization (ELOT) 313, Acharnon Street, GR-111 45 Athens
Hungary	Magyar Szabványügyi Testület (MSZT) Üllöi út 25, Pf. 24., H-1450 Budapest 9
Iceland	Icelandic Council for Standardization (STRI) Keldnaholt, IS-112 Reykjavik
India	Bureau of Indian Standards (BIS) Manak Bhavan, 9 Bahadur Shah Zafar Marg, New Delhi 110002
Indonesia	Badan Standardisasi Nasional (BSN) c/o Pusat Standardisasi—LIPI, Jalan Jend. Gatot Subroto 10, Jakarta 12710
Iran	Institute of Standards and Industrial Research of Iran (ISIRI) P.O. Box 31585-163, Karaj
Ireland	National Standards Authority of Ireland (NSAI) Glasnevin, Dublin-9
Israel	Standards Institution of Israel (SII) 42 Chaim Levanon Street, Tel Aviv 69977
Italy	Ente Nazionale Italiano di Unificazione (UNI) Via Battistotti Sassi 11/b, I-20133 Milano
Jamaica	Jamaica Bureau of Standards (JBS) 6 Winchester Road, P.O. Box 113, Kingston 10
Japan	Japanese Industrial Standards Committee (JISC) c/o Standards Department, Ministry of International Trade and Industry 1-3-1, Kasumigaseki, Chiyoda-ku, Tokyo 100

Kazakhstan	Committee for Standardization, Metrology and Certification (KAZMEMST) Pushkin Str 166/5, KZ-473000 Astana
Kenya	Kenya Bureau of Standards (KEBS) P.O. Box 54974, Nairobi
Korean, Dem. P. Rep. of	Committee for Standardization of the Democratic People's Republic of Korea (CSK) Zung Gu Yok Seungli-Street, Pyongyang
Korea, Republic of	Korean National Institute of Technology and Quality (KNITQ) 2, Joongang-dong, Kwachon, Kyunggi-do 427-010
Kuwait	Public Authority for Industry, Standards and Industrial Services Affairs, Standards and Metrology Department (KOWSMD) Post Box 4690, Safat, KW- 13047
Libyan Arab Jamahiriya	Libyan National Centre for Standardization and Metrology (LNCSM) Industrial Research Centre Building, P.O. Box 5178, Tripoli
Luxembourg	Service de l'Energie de l'Etat, Department Normalisation (SEE) 34 avenue de la Porte-Neuve, B.P. 10, L-2010 Luxembourg
Malaysia	Department of Standards Malaysia (DSM) 21st Floor, Wisma MPSA, Persiaran Perbandaran, 40675 Shah Alam Selangor Darul Ehsan
Malta	Malta Standards Authority (MSA) 2nd Floor, Evans Bldg., Merchants Street, MT-Valletta VLT 03
Mauritius	Mauritius Standards Bureau (MSB) Moka
Mexico	Dirección General do Normas (DGN) Calle Puente de Tecamachalco No 6, Lomas de Tecamachalco Sección Fuentes, Naucalpan de Juárez, 53 950 Mexico
Mongolia	Mongolian National Centre for Standardization and Metrology (MNCSM) P.O. Box 48, Ulaanbaatar 211051

Morocco	Service de normalisation industrielle marocaine (SNIMA) Ministère do commerce, de l'industrie et l'artisanat Angle Avenue Al Filao, et Rue Dadi Secteur 21 Hay Riad, 10100 Rabat
Netherlands	Nederlands Normalisatie-instituut (NNI) Kalfjeslaan 2, P.O. Box 5059, NL-2600 GB Delft
New Zealand	Standards New Zealand (SNZ) Radio New Zealand House, 155 the Terrace, Wellington 6001
Nigeria	Standards Organisation of Nigeria (SON) Federal Secretariat, Phase 1, 9th Floor, Ikoyi, Lagos
Norway	Norges Standardiseringsforbund (NSF) Drammensveien 145 A, Postboks 353 Skoyen, N-0212 Oslo
Pakistan	Pakistan Standards Institution (PSI) 39 Garden Road, Saddar, Karachi-74400
Panama	Comisión Panameña de Normas Industriales y Técnicas (COPANIT) Ministerio de Comercio y Industrias, Apartado Postal 9658, Panama, Zona 4
Philippines	Bureau of Product Standards (BPS) Dept. of Trade and Industry, 361 Sen. Gil J. Puyat Avenue, Makati Metro Manila 1200
Poland	Polish Committee for Standardization (PKN) ul. Elektoralna 2, P.O. Box 411, PL-00-950 Warszawa
Portugal	Insituto Português da Qualidade (IPQ) Rua C à Avenida dos Três Vales, P-2825 Monte de Caparica
Romania	Institutul Român de Standardizare (IRS) Str. Jean-Louis Calderon Nr. 13, Cod 70201, R-Bucuresti 2
Russian Federation	State Committee of the Russian Federation for Standardization, Metrology and Certification (GOST R) Leninsky Prospekt 9, Moskva 117049
Saudi Arabia	Saudi Arabian Standards Organization (SASO) Imam Saud Bin Abdul Aziz Bin Mohammed Road (West End) P.O. Box 3437, Riyadh 11471
Singapore	Singapore Productivity and Standards Board (PSB) 1 Science Park Drive, Singapore 118221

Slovakia	Slovak Institute for Standardization (SUTN) P.O. Box 246, Karloveska 63, SK-840 00 Bratislava 4
Slovenia	Standards and Metrology Institute of the Republic of Slovenia (SMIS) Kotnikova 6, SI-1000 Ljubljana
South Africa	South African Bureau of Standards (SABS) 1 Dr Lategan Rd, Groenkloof, Private Bag X191, Pretoria 0001
Spain	Asociación Española de Normalización y Certificación (AENOR) Génova, 6, E-28004 Madrid
Sri Lanka	Sri Lanka Standards Institution (SLSI) 53 Dharmapala Mawatha, P.O. Box 17, Colombo 3
Sweden	Standardiseringen I Sverige (SIS) St Eriksgatan 115, Box 6455, S-113 82 Stockholm
Switzerland	Swiss Association for Standardization (SNV) Mühlebachstrasse 54, CH-8008 Zurich
Syrian Arab Republic	Syrian Arab Organization for Standardization and Metrology (SASMO) P.O. Box 11836, Damascus
Tanzania	Tanzania Bureau of Standards (TBS) Ubungo Area, Morogoro Road/Sam Nujoma Road, Dar es Salaam
Thailand	Thai Industrial Standards Institute (TISI) Ministry of Industry, Rama VI Street, Bangkok 10400
Trinidad and Tobago	Trinidad and Tobago Bureau of Standards (TTBS) P.O. Box 467, Port of Spain
Tunisia	Institut national de la normalisation et de le propriété industrielle (INNORPI) B.P. 23, 1012 Tunis-Belvédère
Turkey	Türk Standardlari Enstitüsü (TSE) Necatibey Cad. 112, Bakanliklar, TR-06100 Ankara
Ukraine	State Committee of Ukraine for Standardization, Metrology and Certification (DSTU) 174 Gorky Street, GSP, Kyiv-6, 252650
United Arab Emirates	Directorate of Standardization and Metrology (SSUAE) Ministry of Finance and Industry El Falah St., P O Box 433, AE-Abu Dhabi
United Kingdom	British Standards Institution (BSI) 389 Chiswick High Road, GB-London W4 4AL

USA	American National Standards Institute (ANSI) 1819 "L" Street NW, Washington, DC 20036
Uruguay	Instituto Uruguayo do Normas Técnicas (UNIT) Pza. Independencia 812, Piso 2, Montevideo
Uzbekistan	Uzbek State Centre for Standardization, Metrology and Certification (UZGOST) Ulitsa Farobi, 333-A, 700049 Tachkent
Venezuela	Fondo para la Normalización y Certificación de la Calidad (FONDONORMA) Av. Andrés Bello-Edf Torre Fondo Común, Pisos 11 y 12 - Apartado Postal 51116, Caracas 1050-A
Viet Nam	Directorate for Standards and Quality (TCVN) 70, Tran Hung Dao Street, Hanoi
Yugoslav Republic of Macedonia, former	Zavod za standardizacija I metrologija (ZSM) Ministry of Economy, Samoilova 10, 91000 Skopje
Yugoslavia	Savezni zavod za standardizaciju (SZS) Kneza Milosa 20, Post Pregr. 933, YU-11000 Beograd
Zimbabwe	Standards Association of Zimbabwe (SAZ) P.O. Box 2259, Harare

APPENDIX C

TC 176 Membership

Membership of all ISO technical committees is made up of three categories of members:

- Participating members, representing countries wishing to vote, participate in discussions, and have access to all relevant documentation.
- Observing members, representing countries not wishing to vote, but who want to participate in discussions and receive all relevant information.
- Liaison organizations, representing international or broadly based regional organizations, that may take part in discussions and receive all relevant information, but may not vote.

The following nations are participating members of TC 176:

Algeria
Argentina
Australia
Austria
Belgium
Bulgaria
Brazil
Canada
Chile
China
Columbia
Cuba
Czech Republic
Denmark
Ecuador
Egypt
Finland
France
Germany
Greece
Hungary
India
Indonesia
Iran
Ireland
Israel
Italy
Jamaica
Japan
Korea, Republic of
Malaysia
Mauritius
Mexico
Netherlands
New Zealand
Norway
Philippines
Poland
Portugal
Russian Federation
Singapore
Slovakia
South Africa
Spain
Sri Lanka
Sweden
Switzerland
Tanzania

Thailand	United Kingdom	Yugoslavia
Trinidad/Tobago	Uruguay	Zimbabwe
Turkey	USA	
Ukraine	Venezuela	

The following nations are observing members of TC 176:

Armenia	Ethiopia	Saudi Arabia
Botswana	Hong Kong	Slovenia
Costa Rica	Iceland	Syria
Croatia	Lithuania	Tunisia
Cyprus	Moldova, Republic of	Viet Nam
Estonia	Romania	

The following are liaison organizations to TC 176:

Arab Management Society
Asian Productivity Organization
American Society for Quality
European Chemical Industry Council
Consumers International
European Association of Automotive Supplies
European Commission DG III EC-DG III
European Foundation for Quality Management
European Organization for Quality
International Automotive Task Force
International Lab Accreditation Conference
International Information Centre for Terminology
International Federation of Consulting Engineers
International Atomic Energy Agency
International Accreditation Forum
International Academy for Quality
International Institute of Synthetic Rubber Procedures, Inc.
International Project Management Association
International Organization of Legal Metrology
International Union of Testing and Research Laboratories for Materials and Structures
The Latinoamerican Institute for Quality Assurance
International Telecom Union
World Association of Societies of Pathology (Anatomic and Clinical)

APPENDIX D

Relationship Between ISO 9000 and ISO 14000

ISO has made the decision, for now at least, not to merge the two standards into one. However, ISO has taken on the task of making it easier for organizations to implement both standards by making them more compatible. "More compatible" means:

- Common elements of the standards can be implemented in a shared manner.
- Relevant terms and definitions should be identical.
- The use of terminology should be consistent.
- The management system standards should be compatible and, as far as possible, aligned.
- Auditing standards should be integrated.

Although this remains a work-in-progress, much has already been accomplished. ISO 9000:2000 has made a leap forward in compatibility. ISO 19011 accomplishes the integration of the auditing standards. The next revision of ISO 14000 is expected to continue the movement toward compatibility.

ISO has provided a comprehensive table in ISO 9001:2000, Annex A that can assist organizations in the following circumstances:

- Organizations currently operating under both ISO 9000 and ISO 14000, and seeking ways to simplify and eliminate duplication.
- Organizations currently operating under one or the other of the standards and contemplating registration to the second, but seeking some integration.
- Organizations preparing for initial registration to both ISO 9000 and 14000, seeking efficiency and elimination of duplication in the process.

The table is intended to indicate correspondence of the clauses with similar purposes from the ISO 9000 and ISO 14000 standards. It is not intended to imply that one clause will serve for the other. However, it may help the organization find opportunities to harmonize its management system policies, procedures, resources, and the like while satisfying both clauses.

APPENDIX E

ISO 9001 Checklist

Purpose of the Checklist

The ISO 9001 Checklist is intended to assist organizations in self-assessing their readiness for registration. It is not a shortcut to conformance. The organization must develop and implement its conforming quality management system according to ISO 9001. The checklist will, however, assist organizations in keeping track of QMS development and implementation by providing a convenient means of tracking progress. It also may reveal more clearly than the standard itself what remains to be done.

Checklist Format

The checklist corresponds directly with the requirements of the ISO 9001 standard. It is organized by the same headings and subheadings of the standard, and all requirements in the standard are listed.

How to Use the Checklist

The checklist can be used as an internal tool for preparing an organization for ISO 9000 registration. For each question an affirmative answer indicates, but does not prove, that the organization has satisfied the requirement in question.

While using this checklist, refer to Chapters 3 through 6 (especially the Application Information in Chapter 4) for clarification.

ISO 9001 Checklist

Note: As stated earlier, each checklist item relates directly to the specified clause of ISO 9001:2000.

QUALITY MANAGEMENT SYSTEM—REQUIREMENTS OF CLAUSE 4

4.1 General Requirements

1. Has the organization established a quality management system to meet the requirements of clause 4?

2. Has the QMS been appropriately documented?
3. Is the QMS implemented and maintained?
4. Is the effectiveness of the QMS continually improved?

4.1 a)

5. Has the organization identified the processes used in the development and production of products or services and any other processes relating to or interacting with them (including relevant outsourced processes)?
6. Has the organization identified the processes necessary to support development and production processes?
7. Has the organization determined the organizational activities responsible for operating the processes, and providing input to and taking output from them?
8. Does the organization fully understand the processes and their functions?

4.1 b)

9. Has the organization determined the sequence and interactions of the processes identified under clause 4.1a?

4.1 c)

10. Has the organization established the criteria and methods for operation of the QMS processes identified above?
11. Has the organization established controls to ensure that QMS processes are always operated according to the established criteria and methods?

4.1 d)

12. Has the organization ensured the availability of the physical resources needed by the QMS (including people, tools and equipment, materials, environment)?
13. Has the organization ensured the availability of information resources needed by the QMS (including quality policy, process specifications and operating parameters, product specifications, drawings, procedures and work instructions, test instructions)?

4.1 e)

14. Does the organization monitor its processes to ensure operation within intended specifications and parameters?
15. Has the organization implemented appropriate measurements for its processes to ensure operation within specified tolerance ranges?
16. Does the organization analyze the process data collected to determine corrective and preventive action and potential process improvement?

4.1 f)

17. Has the organization implemented the actions necessary to achieve planned results from the QMS processes?
18. Has the organization implemented the actions necessary to achieve continual improvement of the QMS processes?

4.2 Documentation Requirements

4.2.1 General

4.2.1 a)

19. Has the organization documented a quality policy?
20. Does the organization have documented relevant (up-to-date) quality objectives?

4.2.1 b)

21. Does the organization have a documented quality manual that conforms to the standard? (See 4.2.2.)

4.2.1 c)

22. Has the organization documented the procedures that are
    - Explicitly required by the standard, and
    - Required for effective operation and control of processes?

4.2.1 d)

23. Are all documents necessary for the organization to ensure effective planning, operation, and control of its processes included in the QMS documentation?

4.2.1 e)

24. Has the organization made records required by the standard (see Chapter 4, Application Information—Clause 4.2.1 e) part of the QMS documentation?

4.2.2 Quality Manual

25. Has the organization established and does it maintain a quality manual?

4.2.2 a)

26. Does the organization's quality manual describe the scope of the QMS, and does it include details and justification for any Clause 7 exclusions?

4.2.2 b)

27. Does the quality manual include the documented procedures required for the QMS, or does it provide reference to them?

4.2.2 c)

28. Does the quality manual provide descriptions of interaction between QMS processes?

4.2.3 Control of Documents

29. Are all non-record documents required by the QMS subject to a conforming document control system that includes all of the following:
    a) Are all such documents appropriately approved prior to issue?
    b) Are all such documents reviewed, updated as necessary, and reapproved?
    c) Are change and revision status clearly identified on all such documents?
    d) Are relevant versions of applicable documents available where needed?
    e) Does the organization ensure that all QMS documents are legible and identifiable?
    f) Does the organization ensure that documents of external origin are identified and that their distribution is appropriately controlled?
    g) Does the organization prevent unintended use of obsolete documents, and does it identify them as obsolete if they are retained?

4.2.4 Control of Records

30. Does the organization establish and maintain records needed to demonstrate the conformity and effectiveness of the QMS using an organized, structured, orderly, and controlled process that assures records remain legible, are readily identifiable, and easily retrievable?
31. Has the organization documented the records control procedure?
32. Does the documented records control procedure address all of the following:
    a) Identification of records?
    b) Storage (media, location(s), responsibility)?
    c) Protection (from the elements, decay, loss, and unauthorized access)?
    d) Retrieval (process for, responsibility, access authorization)?
    e) Retention time (for the various categories of records)?
    f) Disposition of obsolete records?

MANAGEMENT RESPONSIBILITY—REQUIREMENTS OF CLAUSE 5

5.1 Management Commitment

33. Is top management committed to the development, implementation, and continual improvement of a conforming QMS?
34. Does top management provide evidence of its commitment by
    a) Communicating the importance of meeting customer, and statutory and regulatory requirements?

b) Establishing (developing and adhering to) the quality policy?
c) Ensuring that quality objectives are established (and acted upon)?
d) Conducting management reviews of the QMS?
e) Ensuring the availability of (providing) resources needed for the effective operation of the QMS?

5.2 Customer Focus

35. Does top management ensure that customer requirements are determined?
36. Does top management ensure that customer requirements are met?

5.3 Quality Policy

37. Has top management ensured that the quality policy
    a) Is appropriate to the purpose (vision and mission) of the organization?
    b) Includes a commitment to comply with requirements?
    c) Includes a commitment to continually improve QMS effectiveness?
    d) Provides a framework for establishing and reviewing quality objectives?
    e) Is communicated to all employees and understood by them?
    f) Is periodically reviewed to ensure that it remains suitable over time?

5.4 Planning

5.4.1 Quality Objectives

38. Has top management ensured the establishment of quality objectives (including those needed to meet product requirements) at relevant organizational departments and levels?
39. Are quality objectives measurable? (See clause 5.4.1, Application Information, *Requirement 2.*)
40. Are quality objectives always consistent with the quality policy (and the organization's strategic plans)?

5.4.2 Quality Management System Planning

41. Has top management ensured that
    a) Quality planning is carried out in a manner to effectively and efficiently meet the requirements of clause 4.1, and its quality objectives?
    b) Whenever changes to the QMS are considered, is planning directed at ensuring the continued integrity of the QMS?

5.5 Responsibility, Authority and Communication

5.5.1 Responsibility and Authority

42. Has top management defined organizational responsibilities and authorities?
43. Has top management communicated those responsibilities and authorities within the organization?

5.5.2 Management Representative

44. Has top management created a position of management representative whose duties include
    a) Ensuring that QMS processes are identified (established), implemented, and maintained?
    b) Reporting to top management on QMS performance and needs for improvement?
    c) Ensuring the promotion of customer requirements awareness throughout the organization?

5.5.3 Internal Communication

45. Has top management ensured the establishment of appropriate internal communication processes?
46. Has top management ensured that these internal communication processes are used for the effectiveness of the QMS?

5.6 Management Review

5.6.1 General

47. Does top management conduct periodic reviews of the QMS to ensure that it remains suitable, adequate, and effective?
48. Do these management reviews include an assessment of continual improvement opportunities?
49. Do these management reviews include an assessment of the need for change to the QMS or its constituent parts including the quality policy and quality objectives?
50. Does the organization maintain records of the management reviews in a manner compatible with clause 4.2.4?

5.6.2 Review Input

51. Do inputs to the periodic management reviews of the QMS include data or information concerning
    a) Internal and external audits?
    b) (Internal and external) customer feedback?

c) Process performance and product conformity measurement?
d) Status of preventive and corrective (and continual improvement) actions underway and completed during the review period?
e) Follow-up reports on action items continuing from previous reviews?
f) Changes that could affect the QMS?
g) Recommendations for improvement?

5.6.3 Review Output

52. Does the output from periodic management reviews include decisions and actions related to
    a) Improving the QMS and its policies, procedures, processes, etc.?
    b) Improvement of products or services related to customer requirements?
    c) Resources of all categories needed by the QMS?

RESOURCE MANAGEMENT—REQUIREMENTS OF CLAUSE 6

6.1 Provision of Resources

53. Has the organization determined the resources needed to
    a) Implement and maintain the QMS and continually improve its effectiveness?
    b) Enhance customer satisfaction by meeting (or exceeding) customer requirements?
54. Has the organization provided these resources?

6.2 Human Resources

6.2.1 General

55. Are all employees whose work can affect product quality competent to perform that work (on the basis of education, training, skills and experience?

6.2.2 Competence, Awareness and Training

56. Has the organization
    a) Determined the level of competence necessary for employees whose work can affect product or service quality?
    b) Provided appropriate training, or taken other action to ensure that employees are competent?
    c) Evaluated the effectiveness of training provided, or of other actions taken?
    d) Ensured that employees are aware of the relevance and importance of their activities, and how they contribute to the achievement of the organization's quality objectives?

e) Maintained appropriate records of employee education, training, skills, and experience?

6.3 Infrastructure

57. Has the organization determined the infrastructure (buildings, workspace, utilities, process equipment, and supporting services) needed to achieve conformity to product requirements?
58. Has the organization provided this infrastructure?
59. Does the organization maintain this infrastructure?

6.4 Work Environment

60. Has the organization determined the work environment necessary to achieve conformity to product requirements?
61. Does the organization manage the work environment on a continuing basis to assure conformity to product requirements?

PRODUCT REALIZATION—REQUIREMENTS OF CLAUSE 7

7.1 Planning of Product Realization

62. Has the organization planned and developed the needed product realization processes?
63. Is that planning of product realization consistent with the requirements of the QMS (clause 4.1)?
64. As part of that planning of product realization, has the organization determined (as appropriate)
    a) Quality objectives and quality requirements for the product?
    b) The need for product-specific new processes, documentation, and resources?
    c) Requirements for verification, validation, monitoring, inspection, and test activities specific to the product, and the product acceptance criteria?
    d) What records are required as evidence that the processes and product meet requirements?
65. Is the output of this planning in a form suitable for use by the organization? (If it needs to be documented, is it documented?)

7.2 Customer Related Processes

7.2.1 Determination of Requirements Related to the Product

66. In relation to its products, does the organization determine
    a) Customer requirements, including delivery and post-delivery activities if applicable?

b) Other requirements necessary for specified or intended use?

c) Any laws or regulations that apply?

d) Any additional internal requirements?

7.2.2 Review of Requirements Related to the Product

67. Before any commitment to produce the product, does the organization review product requirements to ensure that
    a) All necessary product requirements are defined?
    b) Any contract or order requirements that differ from previously expressed requirements are resolved?
    c) The organization has the ability to meet the requirements?
68. Does the organization maintain records of the results of the requirements review, including actions arising from it?
69. In the absence of documented customer requirements, does the organization confirm requirements before order or contract acceptance?
70. If product requirements are changed, does the organization ensure that relevant documents are updated, and that relevant employees are made aware of changes?

7.2.3 Customer Communication

71. Has the organization set up an effective means of communication with its customers in relation to
    a) Product information?
    b) Inquiries, contracts, or order handling, including amendments?
    c) Customer feedback, including complaints?

7.3 Design and Development

7.3.1 Design and Development Planning

72. Does the organization plan and control the design and development of its products?
73. During the design and development planning, does the organization determine the
    a) Design and development stages to be used?
    b) Review, verification, and validation appropriate for each design and development stage?
    c) Responsibility and authority assignments necessary for the design and development processes?
74. Does the organization manage the interfaces between the different groups involved in the design and development activity to ensure effective communication between the groups, and to establish clear assignment of responsibility?

75. Does the organization update planning output as appropriate as the design and development progresses?

7.3.2 Design and Development Inputs

76. Does the organization determine design and development inputs relating to product requirements, including
    a) Functional and performance requirements?
    b) Applicable statutory and regulatory requirements?
    c) Information from previous similar designs where applicable?
    d) Any other essential requirements?
77. Does the organization maintain records of design and development inputs relating to product requirements?
78. Does the organization review design and development inputs to assure that they are complete, unambiguous, and not conflicting?

7.33 Design and Development Outputs

79. Do the organization's design and development outputs
    a) Meet (satisfy) the design and development input requirements?
    b) Provide the information needed for purchasing materials, manufacturing the product, and for service provision?
    c) Either contain or make reference to product acceptance criteria?
    d) Specify the characteristics of the product that are necessary for safe and proper use?
80. Does the organization provide its design and development outputs in a form that enables verification against design and development inputs?
81. Does the organization approve design and development outputs prior to their release for use?

7.3.4 Design and Development Review

82. Does the organization, in accordance with plans, perform systematic reviews of design and development work at suitable stages, to
    a) Evaluate the ability of the design and development results to meet requirements?
    b) Identify any problems encountered, and actions necessary to overcome them?
83. Do the organization's design and development reviews include representatives from each of the functions concerned with the stage(s) being reviewed?
84. Does the organization maintain records of the results of design and development reviews and any necessary actions planned or taken?

7.3.5 Design and Development Verification

85. Does the organization, in accordance with its plans, verify that design and development outputs meet (satisfy) all design and development inputs?

86. Does the organization maintain records of the results of design and development verification and any necessary actions?

7.3.6 Design and Development Validation

87. Does the organization, in accordance with its plans, perform design and development validation to ensure that the product is capable of meeting the requirements for the specified application or intended use (where known)?
88. Does the organization, wherever practicable, complete validation prior to product delivery or implementation?
89. Does the organization maintain records of validation results and any necessary actions?

7.3.7 Control of Design and Development Changes

90. Does the organization identify design and development changes (via engineering change notices or other means)?
91. Does the organization review, verify, and validate design and development changes as appropriate, including evaluation of the change's effect on constituent parts of the product, and on product already delivered?
92. Are design and development changes formally approved before changes are implemented?
93. Does the organization maintain records of design and development change reviews and any necessary actions?

7.4 Purchasing

7.4.1 Purchasing Process

94. Does the organization ensure that purchased product conforms to specified requirements?
95. Has the organization established supplier controls proportionate to the effect the purchased product can have on product realization processes or the final product?
96. Has the organization established a supplier-selection criteria?
97. Does the organization evaluate and select suppliers based on their ability to meet the organization's requirements?
98. Does the organization maintain records of supplier evaluation results and any actions taken?

7.4.2 Purchasing Information

99. Does the organization's purchasing information describe the product to be purchased, including where appropriate
    a) Requirements for approval of product, and procedures, processes and equipment used by the supplier?

b) Requirements for supplier personnel qualification?

c) Quality management system requirements?

100. Does the organization ensure the adequacy of specified purchase requirements prior to transmittal to the supplier?

7.4.3 Verification of Purchased Product

101. Has the organization implemented inspection and/or other activities necessary to ensure that purchased product meets specified requirements?

102. When it is intended that purchased product verification is to take place at the supplier's premises, does the organization include the intended verification arrangements and the method of product release in the purchasing information?

7.5 Production and Service Provision

7.5.1 Control of Production and Service Provision

103. Does the organization plan and carry out production of product and/or provision of service under controlled conditions, including (as applicable)

   a) Availability of all information necessary (specifications, drawings, functional descriptions, etc.) to describe the product's characteristics?

   b) Availability of work instructions needed by the process operators?

   c) Availability and use of suitable production/service provision equipment?

   d) Availability of monitoring and measuring devices?

   e) Implementation of monitoring and measuring of processes and product?

   f) Implementation of product release activities, and delivery and post-delivery activities?

7.5.2 Validation of Processes for Production and Service Provision

104. Does the organization validate processes, the output of which cannot be verified by subsequent testing, including where deficiencies may become apparent only after the product is in use, or the service delivered?

105. Does the validation of such processes demonstrate the ability of the process to achieve conforming product—even though not measurable at that point?

106. Has the organization established arrangements for these processes including, as applicable

   a) A defined process review and approval criteria?

   b) Approval of equipment used and qualification of process operators?

   c) The use of specific process validation methods and procedures?

   d) Requirements for records of process validation?

   e) Revalidation of processes as necessary?